Magnetic Recording

Volume III VIDEO, AUDIO, AND INSTRUMENTATION RECORDING

Library of Congress Cataloging-in-Publication Data
(Revised for vol. 3)

Magnetic recording.

Includes bibliographies and index.
Contents: v. 1. Technology—v. 2. Computer digital
storage—v. 3. Magnetic recording
1. Magnetic recorders and recording.   I. Mee,
C. Denis.   II. Daniel, Eric D.
TK881.6.M24   1987   621.38   86-10432
ISBN 0-07-041271-5 (v. 1)
ISBN 0-07-041272-3 (v. 2)
ISBN 0-07-041273-1 (v. 3)

1234567890 DOC/DOC 89321098

ISBN 0-07-041273-1

*The editors for this book were Theron Shreve and David E.
Fogarty, the designer was Naomi Auerbach, and the production
supervisor was Dianne Walber. It was set in Century
Schoolbook by University Graphics, Inc.*
Printed and bound by R. R. Donnelley & Sons Company.

Magnetic Recording

Volume III VIDEO, AUDIO, AND INSTRUMENTATION RECORDING

EDITORS

C. Denis Mee
IBM Corporation, San Jose, California

Eric D. Daniel
Woodside, California

McGraw-Hill Book Company

New York   St. Louis   San Francisco   Auckland   Bogotá   Caracas
Colorado Springs   Hamburg   Lisbon   London   Madrid   Mexico
Milan   Montreal   New Delhi   Oklahoma City   Panama
Paris   San Juan   São Paulo   Singapore
Sydney   Tokyo   Toronto

Contents

Preface

Information storage applications continue to grow at a rapid rate in the 1980s due to the successful development of business and consumer products for processing data, video, and audio signals. Advances in storage products have occurred largely because of rapid progress in two key storage technologies—semiconductors and magnetic recording. The consequent reductions in cost of storage have all but wiped out competing storage technologies. While some of the alternative static-memory and beam-addressable storage technologies are very successful for specific applications, the versatility of magnetic recording in providing different storage media formats has resulted in its application in the form of tapes, rigid disks, flexible disks, cards, drums, and sheets. This has resulted in the spread of magnetic recording applications to data, video, and audio storage for both professional and consumer applications. Magnetic recording now dominates the video recording industry, and is moving rapidly into new areas such as 8-mm home movies and still-camera photography. All of the many products using magnetic recording utilize the same basic inductive-recording technology which has developed over the last 30 to 50 years. The technology has improved immensely as understanding has been gained of the physics of the recording and reproducing processes. Advances in recording materials and processes for fabricating components have also contributed in a major way to extending the performance of magnetic recording technology into different industries.

Despite the growth of the magnetic recording industry, the early interest of students in the relevant disciplines of solid-state physics, chemistry, mechanical engineering, and information theory has been hampered by the lack of university teaching emphasis on the multidiscipline technology of magnetic recording. In the past few years, new magnetic recording research centers have been established in several universities in the United States, and major recording research groups have emerged overseas, especially in Japan. The primary purpose of these books on magnetic recording is to provide a text for courses in university graduate schools and in industry. The first volume covers the basic technology in depth, and the two companion volumes cover the many and varied applications.

Magnetic Recording, Volume I: *Technology,* is concerned with establishing the underlying technologies that are common to all forms of magnetic recording. Separate chapters treat the processes by which recording and reproduction take place; the materials, design, and fabrication of media; the materials, design, and fabrication of heads; the limits on performance due to noise, interference, and distortion; the key magnetic and recording measurement techniques that have evolved; and, finally, the mechanical interface between the head and medium that is of critical importance in all but optically addressed media.

Magnetic Recording, Volume II: *Computer Data Storage,* and Volume III: *Video, Audio, and Instrumentation Recording* are concerned with the major applications of magnetic recording. These are broadly classified into data or analog recording categories, according to the nature of the signal to be recorded rather than the means of carrying out the recording. The data recording chapters cover the evolution of recording systems using rigid disks, flexible disks, and magnetic tape drives. The analog recording chapters cover video and audio recorders, including those in which the signals are stored in binary digital form. Because of their relevance to all applications, two chapters appear in both Volume II and III: one is devoted to coding and error control, and the other discusses reversible optical recording in which the signals are recorded thermomagnetically then reproduced magnetooptically.

SI units (Système International d'Unités) are used throughout. Where other units are widely used, values expressed in these units are also listed in parentheses.

We are grateful to the authors and to a number of external reviewers in assisting us in our attempts to produce a uniform coverage of the subject matter. Special acknowledgment goes to T. C. Arnoldussen, G. Bate, A. E. Bell, H. N. Bertram, M. O. Felix, R. J. Gambino, J. M. Harker, A. M. Heaslett, M. K. Hill, R. E. Jones, J. G. McKnight, L. Rosier, R. Wood, and J. K. Wolf. We are also grateful to others in the recording industry who have kindly provided original materials to us. Finally, we wish to thank IBM for providing the environment and support necessary to make these volumes a reality, with special thanks to G. L. Cavagnaro, J. A. Harriss, and the members of the Technical Document Service for their patience and diligence in working with the manuscripts.

Contents of Other Volumes

Lists of Abbreviations and Symbols

Abbreviations

ADC	analog-to-digital converter
AGC	automatic gain control
ANSI	American National Standards Institute
APC	automatic phase control
APD	avalanche photo diode
AXP	adaptive cross-parity code
BCH	binary-coded hexadecimal
CCIR	Comité Consultatif Internationale des Radio-Communications
CIRC	cross-interleaved Reed-Solomon code
CNR	carrier-to-noise ratio
CRC	cyclic redundancy code
DAC	digital-to-analog converter
DASD	direct access storage device
DASH	digital audio stationary head
DR	density ratio
DSL	digital simulation language
ECC	error correction coding
EIA-J	Electronic Industry Association of Japan
ELP	extra long play
ERP	error recovery procedures
FDD	floppy disk drive
FM	frequency modulation
GCR	group-coded recording
GF	Galois field
HDA	head-disk assembly
HDDR	high-density digital recording
HIP	hot isostatic pressed
HPF	hot pressed ferrite

I/O	input/output
IEC	International Electrotechnical Commission
IRIG	Inter-Range Instrumentation Group
ISI	intersymbol interference
LQR	linear quadratic regulator
LRC	longitudinal redundancy code
LSB	least significant bit
M^2	Miller squared
ME	metal evaporated
MFM	modified frequency modulation
MOL	maximum output level
MP	metal particle
MSB	most significant bit
MSS	mass storage system
MTBF	mean time between failures
MTF	modulation transfer function
MUSE	multiple sub-Nyquist encoding
MZM	modified zero modulation
NEP	noise-equivalent power
NPR	noise-power ratio
NRZ	non-return to zero
NRZ-ASE	non-return to zero with adapted spectral energy
NRZI	modified non-return to zero
NRZI-S	synchronized non-return to zero
NTSC	National Television System Committee
ORC	orthogonal rectangular code
PAL	phase alternation line
PAM	pulse amplitude modulation
PBS	polarizing beam splitter
PCM	pulse-code modulation
PE	phase encoding
PES	position error signal
PET	polyethylene teraphthalate
PIN	positive intrinsic negative
PLL	phase-locked loop
PMMA	polymethyl methacrylate
R-DAT	rotating-head digital audio tape recorder
RLL	run-length-limited

S-I-NRZI	scrambled interleaved modified non-return to zero
SCF	single crystal ferrite
SECAM	sequential color and memory
SER	soft error rate
SFD	switching field distribution
SNR	signal-to-noise ratio
SP	standard play
SR	shift register
TBC	time base correction
THD	third harmonic distortion
TMR	track misregistration
VCM	voice coil motor
VFO	variable frequency oscillator
VHS	video home system
VRC	vertical redundancy code
VSM	vibrating sample magnetometer
VTR	video tape recorder
XOR	exclusive or
ZM	zero modulation
3PM	three-position modulation

Symbols

A	area; exchange energy constant; exchange stiffness; an element of $GF(2^8)$
a	acceleration; arctangent transition parameter; lattice constant
a_i	first digit of ith digit-pair
$A_m(t)$	mth bit in track t in set A
$A(x)$	polynomial representation of A
B	magnetic induction (flux density)
b	base-film thickness; bit length; length of burst-error; width of guard band
B_g	air-gap flux density
B_i	written data byte in ith position
$\hat{B}_i$	read data byte in ith position
$B_i(T)$	Brillouin function for atom of type i
b_i	second digit of ith digit-pair
B_m	maximum induction (flux density)
$B_m(t)$	mth bit in track t in set B

B_r	remanent induction (flux density)
B_s	saturation induction (flux density)
$B(x)$	data polynomial
C	capacitance; capacity (bytes)
c	accumulated dc charge at any digit position in a binary sequence; average length per turn in magnet coil
C_i	specific heat per unit volume of layer i
C_R	damping constant for rotary actuator
C_r	crosstalk
$C(s)$	compensator transfer function
$C(x)$	check polynomial
D	data rate; delay factor; recording density
d	head-drum diameter; head-to-medium spacing; minimum run-length of zeros in a binary sequence; RLL code constraint
D_a	areal density (b/mm^2)
D_c	recording density for maximum resolution
d_{eff}	effective head-to-medium spacing
d_i	data bit in the ith position in a data stream
D_l	linear (bit) density (b/mm)
d_l	lens diameter
D_r	recording density (fr/mm)
D_t	track density (t/mm)
D_u	user density (bytes/mm)
d_0	minimum stable domain diameter; spacing corresponding to nominal "in-contact" conditions
D_{50}	linear density at which the output falls 50%
$D_{(i)}$	thermal diffusivity of magnetooptical film ($\{i\} = f$) and dielectric ($\{i\} = d$)
E	energy; Young's modulus
e	charge of electron; cycle length of modulo-$G(x)$ shift register; exponent of a polynomial $G(x)$; head output voltage
E_a	anisotropy energy
e_b	back emf
E_d	demagnetizing energy
E_e	exchange energy
E_f	Fermi energy
E_h	external magnetic field energy
E_i	error pattern in ith position
E_k	magnetocrystalline energy

E_t	total energy
$E(x)$	error polynomial
F	force; magnet filling factor; system matrix used in servo design; video field frequency
f	focal length; frequency; number of data bits in a section
f_c	carrier frequency
F_f	field frequency
f_H	horizontal sync frequency
f_{max}	maximum recorded frequency
f_s	signal frequency; sampling frequency
f_{sc}	color subcarrier frequency
f_u	color-under subcarrier frequency
F_r	radial force
f_0	Larmor frequency
$F(\theta)$	particle orientation factor
G	gain
g	gap length
g_{eff}	effective gap length
g_i	ith coefficient in the generator polynomial $G(x)$
$G(x)$	generator polynomial
$G_{\{i\}}$	intensity enhancement of bilayer ($\{i\} = b$) and quadrilayer ($\{i\} = q$) over bare thick medium
gb	guard band between tracks
$GF(2^m)$	Galois field of 2^m elements
H	magnetic field
h	magnet coil efficiency; linear system output matrix
H_a	total anisotropy field
H_b	bias field
H_c	coercivity
H_d	demagnetizing field
H_p	print field
H_u	effective field seen by spin-up electrons due to spin-orbit effect
H_0	particle switching (nucleation) field
I	intensity of electron beam; Laplace transform of current
i	current
i_n	total noise current
i_{ne}	electronic noise current
i_{nt}	thermal noise current

$I(r)$	light intensity at radius r
$I_{\{i\}}$	intensity of magnetooptical radiation from semi-infinite ($\{i\} = s$) or quadrilayer ($\{i\} = q$) sample
J	exchange integral; moment of inertia; total angular momentum quantum number
j	$\sqrt{-1}$
J_{ij}	exchange integral between nearest neighbor atoms i and j
K	magnetic anisotropy constant
k	Boltzmann constant; maximum runlength of zeros in a binary sequence; number of information bits in a code word; number of message symbols in a block; ratio of read-to-write widths (w_r/w_w); RLL code constraint; wave number ($2\pi/\lambda$)
k_b	back emf coefficient
k_d	damping coefficient
K_{eff}	effective uniaxial anisotropy constant
k_f	actuator force factor
K_p	figure of merit including laser power tolerance
K_r	figure of merit for rotating power
K_u	uniaxial anisotropy constant
k_x	position gain constant
L	inductance; inductance of voice coil; orbital angular momentum quantum number
l	length of head magnetic circuit; skid length
l_c	length of head core
l_{coil}	coil length (actuator)
$L_{\text{dif}}(t)$	approximate thickness of dielectric layer heated in time t
L_1, L_2	inductance components of voice coil with shorted turns
$L_{1,2}$	coupling inductance of voice coil with shorted turn
$L_{\{i\}}$	thickness of magnetic layer ($\{i\} = f$), overlayer ($\{i\} = o$), and intermediate layer ($\{i\} = i$)
M	magnetization; mass; moving mass of actuator
m	mass; mass of electron; number of data bits
m_B	Bohr magneton
M_p	printed magnetization
M_r	remanent magnetization (remanence); modulation ratio (b/fr)
M_s	saturation magnetization
N	number of atoms per unit volume; number of particles per unit volume; seek distance in units of tracks
n	index of number of tracks; index of mode of vibration; number of

	bits in a code word; number of message bits; number of symbols in a block; number of turns; refractive index
N_A	numerical aperture
N_b	number of bits per byte
N_d	noise density
N_e	tracks remaining to target track
N_h	number of turns on head
N_l	number of turns on coil
N_s	total number of sectors
N_t	total number of tracks
$N_u(E)$	density of spin-up states at energy E
$N_d(E)$	density of spin-down states at energy E
$\mathbf{n}^+$	complex refractive index for right circularly polarized light
$\mathbf{n}^-$	complex refractive index for left circularly polarized light
$n_{\{i\}}$	refractive index of overlayer ($\{i\} = o$), or intermediate layer ($\{i\} = i$)
P	average laser power on each detector; power; print-to-signal ratio
p	number of parity symbols in a block; location of error in a code word; pole-tip length; pressure; volumetric packing density
P, P_e	power dissipated
P_i	laser power incident on medium
p_i	coded bit in the ith position in a data word
P_m	maximum print-to-signal ratio
P_s	optical signal power on each detector
P_{shot}	optical shot noise power on each detector
p_{50}	pulse width at 50% amplitude
$P(A)$	look-ahead parity in zero modulation
$P(B)$	look-back parity in zero modulation
$P(x)$	prime polynomial
Q	dimensionless function; head parameter in calculating transition length; information bit redundancy; quality factor in servo design
q	factor of order unity
Q'	dimensionless function
R	data rate; reflectivity; resolution; resistance
r	radial coordinate; radius; number of check bits in a code word
r_b	e^{-1} intensity radius of laser beam
r_c	critical radius
r_d	disk radius

R_e	resolution
r_h	radius of head
r_i	inner radius
r_{ij}	distance between nearest neighbor atoms i and j
R_m	modulation code rate
r_n	normalized "tent" radius (r/r_t)
r_o	outer radius
R_{sense}	sense resistor
r_t	radial extention of spacing "tent"
r_x	ordinary (unconverted) reflected amplitude for linear polarization x
r_y	magnetooptical (converted) reflected amplitude for linear polarization y for incident polarization x
r_+	reflected amplitude for right circularly polarized light
r_-	reflected amplitude for left circularly polarized light
$r_{\langle i \rangle}$	complex Fresnel coefficient ($\langle i \rangle = x, y, +$ or $-$)
S	element of surface; signal normalized to zero-peak signal from saturated region; spin angular momentum quantum number; synchronization pattern
s	number of message bits in a symbol
S_{MTF}	normalized zero-to-peak signal from an NRZ 1010 ... pattern
S_{sat}	zero-to-peak signal from a saturated region
S_{trans}	normalized zero-to-peak signal from an isolated transition
S^*	coercivity squareness factor
$S(x)$	syndrome polynomial
Sd_m^a	mth cross-parity syndrome in set A
Sd_m^b	mth cross-parity syndrome in set B
Sv_m^a	mth vertical parity syndrome in set A
Sv_m^b	mth vertical parity syndrome in set B
T	companion matrix of a polynomial; data period; temperature; tension; torque
t	time; total thickness of medium including substrate; track spacing
T_A	multiplier matrix corresponding to element A
t_a	audio track width
T_c	Curie temperature
t_c	clock window; control track width
T_{comp}	compensation temperature
t_g	guard band track width
t_{min}	minimum track spacing

t_v	video track width
$\text{TMR}_{W,R}$	write-to-read track misregistration
$\text{TMR}_{W,W}$	write-to-write track misregistration
$\text{TMR}_{3\sigma}$	3σ value of track misregistration
t_p	track pitch
t_r	response time of rigid-disk file
T_s	sampling time
t_{se}	service time of rigid-disk file
t_w	wait time of rigid-disk file
$T_{\{i\}}$	timing window for $\{i\}$ = RLL and $\{i\}$ = NRZ codes
V	head-to-medium velocity; output voltage
v	data word corresponding to a code word w; particle volume; velocity of actuator; mean particle volume
V_c	clip level
V_m	medium velocity (when different from head-to-medium velocity)
$v_{\max}$	maximum velocity of actuator
v_r	velocity of disk
W	code word; full-width-at-half-maximum of Gaussian read beam
w	track width; written mark width
w_a	audio track width
w_c	control track width
w_{eff}	effective track width
w_h	width of head
W_i	Strength of single ion anisotropy of atom of type i
w_r	read track width
w_t	tape width
w_v	video track width
w_w	write track width
$W(x)$	code-word polynomial
X	stroke length
x_d	command to servo
x_e	position error
X_i	atomic fraction of atom of type i
x_i	data bit in the ith position in a data word; first digit of the ith group of three digits
x/y	rate of a run-length limited code
Y_{coil}	admittance of voice coil
y_i	second digit of the ith group of three digits

Z	atomic number; impedance
z	number of balls in a bearing
z_i	third digit of the ith group of three digits
Z_{ij}	number of j atoms that are nearest neighbors of atom i
α	absorption coefficient; angle of polarizer; coefficient of thermal expansion
α_a	coefficient of thermal expansion of arm
α_s	turns ratio for shorted turn
α_{OT}	off-track constant
β	bandwidth
Γ	discrete time system input matrix
γ	fraction of NRZ timing window used by noise
Δ	differential phase angle; parameter of order of d-bandwidth
Δf	frequency deviation
ΔT	change in temperature
Δw	side fringe width; discrepancy between track and playback head
δ	depth of focus; thickness of a magnetic medium
δ_f	thickness of base film
δ_w	domain wall width parameter; side fringe width of read head
δw_w	side fringe width of write head
$\varepsilon(\omega)$	complex dielectric tensor (at frequency ω)
ζ	fraction of heat residing in magnetic film
η	efficiency; quantum efficiency
η_k	polar Kerr ellipticity
Θ	angle of rotary actuator
θ	angular coordinate; video track inclined angle
θ_h	half-angle of lens
θ_k	polar Kerr rotation
θ_s	maximum Kerr rotation in system
θ_0	video track inclined angle when tape is stationary
κ	thermal conductivity
Λ_{trans}	characteristic length of readback from isolated transition
λ	integer constant; wavelength
λ_c	critical wavelength for maximum resolution
$\lambda_{\mathrm{cutoff}}$	threshold for modulation transfer function
λ_m	wavelength for maximum print-through
λ_{min}	minimum recordable wavelength
λ_s	magnetostriction coefficient

μ	relative magnetic permeability
μ_0	magnetic permeability of vacuum
ν	frequency
ξ	polarizing beam splitter efficiency
ρ	density; resistivity
ρ_m	magnetic pole volume density
σ	standard deviation; stress
$\sigma(\omega)$	complex conductivity tensor (at angular frequency ω)
τ	time constant; dibit peak shift
$\bar{\tau}$	average access time
τ_d	penetration diffusion time
τ_e	electrical time constant
τ_{eq}	time constant to reach uniform film temperature
τ_l	mechanical time constant
τ_{lat}	lateral diffusion time
τ_m	time constant of magnet
τ_s	settling time
τ_{sk}	seek time
$\bar{\tau}_{sk}$	average seek time
τ_v	dominant time constant of actuator
$\tau(n)$	time to access n tracks
Φ	state transition matrix
ϕ	angular coordinate; gap azimuth angle; magnetic flux; physical rotation of polarizing beam splitter
χ	constant relating fluctuations in anisotropy to average anisotropy; magnetic susceptibility
χ_p	print susceptibility
ψ	azimuthal angle
Ω	rotational rate
ω	angular frequency $(2\pi f)$
ω_f	filter frequency

Magnetic Recording

Volume III VIDEO, AUDIO, AND INSTRUMENTATION RECORDING

Introduction

C. Denis Mee

IBM Corporation, San Jose, California

Eric D. Daniel

Woodside, California

Magnetic recording applications can logically be classified in several ways according to the type of signal to be recorded, the geometric shape of the recording medium, or the type of signal encoding used. Another classification recognizes that the interests of those working in the magnetic recording industry and, presumably, of the readers of these volumes fall into certain well-defined categories. Classification by these categories has been the primary guide in organizing the contents of Volumes II and III.

Chapters 2 through 4 in Volume II are concerned with the storage of computer data on rigid disks, flexible disks, and tapes. These applications all use binary digital signals and are usually referred to collectively as *digital recording applications*. Chapters 2 through 4 in Volume III deal with the recording of video, audio, and instrumentation signals. Each of these applications may use frequency modulation or digital encoding, as well as linear analog recording techniques. Nevertheless, they are usually referred to collectively as *analog recording applications* since they ideally provide an output that is an exact replica of the input.

Digital and analog applications differ mainly in terms of the requirements placed upon system performance. The primary requirements of a digital application are high data reliability, fast access to stored data, and low cost per bit. The primary requirements of an analog application are high signal-to-noise ratio, low distortion, and low cost per unit of playing time. Of these differences, the most fundamental is with respect to access time because this largely dictates the mechanical systems and media configurations that are used. In analog applications, the signal is usually in the form of large blocks of serial information—a musical composition, a television show, a transmission of data from a satellite. In such cases, a recording system using a medium in the form of a long length of tape is very suitable. However, in the words of Oberlin Smith, one of the early pioneers of magnetic recording, one disadvantage of a tape is "that if some small portion of the record near the middle has to be repeated there is a good deal of unwinding to do to get at it" (Smith, 1888). This difficulty of achieving quick access to stored data rules out tape media for on-line computer usage. Instead, magnetic media are more attractive in such forms as cards, strips, drums, and disks.

Despite the differences outlined above, the two types of applications have much in common. The various species of digital and analog systems evolved over the years from the same origin and still share the same underlying technology—the subject matter of Volume I of this three-volume work. Unfortunately, during the course of this evolutionary process, the technical community working in magnetic recording have become more firmly divided than seems reasonable in view of their common ground. Such polarization is fundamentally undesirable because it inhibits communication and slows down (even prevents) the transfer of technological advances from one area to another. Volumes II and III bring together detailed discussions of all the major magnetic recording applications in one work. By so doing, it is hoped that some of the barriers that exist between workers in different areas will be reduced.

The depth of coverage of recording systems for different applications is weighted by the individual experiences of the authors. Nevertheless, in these volumes there is collectively a detailed description of the generic technologies used in today's products. For instance, the designs of digital recording channels described for rigid disks and tapes are largely applicable to flexible disks and magnetooptical disks. Likewise, the servo technology discussion on rigid disks applies to future flexible and magnetooptical disk systems. While signal coding and error correction coding are mentioned in each of the digital recording chapters, the detailed coverage of the subject in Chap. 5 is relevant to all applications. Thus, it is hoped that readers will profit from studying chapters that are beyond but related to their specific fields of interest.

The following description of magnetic recording systems will serve as an introduction to the body of Volumes II and III for those not familiar with magnetic recording. Three introductory books have been published recently (White, 1985; Yokoyama, 1985; Jorgensen, 1986), and a selection of important papers and patents is also available in book form (Camras, 1985). At the end of this chapter recent review publications are listed for those interested in further general reading.

1.1   Evolution of Magnetic Recording Systems

Every magnetic recording system comprises a means for mechanically transporting recording media and heads with respect to each other, together with a package for the recording medium appropriate to the application. Also included is an electronic signal processor to deliver the signal to the recording head in a form suited to the recording method chosen. Another signal processor reconstructs the reproduced signal into an undistorted replica of the original or into a form that retains certain significant characteristics of the original.

The traditional systems using magnetically coated tape, ring heads, and longitudinal magnetization have dominated product applications since the late 1940s. The first major product application was for audio recording, which was founded on, and continues to use, linear analog methods based on ac biasing. Many instrumentation applications have also used ac bias. The requirements for providing high signal-to-noise ratios while increasing recording densities have kept audio and instrumentation applications at the leading edge of high-density magnetic recording technology.

Extensions of tape recording to the storage of digital data were developed using unbiased nonlinear recording (Harris et al., 1981). New tape-drive devices were required to operate at high tape speeds and provide fast access to the stored data. The requirement for fast access spurred the development of magnetic strips and loops, but these approaches suffered from poor reliability. After some 30 years, tape-drive systems continue to be the primary removable storage technology for digital data. Attempts to extend stationary-head tape machines to video recording failed because they required excessive tape speed to record the very high frequencies involved. This problem was solved in 1956 with the introduction of a scanning-head machine. This major innovation was initially applied to professional video recording using high-speed transverse scanning of a slowly moving tape. Subsequent evolution of helical scanning techniques expanded the application to lower-cost drives and ushered in the consumer video-recording market (Kihara et al., 1976; Shiraishi and Hirota, 1978).

Another significant innovation was the introduction of the rotating rigid disk for digital data storage in 1957 (Lesser and Haanstra, 1957;

Noyes and Dickinson, 1957). Using an air bearing to support a read-write head allowed a substantial increase in head-media velocity. Not only did this provide high reliability and high data rate, but it enabled fast radial accessing over the disk. The rigid-disk system has become the ubiquitous design of choice for on-line storage of computer data. The related flexible-disk system using in-contact recording became a major product with the advent of the personal computer (Noble, 1974; Engh, 1981). This inexpensive storage technology uses a single coated plastic disk inside a protective jacket. The removable, lightweight "diskette" has become the primary low-cost storage medium for small computers.

The success of magnetic recording products in penetrating the markets for data storage and video and audio applications has relied on a sustained upgrading of heads, media and electronics. Improvements in linear recording density resulted from advances in the magnetic and mechanical properties of head and media materials, described in detail in Volume I. The increases in areal density for data, audio, and video recording for the last 30 years are shown in Fig. 1.1a. Areal density has been increased by improvements in both linear and track densities, as shown in Fig. 1.1b, for the major applications of disk and tape recording. There are always some technical factors which limit the relative rates of linear and track density improvement, but the major impediments in some applications are the constraints of standardization. This is particularly true of tape recording where, for example, compatibility requirements in computer data tapes prohibited track density advances until recently. The 1985 introduction of an alternative to the nine-track open tape reel breaks with the data storage standard and advances linear and track densities substantially. In audio recording, the introduction of digital audio recording was accompanied by very high linear and track densities using either stationary multitrack heads or rotating single-track heads (Nakajima et al., 1983; Mukasa, 1985).

The combination of higher linear density, greater head-medium velocity, and multiple-track recording has provided an increase in the upper limit of recording bandwidth of nine orders of magnitude over the last 30 years. Progress in providing faster access to the stored data has also been impressive particularly in rigid-disk systems. Further improvements in accessing by providing more heads, for example a head per track on a disk, are no longer cost competitive with low-cost large-scale integration (LSI) electronic memories. Magnetic bubble stores are superior in access speed to fixed-head disk files but, in general, are also not cost-performance competitive with LSI memory. The combination of high-performance memory and moderate-performance, direct-access disk storage is preferred for on-line data storage.

Some of the difficult challenges for heads and media are circumvented in the emerging optical-beam storage technology (Kryder, 1985). Laser-beam writing and reading on magnetic media, discussed in Chap. 6, provides at once an out-of-contact erasable recording method and the capability for narrow-track operation. This high-density recording technology is currently emerging in its first product forms using rigid disks. While the basic concept for reversible recording was demonstrated about 25 years ago, major product implementations have been slow to emerge. The com-

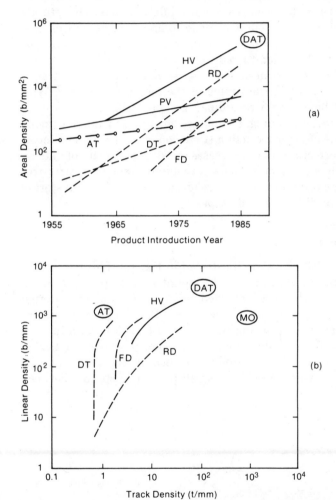

Figure 1.1 (a) Areal density progress for magnetic recording systems; (b) relationships of linear and track densities. The density curves for audio and video recording are obtained from the approximation that the shortest recorded wavelength is equivalent to two recorded bits. HV, home video tape; PV, professional video tape; RD, rigid disk; DT, data tape; FD, flexible disk; AT, audio tape (DAT, digital audio tape); MO, magnetooptical storage projections.

bined performance potential of fast accessing capability and high areal recording density will have to be advanced aggressively to compete with the incumbent storage technology.

1.2 Data Storage Systems

Binary digital signals are particularly well-suited for magnetic recording. Ideally, they are written by reversals of a head field sufficient to saturate the recording medium. No attempt is made to linearize the writing process, and maximum reproduce head flux is obtained on reading the recorded transitions. In order to reduce errors to a minimum, heads, media, and data channels are designed to write and read the data transitions with minimum shift in the timing of the reproduced pulses. With only two states in binary data recording, it is profitable to apply error detection and correction coding techniques in order to provide many orders of magnitude of error reduction (Peterson et al., 1972). The incorporation of error correction codes decreases the storage density of data on the medium, but the net effect is a substantial improvement in error-free storage density. The primary magnetic recording systems for data storage utilize rigid disks, flexible disks, and tapes.

1.2.1 Rigid disk drives

The rudiments of a disk drive are illustrated in Fig. 1.2. This part of the storage system contains a stack of disks mounted on a spindle rotating inside an enclosure which has a controlled air supply to minimize internal contamination. Writing and reading are achieved with a ganged array of heads, each provided with a spring suspension attached to an arm. The head elements are individually mounted on sliders which are loaded against the surface of a disk. A hydrodynamic bearing is generated at the slider-disk interface, which provides a small but stable spacing between

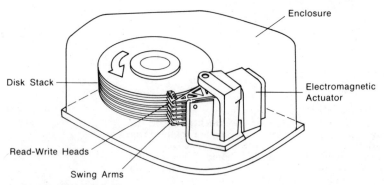

Figure 1.2 Rigid disk file components.

the heads and the disk. The head arms are connected to a common spindle which is positioned by an electromagnetic actuator to provide random access of the heads to any desired track on the disk. Usually, one head near the center of the stack reads a control track, and the signal from this head is used to control the swing-arm electromagnetic actuator. In other designs, a linear electromagnetic actuator is used and provides somewhat faster accessing speeds. Disk files based on these general designs have been developed to provide a range of storage capacities of over two orders of magnitude. The largest capacity files may use up to about ten 14-in-diameter disks, while the small-capacity files use fewer and smaller-diameter disks, down to 3.5-in diameter.

The accessing performance of different designs of large and small files varies because of the ranges of the track-seeking time, the disk rotational delay time, and the channel-busy time. Typically, these three times are about equal to each other. Their sum leads to the average time to complete a request for data, known as the *service time*. The total response time comprises the service time plus the data queueing time, which depends on the number of access paths into the data. As storage densities go up, there is a need to increase the number of access arms and heads in order to contain the queueing time within a reasonable fraction of the total response time.

The advances in areal density achieved over the last 30 years have been made possible through a steady introduction of improved design concepts (Harker et al., 1981; Kaneko and Koshimoto, 1982). Early files were designed to use removable disk stacks, but these became impractical as closer head-to-media and track spacings were required. Today, heads are permanently dedicated to a disk surface in order to improve the alignment of heads and recorded tracks. For these designs, there is a complex relationship between the tribological behavior of the heads and disks in the various modes of operation of the disk file, that is, during starting and stopping, during accessing of the head, and during the resting period when the heads and disks are in contact. The environment inside the head-disk enclosure, the formation of debris in the enclosure, and the lubricant on the disk all play a part, as do any changes in these factors with usage of the disk file. Disk coating dispersions have been developed with adequate durability during starting and stopping, but the magnetic layers are required to be progressively thinner in order to improve recording resolution. Thinner coatings, in turn, demand defect-free substrates since they are less effective in masking substrate imperfections. Ferrite heads have been reduced in size, and the materials and processes used have been improved to produce smaller gaps, smaller cores, and narrower tracks. As the trend continues to smaller dimensions, magnetic alloy films have been introduced, both as recording media and head cores. These components are included in some new disk file designs and are expected to be extended to many future designs as costs are reduced and head-medium interfaces are improved. Detailed descriptions of heads and media are given in Chaps. 2 and 3 of Volume I, and their performance in disk files in Chap. 2 of Volume II.

1.2.2 Flexible-disk drives

The key elements of a flexible-disk drive are shown in Fig. 1.3. The drive is designed for easy insertion and removal of the single flexible disk (diskette) in its enclosing jacket. When mounted in the drive, the disk is clamped at its center and rotated at a relatively low speed while the read-write head accesses the disk through a slot in the jacket. In many designs the accessing arms traverse above and below the disk and with read-write head elements mounted on spring suspensions. When reading and writing take place, the suspensions push the heads together with the double-coated disk interposed. In this way writing and reading can be achieved on both sides of the disk. Since the heads are in contact with the recording surface, high linear densities can be achieved. On the other hand, track densities are relatively low compared with rigid-disk files, because of the poor dimensional stability of the plastic disk substrate, the limited positioning accuracy of the disk clamp, and the lack of a track servo system in a low-cost drive. Head positioning is usually accomplished by a stepping motor whose rotational position is translated to a linear position of the head arm via a steel band as seen in Fig. 1.3. Track position inaccuracy is allowed for by providing an erasing head to erase the regions on either

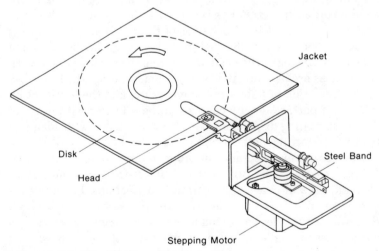

Figure 1.3 Flexible diskette and head accessing system.

side of the read-write track. This "tunnel erase" design is usually a part of the read-write head. As higher track density systems evolve, servos will be required to follow the tracks. One approach is to intersperse servo information at regular intervals around the track between sectors of data.

The trend in flexible-disk designs has been toward smaller disks, protected by a hard-shell jacket, with no exposed recording surfaces when the disk is not in use (Katoh et al., 1981). The constraints imposed by the removable diskette are the unique limitations of this technology. Otherwise, the recording innovations for higher densities follow those developed for tapes and rigid disks; this includes thinner coatings, higher coercivity media, improved wear characteristics of the media and heads, and improved disk lubrication (Bate et al., 1981). The primary applications of flexible disks have been as inexpensive products with high tolerance to handling damage. The track densities used have lagged the achievements of more expensive rigid-disk applications (Fig. 1.1a). More expensive flexible-disk products achieve higher track densities using head servoing technology and faster data rates using high-velocity "flying" disks.

1.2.3   Tape drives

Data recording on magnetic tape involves the application of 0.5-in-wide or narrower tapes in open-reel and single-reel cartridge, or double-reel cassette formats. In the higher-performance systems, the tape-drive mechanism must be capable of accurately guiding the tape at up to 200 in per second in contact with a multitrack read-write head. A requirement for on-line tape drives is the need for very fast starts and stops in order to minimize the wasted time between blocks of data. This requirement led to the use of low-inertia motors under servo control, together with vacuum columns to mechanically decouple the tape from the tape reels in the region of the read-write head. In recent developments, the need for fast starting and stopping is relieved by electronic buffering of the data stream; this allows lower tape acceleration and deceleration without loss of recording area on the tape. Also, interchanging tapes between different drives is helped by the use of electronic buffers to realign the signals for the different tracks of the multitrack head. Thus, the design of a data tape drive is simplified as illustrated in Fig. 1.4, which is a single-cartridge drive using a stationary multitrack read-write head. This drive is designed to match the data rate of an on-line data disk file.

The increases in recording density in 0.5-in data tape have resulted from the development of improved magnetic oxides, reductions in the number of defects in the recording medium, and the introduction of superior head-medium interface designs which maintain a uniformly small spacing between the head and medium with minimum wear. The track

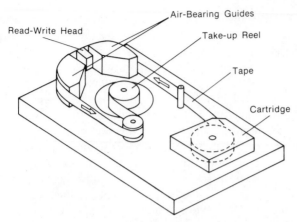

Figure 1.4   Schematic of cartridge drive for data recording.

density for tape recording is relatively low compared with other data recording drive systems and, therefore, a high signal level is achievable; this allows substantial amplitude equalization to be applied to the recording and reading channels with consequent gains in overall recording density. Complementing the equalization, powerful error correction codes have provided many orders of magnitude reduction in errors which would otherwise increase dramatically as the recording density increases.

Several different tape-drive formats have been developed to complement the mainstream evolution of 0.5-in tape drives. One motivation has been to alleviate the high cost of multitrack heads. A promising contending head design is the rotating single-track head developed for video recording (Newby and Yen, 1983). The ability to scan a relatively wide tape shortens the length of tape required for a given data capacity and thereby improves the access time to a given track. High data rate per track is achieved owing to an increase in the head-medium velocity by about an order of magnitude over the stationary-head, 0.5-in tape drives. This approximately offsets the data rate reduction due to using a single-track head. While the scanning head approach has been applied in video and instrumentation recording, and is emerging in audio recording, the technique has so far received only limited consideration in the computer data field.

Very high density digital tape recording has been developed for recording the wide-bandwidth transmissions encountered in certain instrumentation applications (Chap. 4 of Volume III). One direction of this area of development involves the use of multitrack heads to achieve the required bandwidth. Rotary-head designs have also been developed with higher recording density capability. Magnetic recording still remains the most cost-effective storage and retrieval technology for this type of data (Kalil, 1982, 1985).

1.3   Image Recording Systems

1.3.1   FM video recording

As mentioned earlier, stationary-head tape machines proved impractical to record the large bandwidth of video information, which stretches from dc to over 4 MHz. The dc requirement can be overcome by using FM, but the high-frequency requirement leads to either excessive tape speeds or uneconomical multiple-track multiplexing schemes. The breakthrough occurred in 1956 when Ampex introduced the Quadruplex video recorder (Ginsburg, 1986; Kirk, 1981). This uses a head-scanning mechanism in which four heads, mounted on a rotating wheel, scan successively across the width of a 2-in-wide tape at about 40 m/s (1600 in/s), while the tape moves by at a leisurely 38 cm/s (15 in/s). The high head-to-tape speed allows the whole video signal to be recorded using a specially developed low-modulation-index FM (Robinson, 1981; Benson, 1986).

The transverse-scanning machines, in a series of enhanced versions, remained the standard recorders for broadcast use for over two decades and are still in use. The alternative is to scan heads (usually two) across a narrower (usually 1-in or less) tape, forming long tracks at an acute angle to the tape length. This means that the tape moves in a helical path around a drum containing the rotating heads. A home video tape recorder using a helical-scanning head and a two-reel cassette is shown in Fig. 1.5. The helical-scanning method simplifies the electronics in that a whole video field is recorded by one head scan. Without correction, the time-base stability associated with the long track is inherently insufficient to meet broadcasting standards but may be adequate to meet the requirements of many closed-circuit television applications in industry and other areas. The development of electronic means of time-base correction has allowed the helical-scan method to be developed for broadcasting use where, compared with transverse-scan, it offers economies in equipment and tape costs. The helical-scan method has also become the basis of the huge market for home video recording. Modifications are required in

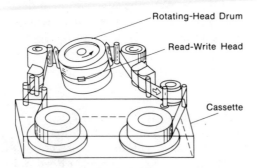

Rotating-Head Drum

Read-Write Head

Cassette

Figure 1.5 Helical-scan video recorder.

order to make the necessary dramatic reductions in the cost and size of the equipment and in the tape consumption per hour of playing time. First, at some small expense in color picture quality, the high-frequency color carrier is placed at a lower frequency than the luminance signal ("color-under"), thus compressing the overall bandwidth. Second, the guard bands, normally required between successive scans of the tape, are eliminated by using different head gap azimuth angles for adjacent recorded tracks. The consequent misalignment of the recorded signals reduces the pickup of the adjacent-track signal, should the head wander off the desired track. The scheme fits well with the convention, convenient for other reasons, of using two head elements on the rotating drum.

Head improvements have been required to achieve shorter wavelength recording as well as narrower tracking capability. This challenge has been met, for the most part, by improving or modifying ferrite heads. In particular, single-crystal ferrites are used for combined mechanical and magnetic performance reasons. More recently, alloy pole tips are added to provide sufficient recording field for the high-coercivity particles required for very short wavelength recording. Track-following servos have also been introduced to achieve better tracking of the recorded media. Piezoelectric mounting arms for the heads provide up to 100 μin of lateral head motion for tracking purposes.

Additional signal manipulation is being applied to different applications for video recording. In the consumer recorder, for instance, the repetitious nature of the video signal is used to conceal the signal degradation due to imperfect media. When loss of signal occurs, the error is concealed by repeating the previous television field.

As described in Chap. 2 of Volume III, 0.5-in and 8-mm video recording are now available in small, hand-held camera-recorder systems, continuing the preference for magnetic tape as the leading technology for video recording. Magnetic disks provide more limited storage capability but are a candidate for a magnetic-recording still-picture camera. Perhaps the major challenge for video tape recording is the lack of a fast, low-cost duplication process. Today, copies are made by rerecording at normal operating tape speeds. Replication of disks is a low-cost advantage for high-density, read-only optical disk storage. With the advent of high-density magnetooptical storage technology (Chap. 6), there may emerge another magnetic disk storage technology suitable for video recording and replication.

1.3.2 Digital video recording

In professional video recording, the significant advantages of digital recording—higher signal-to-noise ratio, no copying degradation, fewer uncorrectable errors—are receiving increased attention, and international standards are being promulgated. The application of powerful Reed-Solomon error correction plus interpolation of picture elements should allow digital recording densities to be achieved with similar tape consumption to analog recording.

1.4   Audio Recording Systems

1.4.1   Linear analog audio recording

The recording of audio signals, using linear analog techniques, is the oldest application of magnetic recording and is still one of the largest. It is also one of the most demanding in terms of signal-to-noise ratio and bandwidth (10 octaves) and the requirements these place upon the properties of the magnetic medium.

The elements of the generic audio recording have changed little since it was discovered that ac bias could be used to linearize the recording process. Separate recording, reproducing, and erasing heads are used on all but the least expensive recorders (Fig. 1.6). The original single-track, quarter-inch tape recorder has been expanded or contracted to form many configurations aimed at meeting the special needs of the recording studio, the home, industry, and school. Thus, at one extreme, there are machines for the professional recording engineer that use up to 32 tracks on 2-in-wide tape for program editing and mastering. At the other extreme, there are machines designed to fulfill special needs such as portable headphone recorders, microcassette dictating machines, and telephone-answering devices.

At the consumer level, the quarter-inch tape recorder has all but disappeared, and the two-reel "compact cassette," has become the worldwide standard. This format uses two pairs of stereo tracks on 3.8-mm-wide tape running at 48 mm/s. Despite the narrow tracks and slow tape speed, a good-quality cassette recording delivers an audio performance that rivals that of its quarter-inch predecessors running at four or more times the speed. This performance has been achieved by making improvements in

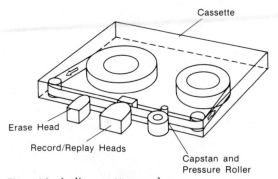

Figure 1.6 Audio cassette recorder.

all the components of the system: record, replay, and erase heads; mechanical drives; magnetic media; and signal processing. Of these, the last two deserve brief mention here.

Traditionally, audio recording, more than any other form of recording, provided the stimulus for exploring new materials for magnetic media. In part, this was because there was always a need to increase signal-to-noise ratio while maintaining (or increasing) bandwidth. Also, the volume of material involved in audio applications is large enough to provide the incentive to the independent manufacturers of such materials. Finally, although governed by strict international standards, the audio recording industry is not loath to try something new, provided the advantages offered are significant. Thus, audio recording has been the proving ground for many new materials—fine-particle oxides, chromium dioxide, cobalt-modified oxides, metal particles, and even evaporated metal—before these materials found application in video or data recording.

Another way of increasing signal-to-noise ratio is to introduce complementary signal-processing techniques into the recording and reproducing channels which have the net effect of reducing the noise that is not already masked by a high signal level (Dolby, 1967). These noise-reduction schemes, described in Chap. 3 of Volume III, have had a major impact on audio recording performance, both at the professional and the consumer level.

1.4.2 Digital audio recording

Digital audio recording has the potential of producing a high signal-to-noise ratio while avoiding virtually all of the other signal degradations associated with the analog approach. Digital recording is increasingly used in producing master tapes for high-quality program duplication when the program production does not demand complex editing and mixing between a large number of channels. When such mixing is required, as it is for much popular music, ac-biased master recording is still employed. In order to record at the approximate data rate of 1 Mb/s per channel, professional mastering recorders use either a stationary-head recorder with the digital audio signal divided between multiple tracks or a modified video helical-scan recorder.

For the consumer, audio recording using either FM pulse code modulation or digital (PCM) encoding is available on the higher quality home video recorders. Also, dedicated digital audio technologies have emerged based, again, on either a multiple-track stationary-head approach (S-DAT) or a rotating-head approach (R-DAT). Both are remarkable for the high track density (12.5 and 73.6 tracks per millimeter) and high linear density (2560 and 2440 fr/mm). The rotating-head system probably has the highest areal density of any magnetic-recording product. Both systems use high-coercivity metal-particle tapes.

1.5 Magnetooptical Recording Systems

An alternative type of magnetic-recording system uses an optical-beam read-write device instead of an inductive or flux-sensing head. Writing is achieved via a change in the magnetic properties of the media when heated by the beam. Reading is accomplished with a lower intensity beam using the magnetooptical rotation of the plane of polarization. Very high recording densities are projected as indicated in Fig. 1.1b. This approach, described in Chap. 6, has the major advantage of avoiding close proximity between heads and media, while offering a higher areal density than currently practiced in magnetic recording products.

The rudiments of an optical head and disk are shown in Fig. 1.7. The focusing and tracking lens above the disk can be mechanically accessed with a small electromagnetic actuator of the type used for magnetic recording on disks. Writing is achieved by modulating the beam in the presence of a perpendicular magnetic field applied, for instance, from a current-carrying coil underneath the disk. This magnetic field need not be localized to the recorded track since switching of the magnetization can occur only where the focused spot heats the surface and reduces the coercivity below the applied field magnitude. Recording takes place in physically pregrooved tracks molded into the substrate. Reading of the magnetized track is achieved at reduced beam power and without the field applied. In this mode, the plane of polarization of the reflected beam is changed slightly depending upon the magnetization direction of an element of the recorded track. Detection of this change of polarization is

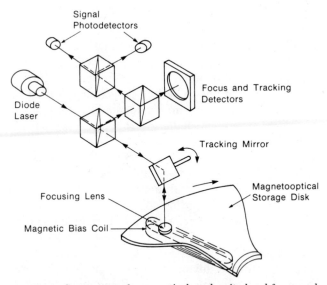

Figure 1.7 Components for an optical read-write head for recording on magnetic media.

achieved with a polarizing beam splitter and photo detector. Overwriting of a previously recorded track is normally achieved by applying an unmodulated writing beam and resetting the magnetization. The drawback of this simple scheme is that it requires two revolutions of the disk to rewrite a new track. A unique advantage of optical storage media is that the recording layer is protected on one side by the transparent substrate and on the other side by a relatively thick overcoat. Thus the surfaces of the disk can be made insensitive to handling damage since the recording beam is focused on the internal recording layer and not on the medium surface. Therefore, under normal handling conditions, optical disks do not require a protective jacket.

The most successful consumer application for optical recording is the read-only compact audio disk. Although this is a different recording technology to magnetooptical recording, it has demonstrated the practicality of recording at the highest areal densities in a commercial product in normal domestic environments including the poor conditions of audio players in automobiles. It is expected that the magnetooptical read-write disks, now emerging, will have similar environmental behavior, providing adequate stability of the recording medium itself is achieved (see Chap. 6). At this stage of development, magnetic coatings for conventional magnetic recording have superior intrinsic stability compared with reversible optical media, and they can be designed with either rigid or flexible substrates. Consequently, although magnetooptical recording will find increasing acceptance, the wide range of product applications described in this book is expected to continue for conventional magnetic recording.

1.6 Outlook

The number of applications for magnetic-recording products has increased impressively over the last 30 years and shows every sign of continuing to advance through the remainder of the twentieth century. In contrast to many predictions that the underlying technology would top out, there has been no slowing down of improvements in recording performance and, as a consequence, little challenge from other technologies for either the consumer or data storage business. In data storage applications the challenges from solid-state storage or from beam-addressable storage have, so far, had little impact on the domination of this business by magnetic recording products. Successful incursion of these new technologies into the realm of magnetic recording applications will depend on their continued rapid advance and a slowing down of magnetic recording advances. In the near future, developments currently underway to improve magnetic recording components and signal processing and control systems will ensure continued advances in recording density, data rates, and access speeds. This will present the challenging technologies with the same kind of moving target they have seen in the past. Any new technology will have to demonstrate the collected set of attributes that have made magnetic recording such a durable technology—that is, reversibility without degradation, environmental stability, and recording performance at an advantageous product cost. The recent successes of read-only optical storage technology in digital-audio applications have spurred a rapid increase in research programs to produce true read-write optical storage. The surge of activity in the 1980s has advanced a number of technologies in this area with magnetooptical recording as a leading contender—but, in effect, magnetooptical recording can be considered to be yet another extension of magnetic recording technology.

We conclude, therefore, that magnetic recording can anticipate a future of increasing applications and unimpeded growth.

References

Bate, G., G. J. Hampton, and B. J. Latta, "A 5-Megabyte Flexible Disk," *IEEE Trans. Magn.*, **MAG-17**, 1408 (1981).

Benson, K. B. (ed.), *Television Engineering Handbook*, McGraw-Hill, New York, 1986.

Camras, M., *Magnetic Tape Recording*, Van Nostrand Reinhold, New York, 1985.

Dolby, R. M., "An Audio Noise Reduction System," *J. Audio Eng. Soc.*, **15**, 383 (1967).

Engh, J. T., "The IBM Diskette and Diskette Drive," *IBM J. Res. Dev.*, **25**, 701 (1981).

Ginsburg, C. P., "Development of the Videotape Recorder," in *Videotape Recording*, Ampex Corporation, Redwood City, Calif., 1986, Chap. 1, p. 3.

Harker, J. M., D. W. Brede, R. E. Pattison, G. R. Santana, and L. G. Taft, "A Quarter Century of Disk File Innovation," *IBM J. Res. Dev.*, **25**, 677 (1981).

Harris, J. P., W. B. Phillips, J. F. Wells, and W. D. Winger, "Innovations in the Design of Magnetic Tape Subsystems," *IBM J. Res. Dev.*, **25**, 691 (1981).

Jorgensen, F., *The Complete Handbook of Magnetic Recording*, 3d ed., TAB Books, Blue Ridge Summit, Penn., 1986.

Kalil, F., *Magnetic Tape Recording for the Eighties*, NASA Reference Publication 1075, U.S. Govt. Printing Office, Washington, D.C., 1982.

Kalil, F., *High-Density Digital Recording*, NASA Reference Publication 1111, U.S. Govt. Printing Office, Washington, D.C., 1985.

Kaneko, R., and Y. Koshimoto, "Technology in Compact and High Recording Density Disk Storage," *IEEE Trans. Magn.*, **MAG-18**, 1221 (1982).

Katoh, Y., M. Nakayama, Y. Tanaka, and K. Takahashi, "Development of a New Compact Floppy Disk Drive System," *IEEE Trans. Magn.*, **MAG-17**, 2742 (1981).

Kihara, N., F. Kohno, and Y. Ishigaki, "Development of a New System of Cassette Tape Consumer VTR," *IEEE Trans. Consum. Electron.*, **22**, 26 (1976).

Kirk, D. (ed.,), *25 Years of Video Tape Recording*, 3M United Kingdom Ltd., Bracknell, England, 1981.

Kryder, M., "Magneto-Optic Recording Technology," *J. Appl. Phys.*, **57**, 3913 (1985).

Lesser, M. L., and J. W. Haanstra, "The Random-Access Memory Accounting Machine. I. System Organization of the IBM 305," *IBM J. Res. Develop.*, **1**, 62 (1957).

Mukasa, K., "Magnetic Heads," *J. Inst, Telev. Eng. Japan.*, **30**, 295 (1985).

Nakajima, H., T. Doi, J. Fukuda, and A. Iga, *Digital Audio Technology*, TAB Books, Blue Ridge Summit, Penn., 1983.

Newby, P. S., and J. L. Yen, "High Density Digital Recording Using Videocassette Recorders," *IEEE Trans. Magn.*, **MAG-19**, 2245 (1983).

Noble, D. L., "Some Design Considerations for an Interchangeable Disk File," *IEEE Trans. Magn.*, **MAG-10**, 571 (1974).

Noyes, T., and W. E. Dickinson, "The Random-Access Memory Accounting Machine. II. The Magnetic-Disk, Random-Access Memory," *IBM J. Res. Develop.*, **1,** 72 (1957).

Peterson, W. W., and E. J. Weldon, *Error Correction Codes*, MIT Press, Cambridge, Mass., 1972.

Robinson, J. F., *Videotape Recording*, Focal Press, London, 1981.

Shiraishi, Y., and A. Hirota, "Video Cassette Recorder Development for Consumers," *IEEE Trans. Consum. Electron.*, **CE-24,** 468 (1978).

Smith, O., "Some Possible Forms of Phonograph," *The Electrical World, 9,* 116 (1888).

White, R. M., *Introduction to Magnetic Recording*, IEEE Press, New York, 1985.

Yokoyama, K. (ed.), *Magnetic Recording: The Latest Technology, Systems and Equipment* (Japanese), General Technology Publishing, Tokyo, 1985.

Video Recording

Hiroshi Sugaya

Matsushita Electric Industrial Company, Limited,
Osaka, Japan

With sections on digital video tape recording by

Katsuya Yokoyama

NHK, Tokyo, Japan

2.1 Historical Development and Background[†]

2.1.1 Television systems

In terms of its impact on our day-to-day lives, television is without a doubt the single most important invention of this century. With the recent successful advent of high-quality color television, the magnitude of this impact has increased several times over; at the present time there are a great many different color television systems on the market (Pritchard and Gibson, 1980).

To understand the problems inherent in video tape recording (VTR), it

---

[†]Important historical developments are summarized in the Appendix, Table 2.9.

is helpful to compare the recording and playback systems used in films and in television broadcasts. In motion pictures, the illusion of continuous motion is created by projecting a series of still-frame pictures onto a screen at such a rate that the eye cannot differentiate between the frames. Owing to the after image phenomenon in the human eye, the segmented pictures are observed as a continuous picture. After a great deal of experimentation, a single international motion picture standard was eventually established: 24 illuminated frames per second, with darkness between each frame.

In order to send a picture as a television signal, the picture is first reconstructed as a series of small dots. In monochrome television, dots of different tone are used—black, white, and gray. The arrangement of these dots constitutes the details of the picture. Although the image is actually constructed from dots, these dots themselves are so small that the human eye blends the individual dots together into an integrated image and perceives them as a complete picture. The definition of the picture will depend upon the total number of lines (series of dots), and the contrast will depend on the number of shades of gray between black and white. The picture is scanned by a television camera, transmitted to the viewer, and displayed on a cathode-ray tube. The camera and the television receiver must be constantly synchronized so that the order in which the camera scans the picture is exactly the same as that on the television display. The camera scans the picture one line at a time and, as each line is completed, a synchronizing pulse is added to the waveform to cause the cathode-ray tube to begin to scan a new line. This continues until the last line of the picture is scanned, at which point a different synchronizing signal is added to the waveform to indicate that not only a new line but a new picture is beginning.

In order to reduce the flicker, which is inherent in a transmission system of this sort, interlace scanning is used in all present-day television systems. As shown in Figure 2.1, the scanning beam moves from line a vertically down the tube, scanning the odd lines to the middle of the bottom line, a'; this is called a *field*. Then the beam traces the even lines beginning at line b, making another field. The two different fields constitute a

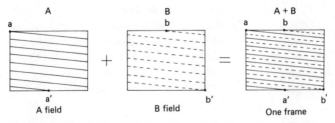

Figure 2.1 Interlace principle and composition of a television picture.

complete scene, which is called a *frame* in the United States and a *picture* in Europe.

Initially, it was necessary for television transmission equipment to be synchronized with the main power frequency. In countries such as the United States and Canada, 30 frames are transmitted each second at the vertical scanning rate of 60 Hz, while 25 frames are transmitted each second in Europe, Australia, China, and other 50-Hz countries. Japan has both 50- and 60-Hz systems. Obviously, the greater the number of frames transmitted per second, the more convincing the picture will appear to the human eye, and 25 frames per second is close to the miminum number that the eye will accept as a continuous picture without flicker. This is the source of a serious dilemma in television broadcasting because this frame rate should be kept as low as possible in order to keep the bandwidth narrow.

The number of horizontal lines determines the bandwidth of the television signal as well as the vertical definition of the television picture. There is a lengthy history behind the selection of the number of horizontal lines. In the United Kingdom, at the very beginning stages of television broadcasting, a standard of 405 lines and 50 fields per second was established. A standard of 819 lines and 50 fields per second was established for monochrome in France and Belgium; eventually a unified standard of 625 lines and 50 fields per second was established for western European color transmissions, including that of the United Kingdom. Meanwhile, the United States, Japan and other countries using 60 Hz power frequency adopted 525 lines. Thus, countries with a 60-Hz electrical standard use 525 horizontal lines and most areas with a 50-Hz standard use 625 lines.

In color television systems, the dots are formed from a trio of the three primary colors, red (R), green (G), and blue (B). In order for color transmission to be compatible with monochrome transmission, color transmission systems broadcast the basic monochrome signal, which is called the *luminance* signal, plus a *chrominance* signal. Color television systems can be classified into three main types: NTSC (National Television System Committee), PAL (phase alternation line), SECAM (sequential color and memory). A unified worldwide standard for color transmission has not yet been achieved, and at the present time there are some thirteen subsystems of these three major standard systems. The major differences among these various NTSC, PAL, and SECAM subsystems are in the specific modulation processes used for encoding and transmitting the chrominance signal (Table 2.1).

2.1.1.1   NTSC.   The basic idea of color television derives from the NTSC system developed by RCA. The standard was established in 1953; it has been used in the United States since 1954 and in Japan since 1960. In

TABLE 2.1   Typical Color Television Systems

	NTSC	PAL	SECAM	High definition†
Fields, per second	60	50	50	60
No. of horizontal lines	525	625	625	1125
Aspect ratio	4:3	4:3	4:3	16:9
Subcarrier, MHz	3.58	4.43	4.25 4.40	
Luminance bandwidth Y, MHz	4.2	5.0	6.0	20
Chrominance bandwidth C, MHz	I: 1.3 Q: 0.4	U: 1.3 V: 1.3	D_R: 1.3 D_B: 1.3	C_W: 7.0 C_N: 5.5
Color system	AM by two-color signals having 90° phase displacement	(Similar to NTSC) U and V color-difference signals reverse in phase on alternate lines	Color-difference signals (R − Y and B − Y) are transmitted by FM on alternate lines	Color-difference signals (C_w and C_N) are processed by component

†Comité Consultatif International des Radio Communications Proposal, 1986.

order to transmit the three primary R, G, B color signals, a frequency bandwidth three times wider than that necessary for monochrome TV signals is required. The NTSC system, however, can transmit the luminance signal Y and the chrominance signals C within the same frequency band as monochrome and is thus perfectly compatible with monochrome TV signals. This composite system uses two carriers of the same frequency (3.58 MHz), having a phase displacement of 90°, amplitude-modulated by color-component signals. These two color signals are represented by two vector signals I and Q having bandwidths of 1.5 and 0.5 MHz. Color hue (tone) is determined by the phase of the combined color subcarrier signal, and color saturation (intensity) is determined by the amplitude of the color subcarrier signal. A color reference signal (color burst signal) is inserted in a horizontal blanking period to synchronize the subcarrier frequency and phase (Fig. 2.2). In color television, therefore, the phase characteristics of the transmission system are critical. This has more important implications when recording magnetically on tape, which has an intrinsic time-base instability due to its flexibility.

2.1.1.2   PAL.   The PAL system, developed by Telefunken in 1962, is similar to the NTSC system, with the exception that the color-difference signals are represented by U and V, which have equal bandwidths and are

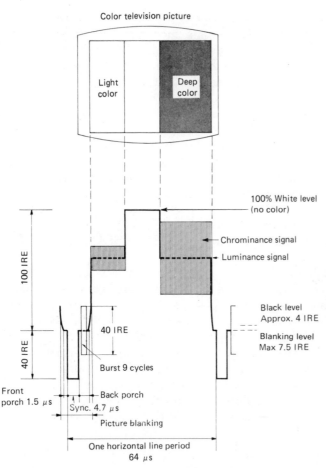

Figure 2.2 One horizontal line of an NTSC color television signal

transmitted on a 4.43-MHz subcarrier, with the V signal reversing in phase on alternate lines. In the NTSC system, phase distortion (color-quadrature distortion) causes an incorrect color transmission, but in the PAL system the V color-difference signal reverses in phase every other line. Thus, the reproduced color signal phase distortion on any two adjacent lines will be in the opposite direction from the true color, and color distortion can be averaged out by means of a single-horizontal-line delay line. Simpler receivers rely on the eye averaging the error on alternate lines.

2.1.1.3 SECAM. The SECAM system was first developed in France in 1967. In this system, the color transmission method is quite different from the NTSC and PAL system. Color-difference signals (R − Y, B − Y) are transformed into frequency modulation signals with different subcarriers and transmitted successively on alternate lines (line sequential transmission). Simultaneous color-difference signals are reproduced using a horizontal scanning period delay line in the TV receiver. This system is, therefore, not influenced by phase distortion in transmitted color signals.

2.1.1.4 High-definition television systems. With the present-day rapid advances in electronic and television technology (including satellite broadcasting), a higher-definition picture quality can be anticipated for many future applications such as broadcasting systems, movie production (including electronic movie theaters), and a variety of industrial information transmission systems. Moreover, the amount of information which can be transmitted over the TV screen has increased dramatically in recent years. Future needs will dictate the necessity for pictures of higher definition than the present capabilities of the NTSC, PAL, and SECAM systems.

The development of high-definition television has been guided by studies of the relationships among three main parameters: picture size, number of scanning lines, and visual impact. Both a film simulation method and a real television system have been used (Hayashi, 1981). One proposed future system uses 1125 scanning lines, 60 fields per second, and an aspect ratio of 16:9. The bandwidth of the color-component signal of this 1125-line system is more than 5 MHz, and no recognizable deterioration in line-sequential transmission of $R - Y$ or $B - Y$ is noted. The signals C_W and C_N are obtained by slightly amending $R - Y$ and $B - Y$, and are used as color-component signals (Fujio, 1978). The major characteristics are listed in Table 2.1.

2.1.2 Broadcast video tape recorders†

As the television networks in the United states expanded, the handling of time differences from the east to the west coast of the American continent emerged as a serious problem. The development of a television recording device would alleviate this problem, and several such developments were undertaken using fixed-head mechanisms in the 1950s (Abramgon, 1973; Sugaya, 1986). These development activities were directed specifically toward the recording of video signals using a type of tape-transport mechanism similar to that used in audio tape recorders but with faster tape speeds (Axon, 1958). Multichannel recording was used in order to reduce the recorded frequencies to as low a range as possible (Mullin, 1954). None of these methods produced significant results because the immature technology at that time resulted in a recorded signal that had both too short

---

†For specifications, see the Appendix, Table 2.10.

a wavelength and severe time-base instability; it also had an unrealistically short recording time of at most 15 min (Olson et al., 1954, 1956).

In 1956, Ampex Corporation announced their newly developed rotating head video recorder using 2-in-wide tape (Ginsburg, 1956; Snyder, 1956). Rotating-head technology afforded a high head-to-tape speed of about 38 m/s (1500 in/s), which made it possible to record sufficient bandwidth of the video signal by FM. A longer recording time (over 1 h) was achieved while keeping the tape speed at 38 cm/s (15 in/s), which was the same as that of audio recorders then in broadcast use.

The rotary-head recorder employed a transverse format in which recording was done by four heads mounted on a rotating drum; this was also called the *quadruplex head* (see Fig. 2.3). The original quadruplex system recorded only monochrome. In order to introduce color, it was necessary to shift the FM carrier frequency to a higher range without changing the recorded format on the tape. This became possible with the further development of magnetic recording technology and yielded high-band recording in 1964 and super-high-band recording in 1971.

The quadruplex (or quad) system was a significant innovation in recording technology, but this system as developed at the time was still far from ideal. Each transverse track on the magnetic tape could record only one-sixteenth of one field, and complicated switching was required to reassemble a complete field. Any amplitude or phase differences between the picture segments produced an effect called *banding,* to which the eye is especially sensitive. Moreover, tape consumption was high so that the tape cost per hour, as well as tape storage cost, was correspondingly high (Kazama and Itoh, 1979).

In order to record one field continuously on the tape, it is necessary to lengthen the video track; one way to do this is to record the video track on the tape diagonally. To achieve this, the tape must be physically wrapped around the rotating-head drum in a helical-shaped tape path (Fig. 2.4). This *helical-scan* recording method can record one complete field on a single track, with the result that each field can be switched in

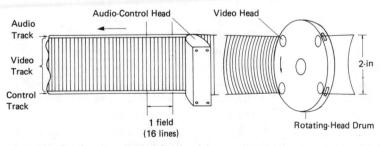

Figure 2.3 Quadruplex rotating-head video recorder and transverse recording format.

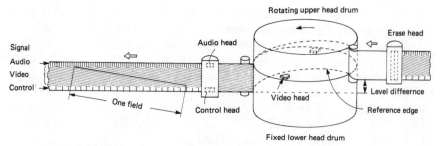

Figure 2.4 Two-head helical-scan video recorder and recording format.

during the vertical blanking period without the necessity for any complicated circuitry. When helical-scan recording was first introduced, the technology of magnetic recording was still primitive, and the helical-scan method needed a head drum of overly large diameter (Sawazaki, et al, 1960). In addition, if a helical-scan recorder designed for the PAL or SECAM system (50 Hz) was adapted for use with the NTSC color system (60 Hz), the head drum diameter had to be increased by 20 percent to achieve the same head-to-tape speed. To solve this problem, the segment sequential-recording format was developed in 1966. In this format, one-fifth of one field of the NTSC signal and one-sixth of one field of the PAL signal (both of which are approximately 52 horizontal lines) are recorded on a single segmented slant track on the tape, without changing the head drum diameter and the tape speed. Thus the head drum rotating speed could be kept constant in order to achieve the same relative head-to-tape speed and could be synchronized to the field rate. This segmented recording method was developed for industrial use with 1-in-wide tape, and the same idea was marketed in 1976 for broadcast use (type B recorders) (Zahn, 1979).

Nonsegmented helical-scan recorders were first used exclusively for industrial and institutional applications but, in 1978, a 1-in-wide tape system for broadcast use was developed (type C recorders) (Alden, 1977). Magnetic recording technology is continuously increasing the recording density and it is now possible to record the PAL and SECAM signals on type C machines without changing the head drum diameter and also to use considerably smaller-diameter head drums and smaller reels.

All commercially marketed recorders used a single track for the video signal until 1982, when the M format was introduced (Arimura and Sadashige, 1983). This uses two video tracks (one for the Y signal and the other for the I and Q signals of NTSC), with the same diameter of head drum and cassette as the half-inch format now in wide use throughout the world (video home system, or VHS). A similar development, known as the L format, also uses two video tracks (one for the Y and one for the R − Y and B − Y signals) and the cassette is mechanically compatible with the Beta half-inch format. The L format, using digital technology, compresses sequentially the R − Y and the B − Y signals to half their length. Other broadcast-use formats have been developed, such as the Quartercam (Reimers et al., 1985) and M-II (Sekimoto et al., 1986).

2.1.3 Home video tape recorders†

Apart from commercial broadcast needs, market demands developed for home, industrial, and institutional video recorders that could utilize narrower bandwidths than those required in broadcasting use. Home video tape recorders were also originally designed on a stationary-head basis, but the recording time, time-base instability, and video bandwidth proved to be insufficient (Fisher, 1963). Endless-loop systems were also tried but without commercial success (Sawazaki et al., 1979; Sadashige, 1980). As a result, rotating-head methods also became the standard in recording television signals for the home.

The history of home video recording has witnessed a series of significant reductions in both head drum diameter and tape width. In 1964, $\frac{1}{2}$-in-wide tape was first used to record a 1-h program on a 7-in-reel tape recorder with a two-head helical-scan method. The particular method used was not commercially acceptable since it relied on the "field-skip" process, which records one field, skips the next field, and then plays back the same field picture twice using the two heads. Even though it was possible with this method to record a full 1-h program, the picture quality deteriorated owing to a staccato picture effect during fast motion. Shortly after this a $\frac{1}{2}$-in recorder was developed that could record a complete-frame picture, although the recording time was limited to 40 min.

In order to solve the playing-time problem, azimuth recording was developed (Okamura, 1964; Sugaya et al., 1979). This technique involves the use of two heads with differently inclined gaps ($\pm\phi$), as indicated in Fig. 2.5. When head A cross-tracks into B, the resultant difference in angle is 14° (2ϕ). This produces an azimuth loss of sufficient magnitude to enable guard bands between tracks to be eliminated. A black-and-white video recorder using the azimuth-recording technique was first developed by Matsushita/Panasonic in 1968 and provided a 90-min playing time on a 7-in reel. The azimuth recording method has been a key element in the growth of home video tape recording.

The Beta format, which utilizes the azimuth recording method and a small cassette, was introduced by Sony in 1975 (Kihara et al., 1976), followed a year later by the VHS format of JVC (Shiraishi and Hirota, 1978),

---

†For specifications, see the Appendix, Table 2.11 (NTSC) and Table 2.12 (PAL/SECAM).

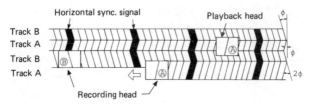

Figure 2.5 Azimuth recording principle (overwrite recording) *(Sugaya, 1986)*.

also using an azimuth-recording cassette system. In 1980, a fourth azimuth recording format, the V-2000, was introduced by Philips mainly for use in the European market. A fifth azimuth recording format, the 8-mm video format, was established by international agreement at the 1984 "8-mm Video Conference."

Based on innovations in magnetic recording technology, which have mainly affected recording heads and recording media, the Beta and VHS formats have been able to double and triple recording time by reducing the video-track pitch to one-half and one-third (Iijima et al., 1977). As a result, the tape cost of video recording per hour has become slightly less than that of audio recording on tape. The development of new recording media and improved heads have led to still higher recording densities; these in turn will make possible still smaller recorders approaching the size and weight of the modern 8-mm movie camera (Morio et al., 1981; Mohri et al., 1981).

2.1.4 Continued increases in recording density

Accompanying innovations and improvements in magnetic tape and heads, the areal recording density has increased and the tape consumption per hour has decreased exponentially over a 25-year period, as shown in Figs. 2.6 and 2.7 (Sugaya, 1982). Tape consumption is calculated simply from tape width and speed, and so includes audio and other signal tracks as well as video. The areal recording density of video disks is shown in Fig. 2.6 for comparison purposes. The 2-in quadruplex recorder was the first successful machine. The major trend of development has been toward helical-scan recorders, first for industrial and institutional use, later for home use. Currently "camera video" captures a portion of the 8-mm movie market. In broadcast recording, type C and type B helical-scan recorders, which have almost the same picture quality as quadruplex super-high-band recorders, have greatly improved tape consumption and are used widely throughout the world today. Quadruplex production stopped in 1980. The M and L formats were developed to meet the requirements of electronic news gathering.

Figure 2.7 shows that, since 1968, azimuth recording has strongly contributed to decreasing the video track pitch. The development of such recording media as higher-coercivity cobalt-absorbed oxide tape and alloy tape has played a significant role in reducing the minimum recordable wavelength. These trends are important in forecasting the future of video recording.

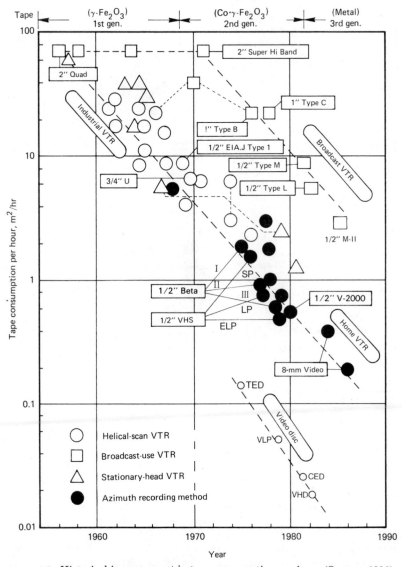

Figure 2.6 Historical improvement in tape consumption per hour *(Sugaya, 1986).*

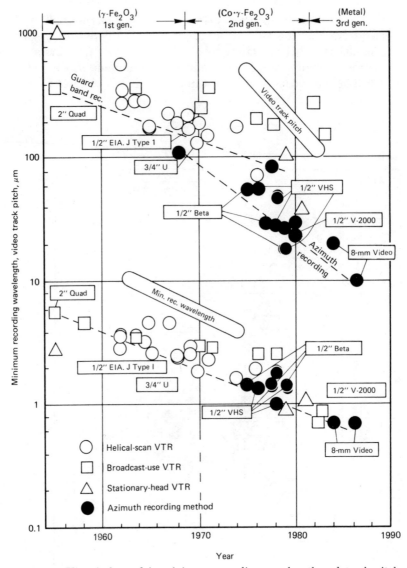

Figure 2.7 Historical trend in minimum recording wavelength and track pitch *(Sugaya, 1986).*

2.2 Principles of Video Recording

2.2.1 Problems in video tape recording

The primary difference between video and audio signals is in the highest frequency required, as shown in Fig. 2.8. Video requires more than 100 times the highest frequency of audio. The other significant difference is the continuity of the signal. An audio signal is continuous, while the video

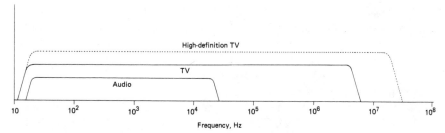

Figure 2.8 Frequency bandwidths of audio, regular television, and high-definition television signals.

signal is segmented into separate fields similar to movie film. In order to record a television signal on tape, it is necessary to achieve continuous recording of at least a single picture field with sufficient bandwidth and time-base stability.

The transverse rotating-head recording method was developed to achieve sufficient relative head-to-tape velocity without increasing the tape speed itself (Marzocchi, 1941). The inertia of the rotating-head drum also alleviates time-base instability problems. The helical-scan rotating-head method is a more attractive approach because it records an entire-field picture signal on tape (Masterson, 1956). There are, however, some specific advantages to the transverse-type rotating-head recording method: the higher rotating speed of the head drum results in less instability; the track pass is shorter so that the recording is affected less by tape flexibility; finally, the tape is supported in a cupped arc by a rigid tape guide, resulting in considerably improved stability and reduction of errors.

There are a number of recording and playback losses related to wavelength, as shown in Fig. 2.9 (for more details refer to Chap. 2 of Volume

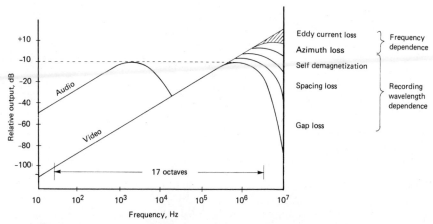

Figure 2.9 Types of losses in magnetic recording and playback with an inductive head.

I). Also, because of the use of inductive heads, the low-frequency signal has poor signal-to-noise ratio, which is a serious problem for video recording with its characteristically wide bandwidths of approximately 17 octaves. The lower frequencies largely determine the contrast of the picture; thus, the signal output must be very stable, otherwise the reproduced picture will be seriously spoiled by streaking noise. Frequency-modulation recording of the video signal was introduced to solve these problems (Anderson, 1957). The audio signal, on the other hand, being restricted to relatively low frequencies, can be easily recorded using ac bias with almost no distortion and can be equalized into a flat frequency response.

2.2.2 Transverse-scan recorders

The mechanical format of the quadruplex system has already been discribed (Fig. 2.3). By rotating the head drum at a speed of 14,400 rpm, an immediate increase of head-to-tape relative speed to 38 m/s (1500 in/s) was achieved, and the time-base instability was reduced by the rotational inertia of the drum. It then remained to develop a suitable signal-modulation scheme. Amplitude modulation was discarded because of large amounts of low-frequency noise. Frequency modulation proved successful because it does not have the same sensitivity to low-frequency noise effects. According to the traditional understanding of frequency modulation, the carrier frequency should be set at a much higher frequency than that of the signal to be transmitted. In magnetic recording, however, the

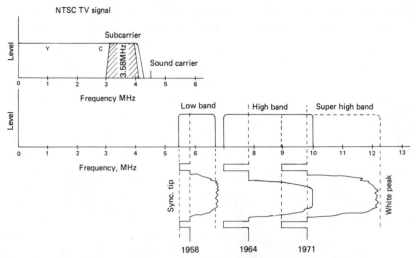

Figure 2.10 FM recording spectrum for NTSC broadcast quadruplex recorders.

highest frequency is limited by both the head-to-tape speed and the minimum recording wavelength, and the highest frequency of modulation is also limited. Thus, the modulation index, which is the modulated frequency deviation divided by the signal frequency, must be less than one. This is called *low-index FM recording.*

At the early stages of transverse-scan recording, the technology available did not permit the recording of the higher frequencies necessary for a FM video signal. As a result, the FM range was located relatively far down in the frequency scale. This was called *low band* and was used until about 1964 (Fig. 2.10). Later, it became possible to realize sufficient FM deviation in higher frequency ranges. This technique, known as *high band,* was responsible for considerable improvement of picture quality. Improvements in head and tape technology and in head drum air-bearing construction, which reduced the high-frequency jitter component, resulted in *super high band* which was introduced in 1971. This made it possible to obtain sufficient deviation without moiré at still higher frequency bands, and a picture of nearly perfect quality could now be recorded.

2.2.3   Helican-scan recording for broadcast use

Experience with the transverse-scanning recording system revealed some difficulties because the process involves splitting a single field into sixteen segments and stringing them out in a line to record them on the magnetic tape. Thus, when the video signal is reproduced, it is essential to keep switching in the horizontal synchronization period so that there is no switching noise in the reproduced picture (Dolby, 1957). It also proved necessary to readjust the distorted picture in order to compensate both for wear on the video heads and for distortion caused by tape interchange (Benson, 1960). On the other hand, if a single-field image is recorded in a single scan, the circuitry can be greatly simplified. As a consequence, helical-scan systems were proposed and developed in the 1950s (Schüller, 1954; Sawazaki, 1955; Masterson, 1956; Sawazaki et al., 1960).

In the beginning stages, the helical-scan technique was conceived of as using a single video head in the head drum and wrapping the tape around the drum in such a manner that the signal is recorded in a slanted or diagonal locus on the surface of the tape. However, since this locus spans the tape surface from edge to edge, no space remains for the recording of the necessary audio and control signals. The two-head helical-scan system (Fig. 2.4) was devised to compensate for this defect (Tomita, 1961). Since the early helical-scan prototototypes were intended for broadcast use, they employed a large head drum in order to obtain a head-to-tape relative speed comparable to that of the transverse-scan recorder. This, in turn, necessitated running the flexible magnetic tape in such a long and complex path that an unacceptable degree of time-base instability and tape skew resulted. Consequently, this system proved not to be competitive with the quadruplex system, which had already firmly established itself in the market.

It is characteristic of the helical-scan system that the diameter of the head drum must be increased in proportion to the number of heads in the drum in order to maintain the same head-to-tape relative speed, as is shown in Fig. 2.11. For this reason, it was generally not considered feasible to increase the number of heads for broadcast use because of the high head-to-tape speed required to ensure sufficient picture quality. In response to this, the 1.5-head system was invented; in this system an auxiliary head is used exclusively to record the synchronizing signal on the extreme edge of the tape (Fig. 2.12). This compensates for the portion formerly lost on both edges of the tape, which was the basic drawback of the single-head system. This 1.5-head system was introduced in 1962 in a 2-in tape format for industrial use (Kihara, 1963; Suzuki et al., 1979). Subsequently, further advances in magnetic recording technology, especially in the area of higher-coercivity tape, made it possible, without any loss in broadcast picture quality, to develop a 1-in tape recorder using this principle; it needed less than one-third the amount of tape that was required in the 2-in transverse-scan recorders.

Advances in semiconductor and large-scale integration (LSI) technology have led to video recording equipment reduced in both size and weight, with a picture quality and time-base stability sufficient for broadcast purposes. As an example, the $\frac{3}{4}$-in U-format recorder began to replace movie film as the medium for electronic news gathering in about 1974. A much smaller and more easily portable recorder, the M-format, which uses VHS cassettes, was developed in 1982 for the same purpose. The U-format transforms the NTSC color subcarrier to low frequencies on the order of 700 kHz by the color-under system (see Sec. 2.3.3). The M-format (component recording) has a two-channel track: one channel is used for

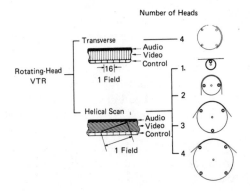

Figure 2.11 Various types of rotating-head video tape recorders (Sugaya, 1986).

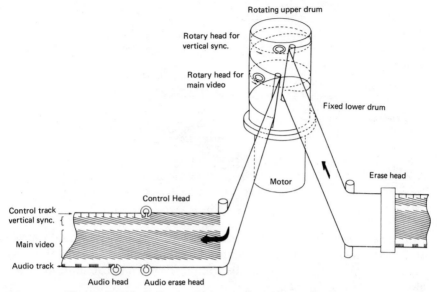

Figure 2.12 The 1.5-head helical-scan video recorder and recording format.

the FM recording of the luminance signal Y; the other is for the FM recording of the chrominance signals I and Q (Fig. 2.13). In this way, not only can an adequate frequency band be assigned to the color signal, but the Y signal can be recorded separately with no disturbance of the chrominance signal. In a variation of the M-format, one track provides for FM recording of the Y signal, and the other employs a system for alternately recording the R − Y and B − Y signals by compressing the frequency band. This system achieves both high resolution and a high signal-to-noise ratio by providing sufficient FM modulation. A comparable recorder, the

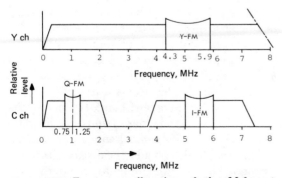

Figure 2.13 Frequency allocation of the M-format recorder for electronic news gathering.

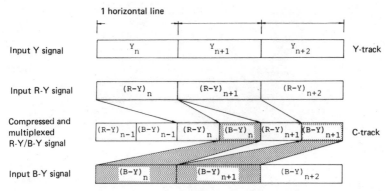

Figure 2.14   Time compression scheme of type L video recorders.

L-format, which uses Beta cassettes, was introduced in 1983, and the M-II format appeared in 1985 (Fig. 2.14). With the M or L formats, the television camera and recorder can be combined into a single unit, ideal for electronic news gathering.

In summary, the helical-scan format has become the mainstream format for broadcasting use. It is anticipated that, even though there will doubtless be further significant advances in signal-recording methods, such as two-channel recording (Y and C) and digital recording, helical-scan recording methods will continue to be relied upon heavily in the future.

2.2.4   Helical-scan recording for home use

One of the most important points to consider in the development of consumer video recorders is that the trade-off between picture quality and tape consumption should be brought to a reasonable compromise. A significant step in decreasing the tape consumption came from the increased track density achieved by eliminating the guard bands. Normally, guard bands are required to protect against disturbances from adjacent tracks. If a video signal is recorded by one machine, and then subsequently played back on a different machine, the video head upon playback does not necessarily follow precisely along the track previously recorded on the tape by the first machine. This is due to deviations arising from both machine and tape tolerances. The wider the guard band, the better the tape interchangeability; but on the other hand, guard bands will increase tape consumption. As a compromise, the guard-band width was conventionally fixed at half of the video track width on all helical-scan recorders.

Azimuth recording was developed in order to eliminate the guard band in two-head helical-scan machines (Fig. 2.15). To prevent the inadvertent reading of adjacent tracks on playback, the head gaps have different angles ($90° \pm \phi$ with respect to the track), such that the azimuth loss is given by

$$L_{az} = 20 \log_{10} \frac{(\pi w/\lambda) \tan 2\phi}{\sin\left[(\pi w/\lambda) \tan 2\phi\right]} \qquad \text{dB} \qquad (2.1)$$

where λ = recorded wavelength
$\quad w$ = video track width
$\quad 2\phi$ = the angular difference between the gaps

When the discrepancy between the video track and playback head is Δw, the ratio between the crosstalk C and the signal S is as follows:

$$\frac{C}{S} = 20 \log_{10} \left\{ \frac{\sin\left[(2\pi \Delta w/\lambda) \tan 2\phi\right]}{(2\pi \Delta w/\lambda) \tan 2\phi} \frac{\Delta w}{w - \Delta w} \right\} \qquad \text{dB} \qquad (2.2)$$

The dependence of crosstalk on frequency for typical values of tape speed, track width, and azimuth angle is shown in Fig. 2.16.

When the first azimuth recording system was developed in 1968, the minimum recordable wavelength was about 2.5 μm, and the azimuth angle was 30°. When ϕ is large, the effective head gap length will increase by $1/\cos \phi$, and the reproduced output will naturally decrease. Moreover, the reproduced signal from the video head will have a time discrepancy of $\Delta w \tan \phi$ if the video head deviates by Δw from the previously recorded video track, and the upper part of each picture played back will be distorted. As the minimum recordable wavelength becomes shorter, smaller azimuth angles can be used, and both sufficient separation and a tolerably low level of crosstalk will still be achieved. Azimuth angles of $\pm 6°$ are used in VHS recorders. With angles this small, the main disadvantage of azimuth recording is eliminated. Additionally, the azimuth recording method easily allows both recording and playback at one-half or one-third speed without the necessity of changing the heads. This is achieved by

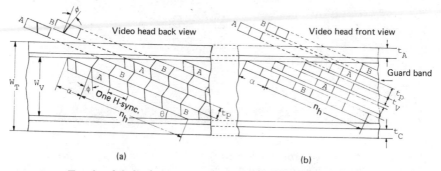

Figure 2.15 Two-head helical-scan tape format. (*a*) Without guard bands (azimuth recording); (*b*) with guard bands.

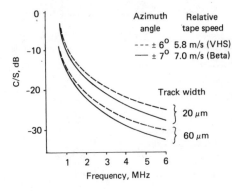

Azimuth   Relative
angle   tape speed

--- ± 6° 5.8 m/s (VHS)
— ± 7° 7.0 m/s (Beta)

Figure 2.16 Frequency characteristics of crosstalk caused by
azimuth loss in the azimuth
recording method (VHS and
Beta).

means of overwrite recording, in which recording by a wider head partly
erases the adjacent track (Fig. 2.5). This was not possible with previous
guard-band recording methods.

2.2.5 Design principles of helical-scan recorders

Since the recorded video track of helical-scan machines is much longer
than that of transverse-track machines, the playback head may inadvertently trace not only the original track but also an adjacent track at the
same time. This is particularly true if there is any nonlinearity in the
video track, as may occur when the record and playback machines are
different or if the playback video head has shifted position from the original track. The horizontal synchronization signal of adjacent tracks thus

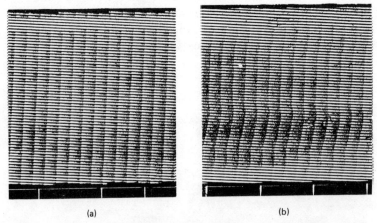

Figure 2.17 EIA-J type I (fixed-head drum) tape format and the jitter pattern
displayed visually by means of fine iron powder. (*a*) Recorded by a standard-
alignment tape recorder (small jitter component); (*b*) recorded by a conventional VTR (significant jitter component).

needs to be aligned in order to eliminate the influence of any interference between the two adjacent tracks. This is called *horizontal sync pulse alignment* and is illustrated diagrammatically in Fig. 2.15. Figure 2.17 shows the recorded tape pattern, made visible by applying very fine, colloidal iron powder directly to the recorded video tape with a special solvent and then evaporating the solvent.

Horizontal sync pulse lineup requires a special relationship between the tape speed V_m, the head-to-tape speed V_h (which sets the drum diameter d), and the helix angle θ_0. The tape speed is given by

$$V_m = \frac{\pi d}{2} \frac{\alpha}{(n_h \pm \alpha)} \frac{F}{\cos \theta_0} \tag{2.3}$$

where F = field frequency (60 or 50 Hz)
    n_h = number of horizontal sync pulses on one video track
    α = number of horizontal sync pulses between the discrepancy of two adjacent track edges

The term n_h is 262.5 for NTSC and 312.5 for PAL and SECAM. The term α is expressed by an integer plus 0.5 of the horizontal sync, as shown in Fig. 2.15.

When the tape is moved, the recorded video track angle θ is the resultant of both head and tape movements:

$$\sin \theta = \frac{\sin \theta_0}{\sqrt{1 \pm 2 \, (2V_m/\pi Fd) \cos \theta_0 + (2V_m/\pi Fd)^2}} \tag{2.4}$$

(Note: If both the rotating head and the tape run in the same direction, − is used; + is used for opposing directions.)

These design principles can be used to establish the tape format. In practice, standard-alignment tapes are required to ensure interchangeability among mass-produced machines (Sugaya et al., 1974).

2.3   Video Recording Technology

2.3.1   Drive mechanisms

2.3.1.1   Helical-scan mechanisms.   The main design difference between video and audio tape recorders is in the use of a rotary head in the video recorders. The essential points to be considered in the mechanical system are thus the construction of the rotary head and the manner in which the tape is wound around the head drum when it is mounted on the machine. As shown in Fig. 2.11, one to four heads are used in rotary-head designs; since there are actually no fundamental differences between any of the various competing systems, the description here will concentrate on the most commonly used two-head system. The signal is recorded on the tape at a small angle by the rotary head and, to achieve this, the tape must be wound at a slant to the head motion around the rotary-head drum. Wrapping the tape in a slanted path means that the takeup and supply reels must be on different levels with respect to the head, as shown in Fig. 2.4. This level difference requirement complicates the construction of the tape-transport mechanism. There are three main types of head drum construction (Fig. 2.18). With the fixed drum shown in Fig. 2.18a there is considerable tape friction, however the tape movement is stable and if alignment of the bottom reference edge of the tape is necessary, a tape guide can be attached to the upper part of the drum. Recently, tape surfaces have become smoother, providing better short-wavelength recording; unfortunately, this often worsens tape-running characteristics (stick-slip), causing tape-running jitter with the fixed drum. The rotating upper drum shown in Fig.2.18b drags in air, forming an air bearing between the tape and half of the upper drum; thus the friction coefficient becomes very low and is unaffected by tape surface characteristics. However, starting up with this format is difficult when the tape is already in the wound condition. Also, if the tape or drum surface is dirty or damp, the tape tends to stick to the drum. This can result in jerking and other problems at high speed.

For the rotating upper drum, the four possible combinations of tape-running direction and the drum-rotating direction are shown in Fig. 2.19. Normally, the type in Fig. 2.19a permits the lightest tape transport. The tape tension in the horizontal direction here is regulated so that the tape runs along the lower fixed-drum reference edge, but this places stress on the tape. In the types in Fig. 2.19c and d, the air intake between the tape and drum tends to be insufficient to cause the necessary tape float, requiring that a fine groove be undercut into the rotating drum surface.

In all versions of the rotating-upper-drum method, it is comparatively difficult to push the tape with precision from the upper to the lower drum

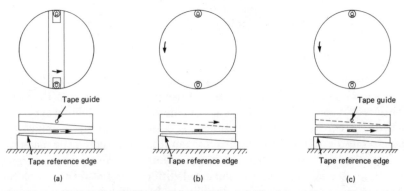

Figure 2.18 Various head drum configurations. (a) Fixed-head drum; (b) rotating upper head drum; (c) rotating intermediate head drum.

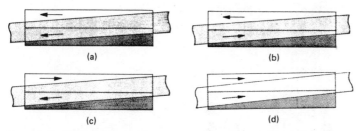

Figure 2.19 Types of upper drum rotation.(*a*) A type ($\frac{3}{4}$-in U, $\frac{1}{2}$-in VHS, $\frac{1}{2}$-in V-2000, and 8-mm video formats); (*b*) B type; (*c*) C type; (*d*) D type ($\frac{1}{2}$-in Philips tandem reel).

reference edge because the upper drum is rotating. A better system is the rotating intermediate drum shown in Fig. 2.18*c*, which combines the best of both the fixed-drum and rotating-drum types. This system is used in both the type B and Beta formats. Since about one-third of the half-inch tape is in contact with the rotating drum in the Beta rotating-intermediate-drum system, there is sufficient air-bearing effect due to the resultant air flow. As the upper drum is fixed, the tape guide can push the upper end of the tape against the reference edge of the lower drum. The disadvantages of this system, however, are its structural complexity and relatively high cost, plus the difficulty of head drum replacement.

2.3.1.2 Tape-loading mechanisms. In general, a container with two reels is called a *cassette* and one with only a single reel, a *cartridge;* video cartridges have more or less disappeared from the market. When the tape is in a cassette (or cartridge), it is necessary that the tape be physically pulled out from the container and wound around the rotary-head drum (Ryan, 1978).

Tape-loading mechanisms can be classified into various types, as shown in Fig. 2.20. In the early years, the large tape reel size prohibited having two reels in a single container. Initially, the one-reel cartridge was used not only for professional but also for home recorders. The two disadvantages of the cartridge are that the tape cannot be removed from the machine until the rewind is completed, and there is a risk of damaging the exposed end of the tape.

As magnetic recording technology developed, reel size diminished, and two reels could be put into a single container. Thus the cassette gradually came into use. There are two types of cassettes: one is the tandem cassette in which the two reels are stacked over each other coaxially; the other is the parallel two-reel cassette. Since the supply and take-up reels in the tandem cassette are on different levels, the tape obviously has to travel in a diagonal path from one reel to another. Although this diagonal tape path is easily adapted to wrapping around the head drum, as required for hel-

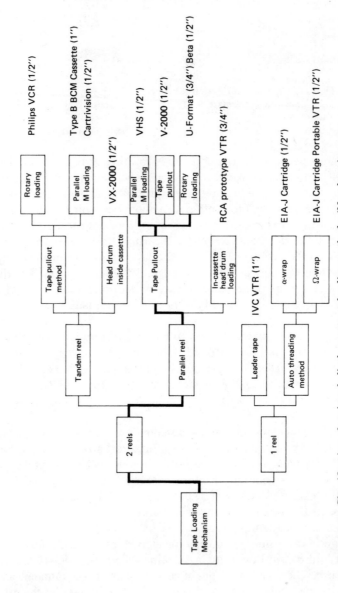

Figure 2.20 Classification of various helical-scan tape-loading methods. (Numbers in parentheses indicate tape widths.)

ical-scan recording techniques, the more complex tape path here necessitates a correspondingly more complex reel drive mechanism. Moreover, the shape of the cassette is rather more bulky than that of the parallel two-reel cassette, which can be manufactured in a convenient size and a shape similar to an ordinary book. The tandem cassette was commercially used for a few models, but is now obsolete in consumer recorders.

The parallel two-reel type (here simply called a cassette) can be classified into two types depending upon whether each reel has flanges or not. The double-flanged reel is somewhat bigger but protects the tape from shock or undue stress. Whichever of these versions is used, a special device is necessary to compensate for the complex path the tape takes as the result of being pulled out of the cassette and wrapped around the head drum in helical-scan recording. As shown in Fig. 2.21, this device consists of cylindrical inclined posts aligned to be perpendicular to the tape-running direction at the position of the posts. In effect, the head drum itself also serves as a kind of inclined post as the tape is wrapped around it. A minimum of two of these inclined posts is necessary to compensate fully for the twisted tape of the helical-scan method. These are in practice realized as fixed posts fabricated from a nonmagnetic material with a smooth surface and a low coefficient of friction. Ideally, they are positioned so that they come in contact with the reverse side of the tape, not with the magnetic recording surface. An inclined post is used for VHS parallel loading. The U- and Beta-formats, on the other hand, use a rotary loading mechanism that compensates for the twisted tape path in a gradual manner by using inclined posts. Several different types of tape-loading mechanisms have been developed. Note the heavy line in Fig. 2.20 showing the mainstream solutions to this problem.

The various tape-loading mechanisms will be explained using concrete examples (Fig. 2.22). The coaxial tandem-reel system is shown in Fig. 2.22a. The reel is driven by a double-core coaxial shaft. Since the level differences inherent in helical scanning can automatically be handled by using the coaxial tandem-reel system, where one reel is above the head

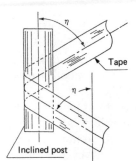

Figure 2.21 Inclined-post tape guide.

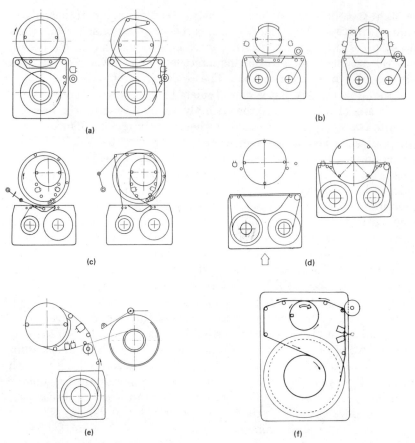

Figure 2.22 Various examples of tape-loading mechanisms using cassettes or cartridges. (*a*) Rotary loading with two tandem reels; (*b*) parallel loading with two parallel reels; (*c*) rotary loading with two parallel reels; (*d*) loading of in-cassette head drum using a cassette with two parallel reels; (*e*) α-wrap leader-tape loading with a single-reel cartridge; (*f*) in-cassette single-head-drum loading with two tandem reels.

and one is below, tape distortion can be minimized more easily in this manner than in parallel-reel types. Another widely used method for wrapping the tape on the drum is that in which the tape is pulled out by two pins, as shown in Fig. 2.22*b*. The correction of the tape twist in such a small space as this is technically difficult to achieve, but this design has the advantages of being able to produce a machine of very small size and of being able to load the tape from the cassette quickly. The rotary-loading methods shown in Fig. 2.22*c* have a longer tape pull-out length; therefore, it is possible to correct any tape twisting, which inevitably accompanies the slanted scanning format, gradually and with no abrupt bends that might place undue stress on the tape. The tape, however, is pulled out to a considerable length from the container in this method; various problems in tape handling can therefore arise, such as requirements for an overly large loading space or slow loading speed. From this point of view, the parallel loading system shown in Fig. 2.22b is a good compromise with respect to tape loading, and the tape-transport unit can be very compact. Sudden bends around the two pins may cause some difficulties during tape transport in the system.

Because of such problems as these, which arise from pulling the tape out of the container and winding it on the drum, various methods for loading the tape on the head drum while it is still in the container have been considered. One of these is the method shown in Fig. 2.22d. To facilitate tape loading, four heads on a large drum are used. The tape wrapping angle is small, therefore the degree to which the head enters the container is also small. Such a system has the advantage that no special mechanism is necessary for winding the tape on the drum, but since it requires a tape-twisting correction mechanism to correct tape level differences inside the tape container, the distance available for correcting the end is even less than is true with the type of mechanism shown in Fig. 2.22b. There are still many difficulties with this approach, such as accuracy problems (due to greater head drum diameter), the problem of minimizing the functional deviations among the four heads, and correction of tape twisting inside the small space of the cassette. Therefore, such systems are not on the market. In contrast, methods have been variously proposed in which the drum diameter is minimized by using only one head and in which the head drum enters the container, as shown in Fig. 2.22f. However, in spite of the simplicity of the one-head design, the container size is too bulky and tape consumption is prohibitively high, and it has been abandoned.

Single-reel cartridge recorders use a method in which only the reel with tape wound on it is placed inside the tape container; the take-up reel is located in the recording machine, and a thick leader tape is used for auto-threading (Fig. 2.22e). This method has the advantage that there are no unnatural twists in the tape, and the optimum take-up-reel location suited to slanted winding is possible, although the mechanism is dimensionally complex and rather large in size (IEC standard, 1983).

An innovative video cassette, developed for use with 8-mm video, has sufficient space to accommodate the guide pins necessary in parallel or rotary loading (Fig. 2.23). It also uses a special design that affords full protection of the tape from the environment, a particularly important feature when high-density tapes, with their extremely smooth surfaces, are used.

In summary, there are two fundamentally different methods of designing a tape transport: one method in which the tape container itself is relatively simple and all of the necessary mechanisms are located on the recorder unit (Fig. 2.23), and another in which the recorder unit has been

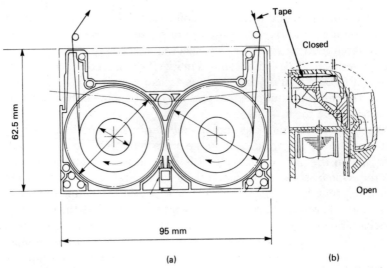

Figure 2.23 The 8-mm video tape cassette. (*a*) Cassette design; (*b*) lid construction.

simplified by including some of the main parts of the mechanism in the tape container itself (Fig. 2.22*d*). In general, the first of these types, having the relatively simple cassette and the correspondingly more complex recorder unit, appears to be more suited for mass production.

2.3.1.3 Miniaturization techniques. An exponential decrease in size and weight of consumer recording machines has occurred without change in the basic design (Fig. 2.24). Miniaturization of the head drum is the key

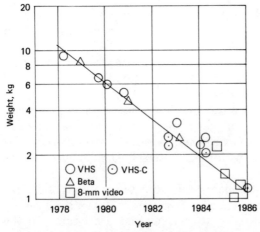

Figure 2.24 Historical decreases in the weight of video recorders (weight of battery included).

problem, since head drum interchangeability must be maintained at all costs. One possible solution to this problem, shown in Fig. 2.25a, is the approach taken by the Beta movie recorder, which uses a double-azimuth head unit (see Fig. 2.43f) with a 300° wrap (Sato et al., 1983). The two heads with different azimuth gap angles are mounted in line. This arrangement reduces the head drum diameter by three-fifths (the ratio of 180°/300°). A drawback to this solution, however, is that recorders based on this principle can record only signals generated by a special television camera designed for this purpose, although tapes recorded with this method can be played back on any regular Beta recorder.

Another approach is the four-head, 270°-wrap method used in the VHS movie recorder (Fig. 2.25b). This device has the advantage of being able to record and play back any type of standard video signal. The construction of the head drum is relatively complex, however, and requires quite fine tolerances. With this method, the head drum diameter can be reduced by a ratio of 180°/270° or two-thirds.

Yet another miniaturization is the one used in the VHS compact (VHS-C) cassette. This method maintains the interchangeability of the original cassette, which can be used with a standard VHS machine, by means of a simple adapter (Fig. 2.25c). The VHS-C cassette can record and play back a 20-min program (standard mode) or a 60-min program (extra-

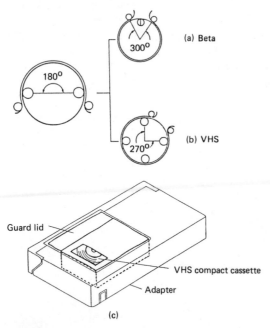

Figure 2.25 Mechanisms for miniaturization: (a) Beta-movie; (b) VHS-movie; (c) VHS-C cassette and adapter.

long-play mode) on VHS-C recorders (or on VHS equipment with an adapter.).

2.3.2 Tracking techniques

With the older transverse-scan recorders, the problems arising from machine-to-machine interchangeability are different from those encountered in helical-scan machines, particularly with respect to head-tip projection (Benson, 1960). The mainstream of video recording research has already shifted to the helical-scan format. For this reason, the tracking techniques discussed in this chapter are limited to those of the nonsegmental type of helical-scan machines.

In NTSC helical-scan recorders, it is necessary to record 60 tracks per second, each containing one field of the video signal. Each of these tracks must be preceded by a vertical synchronization (sync) signal; at the same time, a control signal is recorded on the edge of the tape by a separate stationary head (Fig. 2.4). The control signal is a reference signal used to locate the video head at the proper position on the tape; it is similar in

TABLE 2.2 Tracking-Servo Methods

	Drum servo	Drum-capstan servo
Tape speed	Constant (record and playback)	Variable, according to servo (playback only)
Head drum rotational speed	Variable, according to servo	Constant, according to servo
Jitter (picture)	Slight at low frequency	Small
Wow and flutter (sound)	Small	Slightly less than drum servo
Method	1. Servo directly coupled to head	1. Separate servos directly coupled to each motor
	2. Servo by belt extension	
	3. Servo by stepless speed change	
Tracking	Adjusted by head drum rotation	Adjusted by tape speed
Features	• Single motor possible	• Two motors necessary
	• Low cost	• High-accuracy tracking
	• Picture unstable during starting	• Compact (dc motor)
	• Belt absorbs excessive motor rotation	• Intersync possible
		• Electronic editing possible
		• Small jitter

function to the perforations in movie film. When a video tape is played back, the rotary-head timing is locked to the reference signal, and the tracking position is adjusted by using the recorded control signal as reference. These functions form the tracking servo system. Tracking servo systems can be classified into two types, shown in Table 2.2: drum servo systems and drum-capstan servo systems. Each of these has its own particular advantages.

2.3.2.1 Drum servo systems. Early systems used a single motor drive to both the capstan and the head drum (Figure 2.26). Because the unloaded drum speed was slightly higher than the operational one, a brake could be used to provide first frequency and then phase locking. During recording, the drum was locked to an external vertical sync; in playback, the drum and control-track signals were phase-compared. These systems were inexpensive but were limited in performance and could run at only one speed.

2.3.2.2 Drum-capstan servo systems. As video recorders become more and more miniaturized and are designed to perform an increasing variety

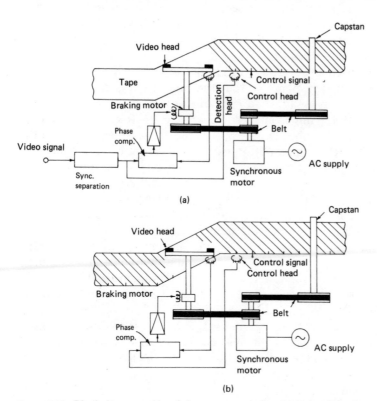

Figure 2.26 Block diagram of head drum servo system. (*a*) Recording; (*b*) playback.

of specialized functions, it is increasingly necessary to design more-precise servo systems to compensate for the lack of inertia of the numerous small motors used. Brushless dc motors and semiconductor devices are better adapted to miniaturization of high-performance recording equipment, especially for such features as perfect still pictures and variable-speed playback; these are becoming comparatively inexpensive.

The drum-capstan servo system has been used in all helical-scan broadcast recorders since about 1971. At present, most consumer video recorders use this type of tracking servo. In this system, the rotating phase is first detected by a pulse generator, which is similar to a detection transducer, and is adjusted so that the switching position in the picture is located at the front of the vertical sync signal (six to eight horizontal lines advanced position) (Fig. 2.27*a*). Then the signal from the frequency generator mounted on the capstan motor is counted and compared with the phase of the vertical sync signal. This causes the capstan motor to rotate at a constant speed. The control signal (30 Hz), which is separated from the vertical sync, is recorded on the edge of the tape, which is in turn driven by the capstan. During playback, the reference sync signal, generated by means of a countdown process from a crystal generator, and the pulse signal of the drum rotation are compared in phase and controlled so that the head drum rotates at a constant speed and in proper phase (Fig. 2.27*b*). At the same time, in order to trace the video head along on the recorded track with no deviation, the reference sync signal and the control signal are compared in phase so that the tape speed is controlled. The phase difference can be adjusted (tracking adjust) to within a very few milliseconds with the important result that machine-to-machine compatibility can be made reliable. Drum-capstan servos are widely used in video recorders both to effect high-performance miniaturization and to increase such special-effect functions as variable-speed play-back, quick search, and still pictures.

The original role of the tracking servo was mainly in phase control of the drum rotation and in tracking adjustment; velocity control was only a supplemental function. Increasing miniaturization has had a significant impact on the design of drum motors and capstan motors—they are rapidly becoming smaller and lighter. Flutter and wow are, therefore, tending to increase owing to the low inertia of these new miniaturized motors. To solve this problem, velocity control by the use of a frequency generator is becoming vital in video recording. (Fig. 2.27).

There are several methods for generating a frequency directly related to the rotation mode. One of the most common methods uses a magnetic head as a detector and a magnetically recorded signal on the rotor. The higher the generated frequency, the more precisely the rotation may be controlled; however, a higher frequency increases production costs. In order to apply velocity control effectively for this purpose, the respective motor must be coupled directly to both the capstan and the drum. This direct-drive capstan, in conjunction with velocity control by a frequency generator, has the additional advantage of being able to produce multi-function devices (stills, slow- and quick-search capabilities, along with good picture quality) as well as being highly reliable and durable. Velocity control by a frequency generator has the following three advantages: (1)

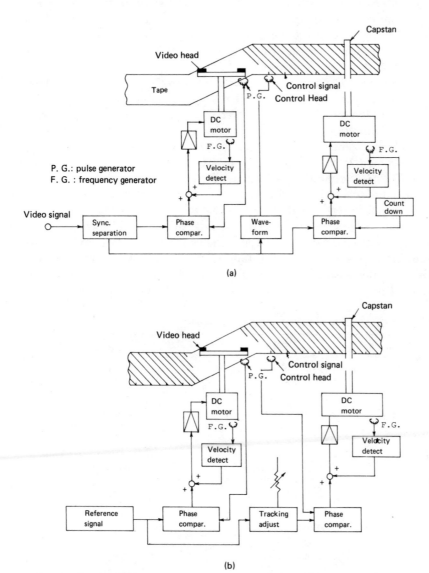

(a)

(b)

Figure 2.27 Block diagram of head drum-capstan servo system. (*a*) Recording; (*b*) playback.

improvement of damping characteristics in the phase-control servo, (2) control of rotating speed within pull-in range of the phase-control servo, (3) control of rotating speeds of the motors, the head drum, and the capstan.

2.3.2.3 Digital servo systems. Accompanying the increase in the numerous functions demanded of the video recorder is the requirement for a great many component parts for the conventional analog servo. This increase in the number of component parts, as well as the adjustments necessary to keep them in order, tends to inhibit the reliability of the machine. With appropriate application of the innovations in semiconductor devices, however, it is possible to realize such digital-process functions as velocity detection, phase comparison, and mono-multi using the clock signal in conjunction with the counter memory (Table 2.3). The fundamental operation of the digital servo process is the counting of a clock signal which is generated from a crystal generator; therefore, there is neither functional deviation nor change over time. Velocity detection is especially effective in digital servo systems. The pull-in range of the horizontal sync signal of television receivers is rather narrow (± 2 percent) although that of the vertical sync signal is somewhat wider. When the recorded picture is played back with a special function (such as still, slow, or quick-search), the drum rotating speed and the capstan rotating speed have to be adjusted so that the head-to-tape relative speed is always constant. This is possible with a digital servo system. The digital servo system was first used only in broadcast systems, but it has since been used in the

TABLE 2.3 Comparison of Digital and Analog Servos

Functions	Digital servo	Analog servo
Velocity detection	Pulse interval measured by clock pulse (no adjustment necessary)	Phase comparison with one-cycle delayed pulse (needs adjustment)
Phase comparison	Phase-difference-signal interval counted by clock pulse	Phase-difference signal stored in a capacitance
Mono-multi	Pulse generated by means of digital counter	Conventional mono-multi by means of RC circuit
Pull-in time	Fast	Slow
Stability	High, no drift	Low
Number of circuit elements	Many	Few
Circuitry	Complicated but flexible	Simple but limited
Influence of environment	Small	Large (power voltage, temperature, and running time)

consumer field, owing both to the rapid progress and cost reductions in LSIs and to the increasing demands for special functions. The quick-search function, especially, is now considered a basic function of home video recorders.

2.3.2.4 Pilot-signal tracking servos.   The control signal discussed so far is recorded on the tape edge by a fixed head which is normally located in the same housing as the audio head, as shown in Fig. 2.4. Instead of this control signal, four pilot signals, recorded on the video track along with the video signal, can be used for a tracking servo (Sanderson, 1981). In 8-mm video, these pilot signals are recorded by the video head, and the signals are located in the lower range of the color-under signal (see Fig. 2.38a). The four pilot signals are these: $f_1 = 102.5$ kHz, $f_2 = 119.0$ kHz, $f_3 = 165.2$ kHz, and $f_4 = 148.7$ kHz. These values provide the following difference frequencies: $f_2 - f_1 = f_3 - f_4 = 16.5$ kHz, and $f_4 - f_1 = f_3 - f_2 = 46.2$ kHz.

As shown in Fig. 2.28, if the video head on track f_1 plays back the signal accurately, only the signal f_1 (102.5 kHz) is reproduced. If the video head deviates to the f_4 track, it will pick up both signals f_1 and f_4 (148.7 kHz). The difference between these two signals $f_4 - f_1$ will give an error signal Δf (46.2 kHz), causing the video head to be moved back to the f_1 track immediately. If the video head shifts to the f_2 track, another error signal, $f_2 - f_1$ (16.5 kHz), will be produced, and the video head will be moved back to the proper f_1 track. Thus, the video head can be made to follow accurately a previously recorded track. This tracking servo has many features: dynamic tracking is easily carried out; no special head is necessary for the tracking signal; a conventional control track can be used for any other signals such as cueing, additional audio, or time-code signals; and finally, no separate fixed audio head is required if the audio signal is also recorded by FM or pulse-code modulation.

If the video head can move rapidly, it can follow a track of any curvature; in other words, with this servo system, equipment can be manufactured having complete machine-to-machine compatibility. In order to ensure that the video head follows very quickly on the track during play-

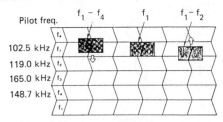

Figure 2.28   Pilot-signal tracking servo for 8-mm video recorder. (See Fig. 2.38 for frequency allocation.)

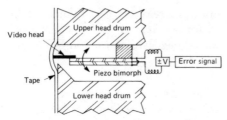

Figure 2.29 Dynamic tracking mechanism.

back, it is mounted on a piezoelectric bimorph element capable of deviat-
ing in shape by a predictable degree corresponding to the (high) voltage
of the error signal, as shown in Fig. 2.29. With this dynamic tracking servo,
the recorder can play back at any tape speed (even during quick search)
with perfect picture quality because the video head follows a complete
track at all times. However, when the quick-search function is used, a
characteristic distortion occurs between tracks in the form of a horizontal
visual noise bar, as discussed in Sec. 2.3.2.5. This can be avoided by using
a combination of the dynamic track-following method with the digital
servo. This results in excellent picture quality and the ability to adjust
the picture speed continuously from slow to fast without any of the noise-
bar characteristics typical of the digital-servo quick search.

2.3.2.5 Special effects. Special effects, such as slow motion, stills,
reverse motion, and high speed (quick search), have been developed primarily for use in helical-scan home recording. The techniques of special
effects can be classified into three generations.

The first generation consists of helical-scan machines with guard bands.
Slow-motion pictures were obtained by reducing the video tape speed, and
still pictures were obtained by stopping the tape movement. These pictures suffered from noise bars caused by the video head crossing the guard
bands. A rather primitive mechanism using two subsidiary heads was
developed to solve this problem. The two main heads for recording and
reproduction and the two subheads are mounted in a staggered position
on the rotary-head drum. In slow motion or still reproduction, the higher
of the reproducing signal levels from the main or subheads is selected in
each field. Thus, a perfect slow-motion effect or still picture can be
obtained without any noise bar.

The second generation refers to the azimuth recording method. If a
wider-track, double-azimuth video head is used for recording and playback, as shown in Fig. 2.30a, a complete video track of a still frame, a slow
frame, or a double-speed frame can be played back. As shown in Fig.
2.30b, the locus of the head does not cross other tracks between reverse
and triple speed, thus a noiseless picture can be played back. At quadru-

ple and higher speeds, other tracks are crossed and horizontal noise still exists at these speeds when the head crosses from one track to another. Both still frame and slow motion have considerable flicker in the playback picture owing to temporal error when the picture moves quickly (a two-field picture is reproduced alternately). This flicker motion can be eliminated by the reproduction of a single-field picture instead of a single-frame picture with the aid of an extra video head having the same azimuth angle as the head used to record this particular field. As long-play recorders developed, including the 6-h-mode VHS, this quick-search capability became a very important feature—so important, in fact, that the occurrence of several noise bars in the picture is acceptable. When quick-search speeds surpass quadruple speed, it is inevitable that several horizontal noise bars occur, as shown in Fig. 2.30b. On the other hand, certain special-effect techniques that yield pictures free from noise bars (such as going directly from reverse motion to triple speed) have been developed where the noise bar is shifted into the vertical blanking area of the reproduced pictures.

The third generation design takes advantage of the dynamic-tracking

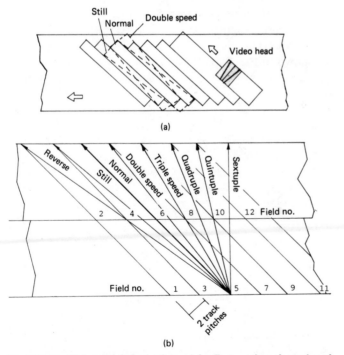

Figure 2.30 Video track loci of special effects using the azimuth recording format. (*a*) Overwrite recording and playback by wider-track double-azimuth video head; (*b*) video track loci.

method already discussed (Fig. 2.29). Since a piezoelectric actuator has several nonlinear characteristics, a closed looped system is used in order to reject these and to obtain perfect track-following. One of the interesting applications of these special effects is time-lapse recording, which is used effectively in security operations involving banks or warehouses. Recording can take place for an extremely long time (the maximum recording time is, for instance, 480 h) at a very low tape speed, either continuously or at intervals, using a conventional video cassette. Two different approaches to the recording method are used in time-lapse systems. In one of these one field is recorded every 1 s in a 120-h recording mode; in the other one frame is recorded every 2 s in the same 120-h recording mode. The recording and reproducing circuitry of the one-frame mode is more straightforward than that of the one-field recording method; however, the one-field recording method avoids the blurring of motion.

2.3.2.6 Electronic editing. With quadruplex video recording, editing was performed by the actual physical splicing of the tape. Mechanical splicing is particularly difficult to perform on helical-scan systems. Electronic editing, has, therefore, completely supplanted mechanical editing in the modern video industry (Anderson, 1969). There are two types of electronic editing: electronic splicing and insert electronic editing.

In electronic splicing, a new program can be recorded immediately after a previously recorded program, on the same tape, without any gaps between the two, as shown in Fig. 2.31. Part *A* is the previously recorded program on recorder A, and part *B* is the new program on recorder B that

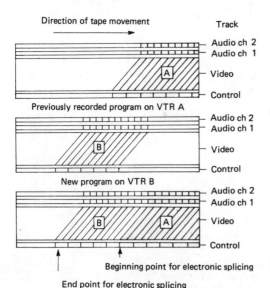

Direction of tape movement

Track

Audio ch 2
Audio ch 1

Video

Control

Previously recorded program on VTR A

Audio ch 2
Audio ch 1

Video

Control

New program on VTR B

Audio ch 2
Audio ch 1

Video

Control

Beginning point for electronic splicing

End point for electronic splicing

Figure 2.31 Electronic splicing.

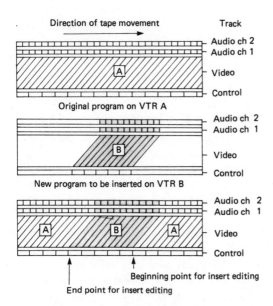

Figure 2.32 Insert-type electronic editing.

is to be added to part A on recorder A. As soon as part A is played back on recorder A, the video playback head of recorder A begins to function as a video recording head, and records the output signal (program B) from recorder B. Recorder B is operating synchronously with recorder A by means of a common sync signal. Any previously recorded video or audio signals on the tape in recorder A following program A are erased in advance by a rotating erase head for the video track and by a separately located audio erase head for the audio track. The control signal on tape A, on the other hand, has been recorded by the saturation recording method; a new signal can be overwritten on the previously recorded signal without any prior erasing. In this type of splicing, the matching of both the control signals and the video signals of A and B on recorder A is done with perfect timing, again as shown in Fig. 2.31. Thus, the program in part B from recorder B can be spliced onto program A on recorder A with perfect synchronization. The tracking on both recorder A and recorder B is done by the head drum-capstan servo method.

In insert-type editing, the new program (part B) on recorder B can be recorded in the middle of the previously recorded program (part A) on recorder A without interruption (Fig. 2.32). In insert-type editing, the video signal of part B (plus the audio signal if necessary), without the control signal, is re-recorded onto the previously recorded program A on recorder A. Thus, the control signal of recorder A, which comes from the previously recorded program (part A), is used for the new program (part B) as well. All other conditions are similar to those employed in electronic splicing.

Both of these electronic editing methods, electronic splicing and insert-type editing, represent highly sophisticated applications of tracking techniques.

2.3.3 FM video recording methods

2.3.3.1 Direct FM recording. As discussed earlier, solving the problem of recording the video signal through frequency modulation was one of the major breakthroughs in video recording technology. In broadcast equipment, picture quality is more important than tape economy, so the FM carrier is located at a high frequency with sufficient deviation so that the NTSC, PAL, or SECAM composite color signals can be recorded directly by FM (See Fig. 2.10).

NTSC and PAL color television signals are extremely sensitive to timing errors; as little as 4 ns can be visually perceived and 20 ns gives a serious change in hue. This is an accuracy impossible to meet in any electromechanical system. Two distinct methods of hue correction are used. The first is to feed the signal through an electronically variable delay as described in Sec. 2.3.3.2. The alternative is the heterodyne process in the color-under system, described later in the same section. During playback, the color signals are heterodyned with the local oscillator signal having the same jitter component as the reproduced color subcarrier. The time "jitter" on the luminance is uncorrected but is visually acceptable. Early quad systems with their high-inertia drums produced relatively small errors (0.5 μs). The helical systems, especially the type C with its 420-mm scan length, had much larger errors that could be corrected only by using digital time-base correctors.

2.3.3.2 Time-base correction. In broadcast equipment, the video signal to be played back must satisfy broadcast standards. The jitter component of the quadruplex recorder, for example, was originally as much as 0.5 μs (Harris, 1961). In order to decrease this jitter to one-hundredth of this amount, a level necessary to meet broadcast standards, the direct color process (Fig. 2.33) was developed. Correction was achieved in two steps—a coarse correction using horizontal sync as reference, followed by a fine correction using the subcarrier. The system was purely analog, with varicaps providing the control (Coleman, 1971).

Digital time-base correction is now universally used in broadcast quality helical-scan recorders (such as type B or type C), which not only have jitter (time-base error) but also skew (time-base discontinuity), a combination which is impossible to correct by analog means. The principles of digital time-base correction are shown in Fig. 2.34. First, the analog playback signal is converted into a digital signal (8 bits or more) with a sampling (clock) frequency of $3f_{sc}$ or $4f_{sc}$ (f_{sc} is the color subcarrier frequency)

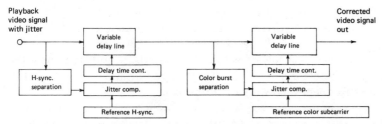

Figure 2.33 Block diagram of direct color process. (*a*) Video signal phase correction (Amtec); (*b*) color phase correction (Colortec).

derived from the color-burst signal; this converted digital signal is then written into the random-access memory (RAM). A stable read-out clock removes the influence of the jitter. One of the most important features of this method is that the error-correction range can be expanded simply by increasing the random-access memory (RAM) capacity. Another important point is that there is absolutely no picture deterioration.

Digital time-base correction is at present limited in use to broadcasting or other professional applications. If an inexpensive type is realized, it will also be used in consumer applications, and home video recording methods will become very much different from what they have been.

2.3.3.3 Color-under method. In designing video recorders for home (or industrial) use, both the relative head-to-tape speed and the actual transport speed of the tape itself should be kept as low as is feasible. Under these conditions, direct color recording is almost impossible, and the color-under system was developed as a solution (Johnson, 1960; Tajiri et al., 1968; Numakura, 1974).

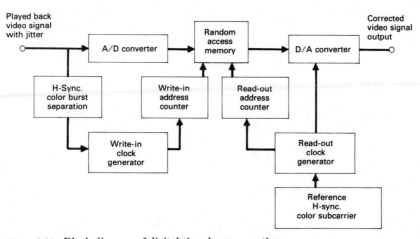

Figure 2.34 Block diagram of digital time-base correction.

When the color signal having a color subcarrier f_{sc} (f_{sc} = 3.58 MHz) is separated from the NTSC composite signal, and converted to the color-under signal (frequency f_u between 630 and 780 kHz), the signal can be recorded easily even with lower head-to-tape speeds. In addition, the jitter component of the color signal will also decrease in accordance with the ratio of the frequency conversion (approximately 5:1). This method, where the color signal is moved to the lower area of the frequency range, is called the *color-under method* and is used throughout the consumer field (e.g., VHS and Beta formats) (Arimura and Taniguchi, 1973). The color-under signal is recorded using frequency division multiplex with the FM luminance signal having carrier frequency 3.4 to 5.4 MHz. Video recording by the color-under method is shown in Fig. 2.35. The frequency

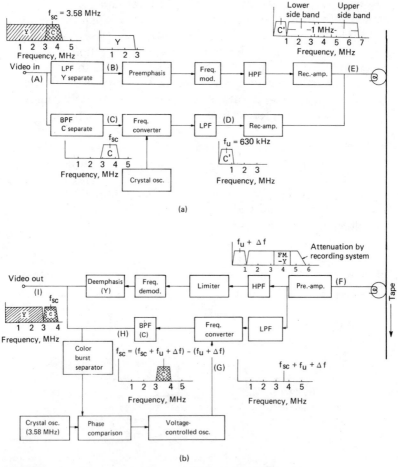

Figure 2.35 Block diagram and spectrum allocations of the automatic phase-control process in the color-under method. (*a*) Recording; (*b*) playback.

allocation of the important points is shown in the figure from (A) to (I). The high-frequency FM luminance signal acts as an ac bias for the low-frequency, low-level chrominance signal (Yokoyama, 1969). The color-under signal f_u can, therefore, be directly recorded on the tape even at low levels with little distortion. When the video signal is played back, the upper sideband of the FM luminance signal is attenuated by the characteristics of the tape and head. The FM deviation of the luminance signal is determined by making an optimum balance of the picture resolution (which requires high-frequency recording) and the signal-to-noise ratio (which requires more output at the carrier frequency). From the reproduced color-under signal, with a substantial jitter component Δf, the color-burst signal is then separated and compared with a reference subcarrier frequency which is, in turn, produced by a crystal oscillator. Thus, an error signal is produced and compensates for the jitter component of the color (chrominance) signal. The color-under frequency f_u is also chosen so that the visibility of undesirable interference beats with the luminance signal is minimized by interleaving them with multiples of the line frequency (Fujita, 1973).

$$f_u = \tfrac{1}{4}(2n + 1)f_H \tag{2.5}$$

where f_H is the horizontal sync frequency and n is an integer.

2.3.3.4 Color-under method for azimuth recording.

In azimuth recording, the FM luminance signal, which has a relatively short wavelength, suffers negligible crosstalk from adjacent video tracks. As the color-under signal has a relatively long wavelength, however, the crosstalk signal from the adjacent video track is not eliminated by the azimuth effect; it spoils the reproduced picture quality by producing beat interference. In order to eliminate this crosstalk component, two methods known as the *phase-invert* (PI) and *phase-shift* (PS) methods were introduced (Amari, 1977; Hirota, 1981).

In the phase-invert method, the phase of the color subcarrier on the A track is inverted 180° every horizontal sync period, but the B track is recorded without phase inversion (Fig. 2.36). When the A track is played back, there is a certain amount of crosstalk from the B track. In order to eliminate this crosstalk component on the A track, the phase is reconverted to its original form (thus inverting by 180° the crosstalk from the B track). A new signal is created with a single delayed horizontal-scanning period which allows the crosstalk to be eliminated.

In the phase-shift method, the phase of the color subcarrier on the A track is shifted by 90° for every horizontal sync period, and on the B track the color phase is also shifted by 90°, but inversely (Fig. 2.37). When the A track is played back with the crosstalk component from the B track,

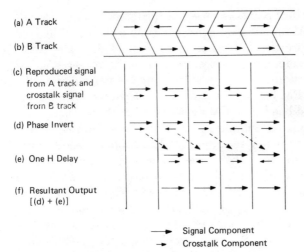

(a) A Track

(b) B Track

(c) Reproduced signal
     from A track and
     crosstalk signal
     from B track

(d) Phase Invert

(e) One H Delay

(f) Resultant Output
     [(d) + (e)]

⟶ Signal Component

⟶ Crosstalk Component

Figure 2.36 Elimination of crosstalk in azimuth recording by the phase-invert method.

the phase is inversely rotated for every single horizontal sync period back to its original form, and the resultant output has no crosstalk component. Thus, the crosstalk from the adjacent *B* track is eliminated by means of line interrelation.

2.3.3.5 Baseband recording by time compression. The color-under recording method is one of the most successful color-signal recording methods for consumer recording. It is quite effective in handling both the

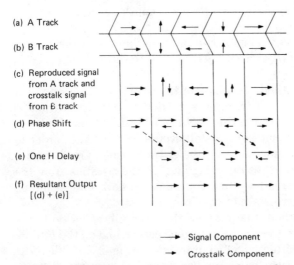

(a) A Track

(b) B Track

(c) Reproduced signal
     from A track and
     crosstalk signal
     from B track

(d) Phase Shift

(e) One H Delay

(f) Resultant Output
     [(d) + (e)]

⟶ Signal Component

⟶ Crosstalk Component

Figure 2.37 Elimination of cross-talk in azimuth recording by phase shift method.

relatively narrow band and the poor time-base instability characteristics of home equipment. Even so, there are some problems with streaking noise in the reproduced color picture caused by the AM recording of the color signal. Another new color-recording method has been proposed in relation to the 8-mm video format. This is termed *baseband recording by time compression* and has been made possible by the rapid development of digital technology and semiconductor devices (Dieter, 1978).

This new method, as used in the type L format, compresses the color-component signals (R − Y, B − Y) into one horizontal line and records them on a separate track (see Fig. 2.14). This type of two-channel recording is too expensive for consumer use and is too wasteful of tape. In an alternative method, the chrominance and luminance signals are recorded on a single track by recording the color-component signals (R − Y, B − Y) sequentially on two horizontal line periods using digital time-compression (Fig. 2.38). The luminance signal is also compressed and recorded along with the color-component signal. When the total signal is reproduced as is, the color-signal pattern is on the left side of the screen and the monochrome picture is on the right side, somewhat compressed. The color-component signal and the luminance signal are then separated and expanded to their original form by digital means. The color-component signals, which are sequentially recorded on two horizontal lines, are formed into a complete color signal by using a single horizontal line delay.

This baseband recording method is an interesting method for recording color signals without any significant deterioration. The jitter component in baseband recording is, however, not reduced as it is in the color-under recording, and commercial use awaits the development of inexpensive LSI circuits, including a digital means for time-base correction.

2.3.4 Audio recording

2.3.4.1 Linear-track recording. When the rotating-head recording method is used for video signals, the tape speed is similar to that of an audio tape recorder. As described earlier, there is also enough space left to record an audio signal on the edge of tape. In these cases, the audio recording in a video recorder is the same in principle as it is in conventional audio tape recorders (Chap. 3). There are, however, some special problems. One problem is caused by the rotating head, which hits the surface of the tape as it rotates and causes flutter in the reproduced sound. This problem has been nearly overcome by reductions in the head drum diameter and by the solution of relevant head-to-tape interface problems. Other problems arise because the tape speed has become very slow (the tape speed of VHS in the extra-long-play mode is, for instance, only 11.1 mm/s), and the space for audio tracks is limited, especially for two-channel audio recording (stereo or bilingual recording). The resulting signal-to-noise ratio is poor, and it is necessary to apply noise-reduction methods

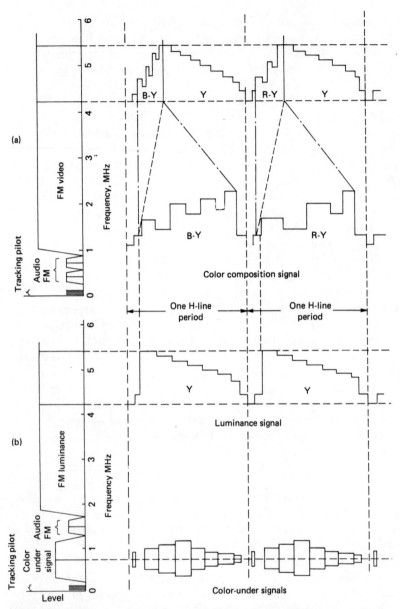

Figure 2.38 Frequency allocation in 8-mm video *(Sugaya, 1986)*. (*a*) Proposed baseband recording by time compression; (*b*) standardized color-under recording.

to these audio recordings (Dolby, 1967). Also, when the tape speed is as slow as this, the rotating speed of the capstan is correspondingly low, and flutter and wow increase owing to the insufficient flywheel effect of the capstan and the motor, even if a separate drive motor is used. For these reasons, FM audio and pulse-code modulation (PCM) audio recording methods have been developed for consumer applications. In broadcast video recorders, the tape speeds are sufficient to record the audio signal at broadcasting standards (the quadruplex recorder uses 381 mm/s; type B, 245 mm/s; type C, 244 mm/s; and types L and M, 118 and 105 mm/s). Therefore, at the present time, most research and development activities to improve the audio characteristics of video recordings are being carried out in the consumer field.

2.3.4.2 FM audio recording. In order to improve the audio quality, while maintaining interchangeability with older machines, the rotating head can be used for audio recording. There are two different approaches: frequency-division multiple recording and double-component multiple recording.

Frequency-division multiple recording. In the frequency allocation of the signals which are recorded by the video head, some usable space can be created for the carrier frequencies of an FM audio signal between the color-under chrominance signal and the FM luminance signals (Fig. 2.39a). In the Beta format, FM audio was developed after the conventional machine using linear audio recording on a separate track was marketed, and space for the new FM audio carrier was created by shifting the FM video range to a 400-kHz higher position. This is believed to be the maximum shift possible while still maintaining interchangeability with conventional machines (Kono, 1983). In order to record the FM audio carrier signal along with the video signal without any interference, an interleaving relation between the video signal and the FM audio frequency needs to be determined. The FM audio carrier frequency for track A and track B (f_A and f_B) must be such that

$$f_A = (n - \tfrac{1}{4}) f_H \tag{2.6}$$
$$f_B = (n - \tfrac{3}{4}) f_H \tag{2.7}$$

where f_H is the horizontal sync frequency (15,734 Hz for NTSC) and n is an integer.

When the FM audio signal is recorded by the video rotating head using the azimuth recording method, the crosstalk from the adjacent track is a serious problem owing to the relatively low frequency range (i.e., relatively long wavelength), and no line interrelation effects can be used to reduce crosstalk as can be done with the chrominance signal. In the Beta format, this problem is solved by making the FM audio have different carrier fre-

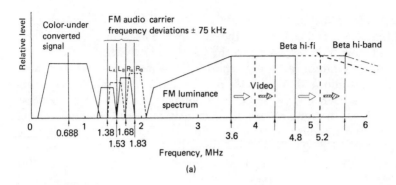

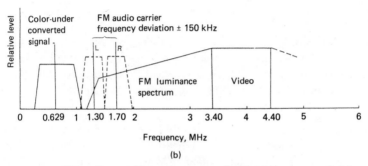

Figure 2.39 FM audio recording using a rotary head. (*a*) Frequency-division multiple recording; (*b*) double-component multiple recording *(Sugaya, 1986)*.

quencies in adjacent tracks, so that the crosstalk can be eliminated by bandpass filters. Two carrier frequencies for each two-channel audio signal are alternately recorded on each video track. Thus, four different carrier frequencies (L_A, L_B, R_A, and R_B), each having a ±75-kHz frequency deviation, are used for FM audio recording and are selected as shown in Fig. 2.39a. In order to avoid interference between the audio and video FM carriers, the audio carrier signal is at a lower level than the video carrier signal.

In 8-mm video, the FM audio signal is limited to a single channel in order to save enough space for the video signal. Use of both a large azimuth angle (±10°) and noise reduction assures that there is little crosstalk between the two tracks in the 8-mm video system (Sugaya, 1986). A single FM audio carrier frequency is therefore used (1.5 MHz), as shown in Fig. 2.38b.

When the lower sideband of the FM video spectrum is replaced by the FM audio spectrum in frequency-division multiple recording, the high-frequency portion of the video spectrum will be eliminated, resulting in a natural decrease in the resolution of the video signal as well. Otherwise, the carrier frequency of FM video has to be shifted to a still higher region. In the frequency-division recording method, a compromise must be made between FM audio (sound quality) and FM video (picture quality). In practice, the FM video signal has been shifted to a range 400 kHz higher in order to maintain interchangeability with previously marketed equipment and tapes; even so, resolution power has been substantially decreased. Thus, the new Hi-Band Beta Hi-Fi has been developed to improve the picture quality by shifting the FM video carrier frequency an additional 800 kHz over the conventional Beta format to 5.6 MHz for peak white frequency. This sacrifices interchangeability, however, and has not been accepted for prerecorded tape use.

Double-component multiple recording. The double-component multiple recording method was developed to improve the sound quality without sacrificing video characteristics while at the same time maintaining interchangeability with conventional equipment (Hitotsumachi, 1983). In this method, the FM audio signal is recorded using a pair of audio rotary heads with a large azimuth angle of $\pm 30°$ in conjunction with a pair of video heads. This azimuth angle is sufficient to eliminate crosstalk of the audio FM signals of 1.3 and 1.7 MHz. The FM audio signal, having a relatively longer wavelength, is recorded first by a rotating audio head with both a large gap length and a large azimuth gap angle. The FM video signal, having a relatively shorter wavelength, is then recorded directly over the same track with a different azimuth angle. This recording is thus called *double-component multiple recording*.

The principles of double-component multiple recording are shown in Fig. 2.40. The longer-wavelength audio signal can be recorded in a deeper portion of the magnetic coating, so the shorter-wavelength video signal can be recorded on the shallower portions of the previously recorded long-wavelength audio signal with a different azimuth angle. The FM audio signal is actually erased to a certain degree (approximately 13 dB in VHS FM audio recording); even so, 30 dB of carrier-to-noise ratio can be maintained. In this recording method, FM audio carrier signals do not disturb the FM video signal, thus the video characteristics are not affected and a wider frequency deviation (± 150 kHz) for FM audio—i.e., a better signal-to-noise ratio—can be obtained. In order to use this FM audio recording method for both standard play and extra-long play, the audio track width needs to be very narrow, as shown in Fig. 2.40, and thus the interchangeability of the FM audio signal is difficult unless the tape transport mechanism is exceptionally rigid. The frequency allocation of the recorded signals of the VHS FM audio system is shown in Fig. 2.39b.

The inertia effect of the rotary-head recording process considerably reduces the flutter and wow components. The FM audio recording process further improves the frequency response, the signal-to-noise ratio, the

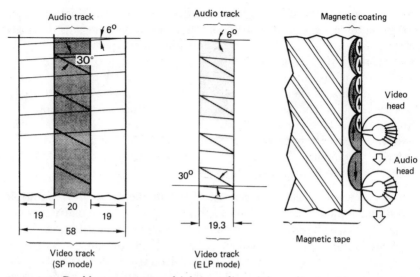

Figure 2.40 Double-component multiple recording of the audio signal by a rotary head *(Sugaya, 1986).*

dynamic range, and the channel separation; dubbing is not possible, however. The frequency-division multiple recording method is used at present only for the NTSC Beta format and the 8-mm video format. The double-component multiple recording method is used for PAL/SECAM Beta format and for all VHS formats.

2.3.4.3 Pulse-code-modulation (PCM) recording.

Recording audio signals using pulse-code modulation has become more attractive both technically and economically, following recent developments in digital and semiconductor technologies (see Chap. 3). One of the advantages of digital processing is that the signal is written into a buffer memory and later can be retrieved from the memory with no deterioration. Thus, sound quality shows no deterioration even after many duplications. The digital process, in other words, permits a discrete, incremental recording of the continuous audio signal.

Track-division multiple recording. In two-head helical-scan recording, the tape must be wrapped around the head slightly more than 180°. In track-division multiple recording, the tape is wrapped an additional 36°, as shown in Fig. 2.41, and the segmented, time-compressed audio signal is written in this region. If the time-base compression rate is high, the recording density and the recording bit frequency will also become high, so that both dropouts and the error rate will increase.

The sampling frequency f_s for PCM recording is determined by the maximum recordable frequency f_{max} ($f_s > 2f_{max}$) and is also related to the video signal ($f_s = n\, f_H$). In the 8-mm video format $f_s = 2f_H = 31.5$ kHz, and the quantization bit number M is 10 bits. In simple terms, the dynamic range D is

$$D = 6M + 1.78 \quad \text{dB} \tag{2.8}$$

Eight bits therefore gives approximately 50 dB of dynamic range, which is insufficient even for home entertainment use. Ten bits are used for analog-to-digital conversion; these are then translated to 8 bits by a nonlinear conversion process. As a result, the dynamic range of 8-mm video PCM audio is approximately 62 dB, increasing to 85 dB with noise reduction. The upper recordable frequency limit is 14 kHz, which is just sufficient for consumer applications (Nakano et al., 1982, and Table 2.4). The most important feature of track-division multiple PCM recording is the possibility of re-recording (dubbing), which FM audio recording does not have. Error compensation of the PCM code in 8-mm video is always completed in each block in order to permit the re-recording of edited signals. (The details of error compensation are discussed in the following section.) Another important feature is simplicity—a single pair of video heads can record both the video signal and the audio signal.

If recording is confined to audio only, another five different channels of PCM audio signal can be recorded by shifting the PCM recording phase (Fig. 2.41). That is, a 12-h signal (2 h × 6) can be recorded on a small 8-mm video cassette, which is almost the same size as an audio compact cassette.

PCM encoder-decoder. Video tape recording technology makes available an extremely large memory capacity. Thus, if an audio signal is encoded onto a composite signal (any of NTSC, PAL, or SECAM) by PCM, the recorder can record the PCM audio signal as an FM video signal without

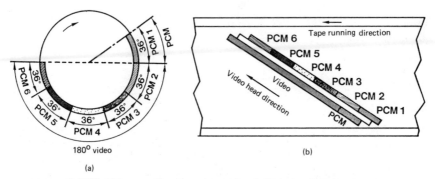

Figure 2.41 PCM audio recording for 8-mm video (video plus PCM audio or 6-channel PCM audio *(Sugaya, 1986)*. (*a*) 8-mm video PCM allocation; (*b*) 8-mm video PCM tape format.

any modification. In this case, only the encoder-decoder (or PCM adapter) has to be standardized. The recordable upper limits of data are approximately 3 Mb/s for both Beta and VHS format recorders. In order to record up to 20 kHz of audio signal, the sampling frequency should be $f_S = 44.1$ kHz. Fourteen bits of linear quantization (dynamic range 86 dB) have been determined to be optimum for a PCM encoder-decoder (see Table 2.4).

Error compensation is one of the most important aspects of the technology of digital magnetic recording, which has intrinsic errors due to dropout and head-to-tape interface problems. Generally, error compensation is accomplished by a combination of four methods: (1) interleaving, in which information is distributed so that the error does not destroy important information; (2) error detection, using a cyclic redundancy check code (CRCC); (3) error correction, using a code such as the Reed-Solomon (Reed and Solomon, 1960); and (4) error concealment, in which missing information is estimated by means of a previous word or by adjacent sequential words. With these error compensation methods, audible errors can be effectively eliminated. A comprehensive discussion of error detection and correction methods is given in Chap. 5.

TABLE 2.4   Typical Two-Channel (Stereo) Sound Characteristics (NTSC)†

	Linear Track		FM	PCM	
	SP	ELP	VHS (Beta)	8-mm video	VHS + EIA-J processor
Tape speed, cm/s	3.33–4.00	1.11–1.33	3.33–(4)	1.43	3.33
Relative tape speed, m/s			5.8–(7)	3.75	5.8
Frequency response, Hz	50–12,000	50–6000	20–20,000	20–14,000	20–20,000
Dynamic range, dB	55‡	55‡	80‡	85‡	86 (linear) [96 (linear)]
Distortion, %	< 1.5	< 1.5	< 0.3	< 0.3	< 0.01
Wow/flutter, %	0.08–0.12	0.22–0.28	< 0.005	< 0.005	0.005
Quantization bits				8 (10/8 conv.)	14 (linear) [16 (linear)]
Sampling frequency, kHz				31.5	44.1

† SP = Standard Play. ELP = Extra Long Play. EIAJ = Electronic Industry Association of Japan.
‡ With noise reduction

2.4   The Video Head

2.4.1   Development of the video head

The video head is one of the most important components in video recording. If an analogy is made with the principles of photography, the video tape corresponds to the photographic film and the video head to the lens. The minimum recordable wavelength will be determined both by the head gap length and by the particle size of the magnetic pigment in the magnetic coating. The corresponding optical properties are the degree of resolution of the lens in a camera and the grain size of the emulsion of the photographic film. One significant difference between the magnetic and optical systems is that there is actual physical contact between the recording medium and the magnetic head, while there is no contact between the film and the lens in a camera. As a consequence, the relationship between the magnetic head and the tape is a more complex one than that between the camera lens and the film (Sugaya, 1985).

The length of the gap is a crucial factor in the design of the video head. In the playback process, the gap length should be less than the minimum recordable wavelength. Theoretically, for maximum output, the gap length should be one-half the signal wavelength. In practice, however, a gap length of one-third the wavelength is preferred to reduce gap loss to a more acceptable level. With modern recording techniques, the recordable wavelength is becoming shorter and shorter, necessitating correspondingly smaller gap lengths. This in turn results in another problem where, if the gap length becomes too small, the main portion of the magnetic flux will be shunted past the gap and insufficient flux will be induced in the core. This is a dilemma which seriously affects the design of the modern video head.

An equally serious problem is that of maintaining the original gap tolerances and the original shape of the gap. The gap dimensions and the shape of the gap itself both tend to be affected adversely over time as the head is subjected to wear from the tape. If the head material is not of sufficient hardness, the head material (as well as the gap material) at the surface can actually be drawn in the direction of the tape movement, resulting in a distortion of the gap shape. This phenomenon of the head and gap material actually moving is called *head-gap smearing;* examples of this are shown in Fig. 2.42.

Alloy materials are relatively soft and are susceptible to this smearing phenomenon. For this reason, head cores and gap materials, which were originally formed from metal alloys, are now mostly formed from ferrite and glass. The adoption of new high-coercivity tapes may present difficulties, in that the limits of magnetic saturation in the poles of the ferrite head are approached. Such developments may demand reversion to the

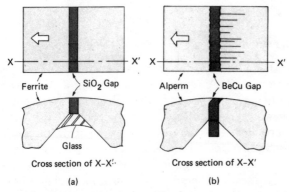

Figure 2.42 Metal-head–gap smearing after passage of tape. (*a*) Typical ferrite head with SiO$_2$ gap and no smearing; (*b*) typical alloy head with Be-Cu gap and smearing.

use of metal for the head pole material or to the use of newly developed amorphous materials. The magnetic materials used in the video head must meet a number of essential criteria:

1. Durable physical properties (low porosity, resistance to chipping and wearing) in conjunction with good machinability (necessary for slicing, forming, finishing, and lapping)

2. High permeability and saturation induction in the MHz range (high specific resistivity)

3. Low coercivity, in order to keep the residual magnetization of the head tip as small as possible

4. Low magnetostriction coefficient, to keep the magnetostriction noise as low as possible (i.e., $\lambda_s \rightarrow 0$)

5. Stability of both magnetic and physical properties after long periods of use in various environments

6. Resistance to possible deterioration of mechanical properties due to assembly processes (e.g., bonding) or machining techniques

7. Reasonable cost

If we consider the conditions imposed by these criteria, ferrite appears to be the optimum material available at present. Ferrite is suited for use in video heads because of its low loss at high frequency and its hardness. On the other hand, ferrite has a rather granular and brittle texture; until the late 1950s it was impossible either to make very fine gaps or to maintain the head surface in good condition (Chynoweth, 1955).

The heads in the first video machines were derived directly from audio-head technology and were constructed out of laminated Permalloy, the standard design at the time. In order to record the very high frequencies necessary for video (3.5 MHz), three strips of extremely thin (50μm) Permalloy were used as shown in Fig. 2.43a, and a rather primitive butt joint was used to form the gap (Olson et al., 1956). Later, as shown in Fig. 2.43b, it was found that a combination of metal alloy and ferrite brought out the desirable qualities of both materials; the alloy could be used to produce an extremely fine gap, and a ferrite body had low loss in the MHz range (Kornei, 1956). Alperm (also known as Alfenol) in 60-μm strips was used as the alloy in the early combination heads because it is much harder than Permalloy and has a higher resistance. The gap of 0.5 μm was formed by silicon monoxide vacuum deposition.

In the quadruplex head design shown in Fig. 2.43c, a similar combina-

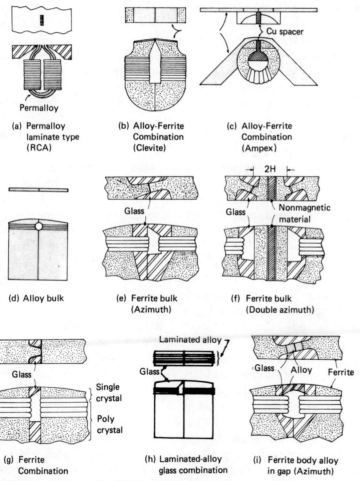

(a) Permalloy laminate type (RCA)

(b) Alloy-Ferrite Combination (Clevite)

(c) Alloy-Ferrite Combination (Ampex)

(d) Alloy bulk

(e) Ferrite bulk (Azimuth)

(f) Ferrite bulk (Double azimuth)

(g) Ferrite Combination

(h) Laminated-alloy glass combination

(i) Ferrite body alloy in gap (Azimuth)

Figure 2.43 Various video head configurations.

tion of alloy and ferrite was utilized, but the design was somewhat different because of the narrow track width (250 μm). A copper spacer, located not only in the Alperm head tips but also in the ferrite core gap, was used as an attempt to enhance the magnetic-flux leakage from the gap. The life of this type of head was extremely short (on the order of 200 h) and it had comparatively poor recording characteristics owing to gap smearing (Sugaya, 1968).

Sendust is another alloy that was considered at an early date for use as a video head material because of its superior physical properties. Sendust is an extremely durable magnetic material, but it is also so brittle that it is not possible to produce it in sheets by rolling (Matsumoto and Yamamoto, 1937). In the early stages of magnetic head development, it was generally believed that the head could not be constructed of a solid block of metal but had to be formed from laminated sheets to reduce eddy currents. In fact, it was subsequently discovered that it was possible to use alloys in bulk form for video heads. This is because eddy currents do not affect the flux at the surface of the head core, thus it is possible to use nonlaminated alloy cores with some loss in efficiency (Sugaya, 1964; Bertram and Mallinson, 1976). Thus bulk-alloy heads made of Sendust were eventually developed (Fig. 2.43d). This design is still used for quadruplex machines because the extremely heavy contact between the heads and the tape does not permit the use of ferrite heads, owing to their fragility and susceptibility to noise induced by mechanical stress.

Ferrite heads can be used with helical-scan recorders because of the relatively light contact between the head and the tape. When helical-scan recorders were first introduced, high-density ferrite was used with bonded glass spacers having the same coefficient of thermal expansion as the ferrite (Duinker, 1960). This type of head was used successfully in audio and instrumentation recorders, but its response proved to be insufficient for video use owing to the porosity (0.1 percent) of the material.

In order to improve the physical properties of ferrite, hot-pressed ferrite (HPF) was developed and successfully used for video heads (Sugaya, 1968). Other approaches taken to improve the porosity characteristics of normal ferrite included the use of single-crystal ferrite (Mizushima, 1971), and hot-isostatic-pressed ferrite (HIP) (Takama and Ito, 1979). Also, various ferrite compositions were tried in the development of the video head. Ni-Zn ferrite was first used for its high resistivity, but Mn-Zn ferrite has now become a common material for video heads because of its higher permeability μ, higher saturation induction B_s, and lower coercivity H_c.

Heads made of hot-pressed ferrite have a certain advantage over single-crystal heads in that the grains are of a uniform size, the minimum mean value of which is approximately 10 μm. When the video track width is wider than 40 μm, several grains of the magnetic hot-pressed ferrite material are located in the gap area. The existence of these grains and the grain boundaries associated with them in sufficient quantity prevents the magnetostriction noise caused by contact shock between the head and the tape (Torii et al., 1980; Kimura et al., 1980). Hot-pressed ferrite is also cheaper to produce than is single-crystal ferrite, which is normally produced by the Bridgeman method (Sugimoto, 1966).

When the track width is less than 20 μm (as in VHS extra-long-play mode), hot-pressed ferrite can no longer be used successfully because only one or two grains are actually located in the head gap area. For this reason, single-crystal ferrite came into general use as a head material for narrow-track video recording. The problem of magnetostriction noise was solved by the use of more-sophisticated and more-accurate machining technology, as well as by the use of smaller head drums, which permitted a smaller head-tip projection, resulting in an extremely light head-to-tape contact pressure.

The problem of high cost was solved by developing techniques to produce ferrite in large ingots of approximately 10 cm in diameter (Torii, 1979). The most common material now in use for video heads in consumer equipment is Mn-Zn single-crystal ferrite. A typical example of such a video head with an azimuth angle gap is shown in Fig. 2.43e. The double-azimuth gap head for special effects and for other applications (such as Beta movies) is an example of extremely sophisticated fabrication techniques (Fig. 2.43f).

In broadcast equipment, on the other hand, the head track width is still quite large (120 μm for type-C recorders), and the contact pressure between the head and the tape is high owing to the large diameter of the head drum and the higher head-to-tape relative speed. Thus, a combination single-crystal and polycrystalline ferrite head was developed for this equipment to take advantage of the strong points of both these materials, as shown in Fig. 2.43g (Camras, 1963a; Tanimura et al., 1983).

With the introduction of alloy magnetic tape, which has a coercivity higher than 80 kA/m (1000 Oe), the low saturation magnetization of ferrite becomes a serious problem, particularly concerning the gap portion of the head where the magnetic flux is concentrated. As a consequence, metal alloy materials, such as Sendust, are once again being investigated as magnetic materials for the video head (Yasuda et al., 1981). Amorphous materials, which have superior physical properties, as shown in Table 2.5, are also used for recording on alloy metal tape (Matsuura et al., 1983).

In conventional ferrite heads, the gap portion of the head, which is in direct contact with the tape, should be wider than the track width in order to prolong head life as much as possible; this is shown in Fig. 2.43e, f, and g. When an alloy material is used for the video head, a construction similar to that used in ferrite heads cannot be used; instead, the alloy head core needs to be reinforced by a stiff nonmagnetic material, such as crystallized glass, for the tape contact portion, as shown in Fig. 2.43h (Shibaya

TABLE 2.5 Physical Properties of Video Head Materials

	Permalloy	Alfenol (Alperm)	Sendust (Alfesil)	Amorphous	High-density ferrite	Hot-pressed ferrite	Single-crystal ferrite	Units
Composition	Ni 79 wt % Mo 4 Fe 17	Al 16 wt % Fe 84	Al 5.4 wt % Si 9.6 Fe 85.0	Fe 4.5 wt % Co 70.5 Si 10 B 15	NiO 1 mol % ZnO 22 Fe_2O_3 67	MnO 28 mol % ZnO 19 Fe_2O_3 53	MnO 29 mol % ZnO 17.5 Fe_2O_3 53.5	
μ/μ_0†								
1 kHz	20000	3000	30000	10000	850	12000	10000	
1 MHz	40	30	60			2100	1540	
10 MHz					550	700	720	
Saturation induction B_s	690 (8700)	1430 (18000)	880 (11000)	680 (8500)	310 (3900)	400 (5000)	400 (5000)	kA/m (G)
Coercivity H_c	4 (0.05)	3 (0.04)	4 (0.05)	1.6 (0.02)	32 (0.4)	3 (0.04)	2.4 (0.03)	A/m (Oe)
Resistivity ρ	55×10^{-8}	140×10^{-8}	80×10^{-8}	150×10^{-8}	10^5	$>5 \times 10^{-2}$	$>10^{-3}$	$\Omega \cdot m$
Curie temp. T_c	460	400	500	420	125	150	167	°C
Vickers hardness	130	350	530	900	600	650	650	
Porosity	0	0	0	0	<1	<0.1	0	%

†μ/μ_0 = relative permeability.

SOURCE: Sugaya (1986).

and Fukuda, 1977). With the development of vacuum-deposited-alloy technology, lamination of alloys is being reconsidered as a means of minimizing the eddy-current loss. Laminated alloy-glass-ferrite head construction is, however, very complicated and is used mainly for broadcast applications. Therefore the metal-in-gap head of the type shown in Fig. 2.43i was developed; in it the gap portion is reinforced by a metal, such as Sendust or amorphous alloy, with high saturation magnetization.

2.4.2 Magnetic materials for video heads

The magnetic materials used in video heads change in the wake of any increase in the recording density of magnetic tape. Gamma ferric oxide of coercivity around 25 kA/m (300 Oe) was used in the early stages of video recording development. The main problem concerning the magnetic material centered on maintaining the head gap in good condition and preventing head wear over time. As a result, as durable a material as possible was desirable for head construction. Sendust was selected as the most suitable alloy material, after experiments with Permalloy and Alfenol (Table 2.5).

The fact that ferrite has a higher resistivity than the alloy materials reduces eddy-current loss and provides a high permeability in the video frequency range; this is an especially important consideration in playback. Figure 2.44 shows the efficiency of the playback head (the ratio of the flux

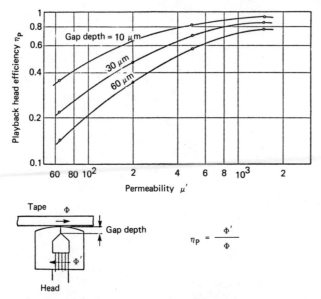

Figure 2.44 Relationship between the playback head efficiency η_p and the permeability μ of the head core.

from the magnetic tape to the induced flux in the head coil) as a function of core permeability. Another very important feature of ferrite material is its superior mechanical properties. Ferrite is a much harder material than conventional metal alloys, yet it is more easily machinable using modern techniques, and machining does not degrade the magnetic properties as it does with alloys.

As mentioned above, single-crystal ferrite is now the most common video head material for consumer equipment, but it does have the draw-back of a strong anisotropy, both magnetically and mechanically (Hirota et al., 1984). Several single-crystal orientations have been tried for use in video head construction, as shown in Fig. 2.45. Type A in this figure is resistant to wear; types B and C are the best selections for output quality.

The first cobalt-modified iron oxide and chromium dioxide video tapes had a coercivity of approximately 40 kA/m (500 Oe), which could be recorded with conventional ferrite heads without saturation. Since then, the coercivity of oxide tapes has increased to around 56 kA/m (700 Oe). Further potential increases in coercivity will be limited by the saturation flux density of the ferrite head material in consumer-oriented equipment, such as VHS and Beta format, which must maintain interchangeability between older equipment and new models. Theoretically, a further increase up to 120 kA/m (1500 Oe) in coercivity is possible if the contact

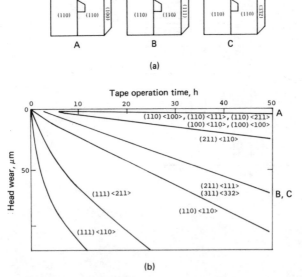

Figure 2.45 Use of single-crystal ferrite for video heads. (*a*) Crystal orientations; (*b*) accelerated wear characteristics. [*Reprinted with permission from Hirota (1984). Copyright 1984 American Chemical Society.*]

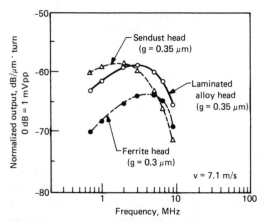

Figure 2.46 Frequency responses of various types of video heads using MP tape. (*Takahashi, 1987*).

between the head and tape can be made extremely close. In practice, it is physically possible but magnetically difficult to achieve such small head-to-tape spacings (Sugaya, 1985).

In order to achieve a major increase in recording density, a completely new recording medium, such as the metal tapes adopted for 8-mm video, will be needed—for instance, metal powder (MP) tape with a coercivity of 120 kA/m (1500 Oe), or metal-evaporated (ME) tape with a coercivity of 72 kA/m (900 Oe). Modern alloy video-head development has progressed from this background (Morio et al., 1981). Sendust is still being used, but new amorphous materials are now being seriously considered because of their durability and superior magnetic performance (Makino et al., 1980). There are two different processes for the production of amorphous metals. One is the liquid-quench method, where the molten magnetic material is quenched suddenly on a roller and is thus cooled before crystallization can take place (Fujimori et al., 1977). The other method is the vacuum-deposition method in which the magnetic material is sputtered or evaporated on a nonmagnetic substrate (Shimada, 1983). In this method, magnetic and nonmagnetic materials are deposited alternately, so that the high-frequency efficiency is considerably enhanced, even at relatively wide track widths (Fig. 2.46) (Takahashi et al., 1987). The head design is similar, as shown in Fig. 2.43*h*. This is another example of an earlier technology reemerging in the development of video recording.

2.4.3 Video head design

All video heads in use at the present time are of the ring type with a very narrow gap and a small coil. The experimental relationship between head gap length and head output is shown in Fig. 2.47; typically, the head gap length is chosen to be one-third of the shortest recordable wavelength. The observed gap length as measured through a microscope is shorter than the effective gap length obtained by measuring the null point (the point at which the effective gap length equals the recorded wavelength). Part of this difference is fundamental; the remainder is the result of what is called the *worked layer* (or the *Beilby layer*). This is a thin, nonmagnetic layer caused by mechanical work or by chemical reaction between the gap surface of the head core and the gap material. In the early stages of ferrite head design, a glass-bonding technique was used to create a strong gap (Duinker, 1960), but this high-temperature bonding process relied on a chemical reaction between the head core and the spacer material to achieve a strong bond. The nonmagnetic portion that was created by the chemical reaction was of no concern when the minimum recordable wavelength was relatively long (longer than 2 μm, for example). In high-density recording, however, the minimum recordable wavelength becomes less than 1 μm, dictating an effective gap length of 0.3 μm. When the effective gap length becomes this small, the presence of this worked layer is no longer negligible. As a consequence, the glass-bonding technique has recently been replaced by the vacuum-sputtering of nonmagnetic materials.

The worked layer on the single-crystal ferrite can be detected by means of x-ray diffraction analysis methods (Sugaya, 1978, 1985). When the surface of a damage-free single-crystal ferrite is observed by x-rays, a regular lattice x-ray diffraction pattern is evident. If there is a worked layer present, this regular lattice pattern will disappear. The pattern will reappear if the worked layer is removed by chemical etching, for example. Use of

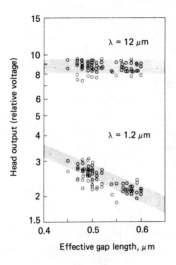

Figure 2.47 Experimental relationship between head output and head gap length.

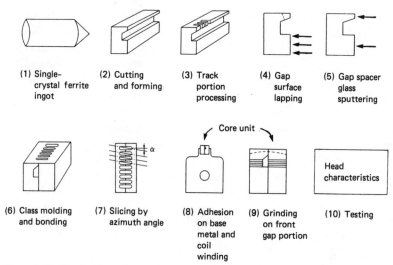

(1) Single-crystal ferrite ingot

(2) Cutting and forming

(3) Track portion processing

(4) Gap surface lapping

(5) Gap spacer glass sputtering

(6) Class molding and bonding

(7) Slicing by azimuth angle

(8) Adhesion on base metal and coil winding

(9) Grinding on front gap portion

(10) Testing

Core unit

Head characteristics

Figure 2.48 Production process for typical ferrite heads shown in Fig. 2.43e.

x-ray diffraction analysis has led to the development of various damage-free lapping techniques as well as to the improvement of the short-wavelength recording performance of video heads. With alloy magnetic materials, the worked layer can be detected by means of the skin effect at frequencies in the MHz range (Sugaya, 1964, 1985). If there is a worked layer present, the magnetic flux, which is concentrated on the surface of the alloy, is disturbed by the nonmagnetic worked layer, and the permeability of the alloy will be decreased. When the worked layer is removed by chemical etching, the original permeability is recovered. In this way, the thickness of the worked layer can be measured; this may lead to the development of improved damage-free lapping techniques.

One of the most important features of the ferrite head is its suitability to mass production techniques. Figure 2.48 shows a typical example of the video head production process. Ferrite heads are normally produced together in core units of several dozen pieces each and are then separated into the individual head units.

Following the increase in video tape coercivity, the saturation flux density of the head material is becoming a critical limitation. There is at present some activity to develop a high-saturation ferrite material, but there are finite upper limits to this, owing to the crystal characteristics of ferrite. The theoretical limit for ferrite is on the order of 480 kA/m (6000 G), while that for alloy materials is far higher, as shown in Table 2.5. For this reason, alloy materials are coming into use once again for video heads. The most serious problems to be faced in the use of alloy materials derive from the complicated design and construction techniques dictated by the nature of the material. These problems arise because the metal alloys, particularly the new amorphous materials, cannot be handled by the same sophisticated processes which are used in manufacturing ferrite heads, except for metal-in-gap heads. Also, in alloy heads the nonmagnetic components, such as glass or ceramics, are usually bonded by an organic adhesive. This adhesive makes contact with and may react with the organic tape binder, triggering head clogging. The development of a direct bonding method or of a suitable inorganic adhesive is anticipated to eliminate the problems of head clogging and other head contamination problems. The various video heads discussed in this section are fabricated mainly through the use of slicing, grinding, and lapping techniques. In the future, however, radically new techniques, such as laser cutting, will be adopted and will, in turn, trigger new designs and construction methods for video heads.

2.5   Video Tape

2.5.1   Development of video tape

As discussed above, the important advances in video recording have centered around the increase of recording density, including the decrease of the minimum recordable wavelength. These advances are strongly dependent on the improvement of magnetic tape materials (Jacobs, 1979). Magnetic tape is discussed in detail in Chap. 3 of Volume I.

In the early stages of development, standard audio tape was used for the recording of video signals. When video recording subsequently became commercially feasible through the introduction of the quadruplex format, it became necessary to develop a special video tape. Standard audio tape was not well adapted either to the high relative head-to-tape speed of rotary-head equipment (38 m/s) or to the necessity of recording short-wavelength signals on an extremely narrow track. A tape designed especially for video use was thus needed to solve these and other conflicting problems; in particular it was necessary to devise a very smooth tape surface in order to maintain the requisite precise contact between the video head and the tape. This must be achieved without undue wear to either the tape surface or the head and without any sacrifice in the tape runability around the head drum.

When further developments led to the adoption of helical scan as the industrywide format, increasingly stringent conditions were added to the requirements of video tape. With helical-scan recording, it now became necessary for this extremely smooth-surfaced tape to be transported smoothly, not only around the rotating head drum, but also through the very complex tape path required to compensate for the angling of the tape as it is wrapped around the drum (more than 180° in the case of a two-head machine). This tape path became even more complex with the use of miniaturized cassette-loading mechanisms. As helical-scan equipment has become more sophisticated, the conditions to which the video tape is subjected have become correspondingly more severe. For example, in the process of playing back a still picture, the video head will, in principle, trace the same track on the tape over and over at a rate as high as 3600 passes per minute; the tape must be extremely well designed to withstand this sort of punishment.

The type of magnetic tape that has eventually been developed for video use is, therefore, the result of many compromises—a well-balanced solution to conflicting demands. The requirements of an ideal video tape can be summarized as follows:

1. *Smoothness of tape surface.* To be able to record signals of extremely short wavelength and play them back with both a high output and a high signal-to-noise ratio, the tape surface should be both very flat and very smooth, with surface irregularities less than 0.1 μm, in order to minimize spacing loss. However, an overly smooth surface will tend to make the tape stick to the fixed-head drum portion and to the erase and audio-control heads.

2. *Optimum head wear.* On one hand, it is desirable to extend video head life as much as possible. On the other hand, a certain amount of carefully planned head wear is vital to maintain the surface of the video head in optimum condition, both magnetically and mechanically, and in order to yield the maximum signal-to-noise ratio. This is true for both alloy video heads and single-crystal ferrite heads (Potgiesser and Koorneef, 1973; Kawamata et al., 1984a).

3. *Hardness of tape surface.* Both a hard tape binder and a hard tape surface are durable and will be resistant to scratches and dropout; the drawback of too hard a tape is that it will cause excessive wear of the video head (Kawamata et al., 1984b).

4. *Coercivity and magnetic flux density.* Both higher coercivity H_c and higher maximum induction B_m are necessary to yield higher output, especially at very short wavelengths; there are limitations to this, however, depending on the saturation magnetization of the head core material.

5. *Tape thickness.* For maximum recording time, the tape should, of course, be as thin as possible; however, very thin tape backings, such as tensilized polyester film, are unstable at high temperatures, in addition to being quite fragile mechanically, particularly on the edge portion of the tape (the stiffness of the tape is proportional to the cube of the tape thickness).

It can be seen, then, that the magnetic requirements for the video tape tend to be in direct conflict with the mechanical requirements for a good video recording; the optimum solution therefore becomes a compromise

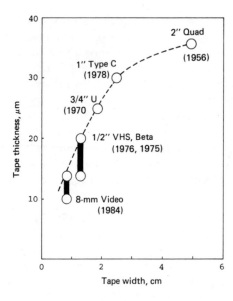

Figure 2.49 Historical relationship between tape thickness and width.

among a number of competing factors. Moreover, in most cases, the video tape is used not only to record the video signal but also to record the audio signal by high-frequency bias recording and to record the control signal by means of saturation recording at the edge of the tape. The magnetic tape for video recording, as a consequence, must meet and satisfy recording conditions which are more stringent and varied than those encountered in magnetic recording of any other form.

The tape width has been decreasing, along with an exponential decrease in tape consumption. Figure 2.49 shows the optimum balance between tape width and tape thickness; this balance also contributes to the decrease in the overall volume of a reel of video cassette tape. The necessity of maintaining sufficient error-free machine-to-machine compatibility has also contributed to very close tape tolerances in modern equipment; this is especially true of machines with extremely narrow-track-pitch recording, such as the Beta III and the VHS 6-h mode, where the tape width tolerance is within $\pm$ 10 μm.

A summary of 30 years of progress in improvement of the properties of video tape is shown in Fig. 2.50.

2.5.2 Video tape materials

In the beginning stages of video recording, gamma ferric oxide (γ-Fe_2O_3), the same material then used in audio recording, was used as the magnetic pigment in the recording tape. Later, a decrease was made in the particle size of the magnetic pigment in order to improve the signal-to-noise ratio while, at the same time, increasing the coercivity from 20 kA/m (250 Oe) to 28 kA/m (350 Oe) (Fig. 2.50). Due to the shape anisotropy of γ-Fe_2O_3, this pigment was a stable and inexpensive magnetic material, but there was an upper limit to the recording density which could be obtained.

Chromium dioxide pigment (CrO_2) was developed and first used for audio magnetic tape in 1961 (Swoboda et al., 1961); subsequently, it was applied to video use in 1970 in the $\frac{3}{4}$-in U-format recorder. Chromium dioxide tape demonstrated improved performance at short wavelengths (i.e., less than 5 μm) owing to its high coercivity of 36 kA/m (450 Oe) and greater acicularity (Darnell, 1961); however, it was both more abrasive and more expensive than the conventional γ-Fe_2O_3 tape (Table 2.6).

The coercivity of a magnetic recording medium can be increased by cobalt doping of the γ-Fe_2O_3 particle, utilizing the strong crystal anisotropy of cobalt. However, the increased coercivity induced in this way by crystal anisotropy is sensitive to the high temperatures and mechanical stresses which are inevitably generated in a video recorder. Eventually, after a great deal of research in this area, cobalt-adsorbed gamma ferric oxide tape (Co–γ-Fe_2O_3) was successfully developed (Umeki et al., 1974). This new tape material combined the crystal anisotropy induced by cobalt

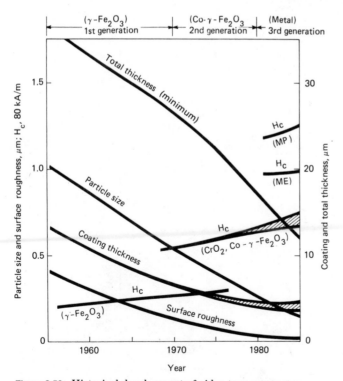

Figure 2.50 Historical development of video tape parameters.

TABLE 2.6 Video Tape Materials

	γ-Fe$_2$O$_3$	CrO$_2$	Co-γ-Fe$_2$O$_3$	Ba-ferrite	Metal		Units
					Fe, Co particle	Co, Ni evap.	
H_c	20–35	36–60	40–80	48–160	56–175	70–100	kA/m (4π Oe)
B_r	80–120	100–130	80–120	70–115	150–260	300–440	kA/m (emu/cm^3)
B_m	110–140	130–150	100–145	120–145	220–300	450	kA/M (emu/cm^3)
S	0.7–0.85	0.8–0.87	0.7–0.85	0.6–0.9	0.7–0.85	0.65–0.8	—
Particle size	0.2–0.8	0.3–0.5	0.2–0.6	0.05–0.1	0.1–0.3	—	μm
T_c	675	125	520	340	1000	1000	°C
Crystal structure	Spinel	Rutile	Spinel	Hexagonal	Body-centered cubic	Hexagonal	—
Shape	Acicular	Acicular	Acicular	Platelet	Acicular	—	—

SOURCE: Sugaya (1986).

with the shape anisotropy of acicular γ-Fe_2O_3. The coercivity can be controlled from 45 kA/m (550 Oe) to 135 kA/m (1700 Oe) with Co–γ-Fe_2O_3 tape; this tape also has the advantage of lower production costs than that of CrO_2 tape and is, at present, the most commonly used material for video tape.

Barium ferrite particles have been developed to effect an increase in the recording density without reference to the coercivity (Fujiwara et al., 1982). Since barium ferrite has a hexagonal-platelet crystal structure and a preferred magnetization along the C axis, it affords, when properly oriented, an optimum axis for magnetization perpendicular to the surface of the coating. There are still unknown areas to be explored in the use of this material, but it appears quite possible that barium ferrite tape may well be situated midway between Co–γ-Fe_2O_3 tape and metal tape.

A coated tape using metal pigment as the recording medium was first reported more than 25 years ago (Nagai, 1961). At that time, however, there was no market demand for such a high-coercivity recording tape, and ferric oxide tapes continued to be dominant in the video recording field for the next 20 years (Sugaya and Tomago, 1983). There are, in addition, other reasons why metal tapes were not readily adopted for practical use; these include chemical instability, noise problems, cost, and incompatibility (see Chap. 3 of Volume I). Metal tape was first introduced commercially in the new tape format of the 8-mm video; that is, an entirely new video recording system was devised to make use of this new tape. There are two types of metal tapes: metal-powder (MP) and metal-evaporated (ME) tape. The metal-powder tape is a coated tape in which the ferric oxide particles are replaced by particles of an acicular alloy, such as Fe-Ni-Co (Kawasaki and Higuchi, 1972; Chubachi and Tamagawa, 1984). The metal-evaporated tape uses an entirely different coating process, shown in Fig. 2.51, based on vacuum evaporation directly onto the surface

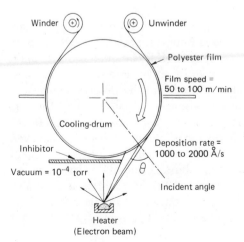

Figure 2.51 Mechanism for continuous vacuum deposition with oblique incidence for metal-evaporated (ME) tape.

of the tape substrate (Maesawa et al., 1982). This results in an anisotropic columnar grain structure which is formed by controlling the angle of incidence and the rate of deposition. The resulting grain structure yields different recording characteristics with different head running directions; it has no influence on either friction or the life of a still frame (Shinohara et al., 1984). Metal-evaporated tape has the following advantages:

1. Higher output at short wavelengths due to the very thin magnetic layer (0.1 μm)

2. Very high saturation magnetization, the result of direct deposition of the Co-Ni magnetic layer without a binder.

Metal-evaporated tape also has some drawbacks. The manufacturing process tape requires an overly long time for the evaporation of the magnetic layer, resulting in low productivity, and the layer thus produced is mechanically fragile and susceptible to cracks. The tape also requires a thin ($<$ 0.02 μm) protective oxidized layer and suffers from the problem of lubrication common to all hard metals. The surface of the evaporated metal alloy material therefore needs to be treated, by various methods, with special lubricants (Iijima and Hanafusa, 1984).

2.5.3 Video tape characteristics

2.5.3.1 Signal output and noise.
In order to increase the signal output at short wavelengths, both the coercivity and the magnetization of the magnetic recording layer are increased. When the coercivity of the video tape is changed, the optimum recording current is also changed, as shown in Fig. 2.52. Because the video signal is recorded without bias, the replay signal goes through a maximum when the recording current is increased; this is due to the recording demagnetization effect and is especially pronounced at high frequencies (see Chap. 2 of Volume I).

Most video signals are recorded by frequency modulation; thus, the carrier-to-noise ratio C/N is one of the most important factors in determining picture quality. Noise induced by the tape itself is classified into two main types: erase noise and modulation noise. In particulate tape, erase noise (the noise remaining after a tape is erased) can be determined by the degree of uniformity of dispersion of the magnetic particles and the surface smoothness of the magnetic layer. Modulation noise is a crucial factor in video recording since a carrier frequency signal is always recorded even if the video signal is zero. Modulation noise is caused by recorded-signal amplitude and frequency deviations attributable mainly to irregularities in the coating and on the tape surface (Fig. 2.53). Modulation noise is also induced by frequency modulation associated with time-base instability resulting from "creaking" (a phenomenon similar to stick-slip, but higher

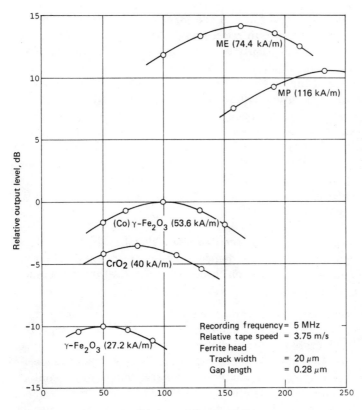

Figure 2.52 Relative output versus relative recording current for various video tapes (noise level: -39 dB).

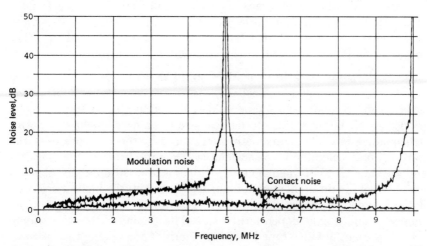

Figure 2.53 Typical modulation-noise spectrum of a video tape (amplifier noise $= 0$ dB).

in frequency) between the head and the tape. There are three essential conditions that must be met to reduce modulation noise: (1) maximum physical and magnetic uniformity of magnetic particles to give increased packing density, (2) maximally uniform particle dispersion, and (3) maximum smoothness of the tape surface.

The carrier-to-noise ratio of video tape has been improving steadily, at a rate of approximately 1 dB per year for the conventional Co-γ-Fe_2O_3 tapes and at an anticipated rate of several decibels per year for the new generation of tapes. It is expected that the next generation of recording media will be based on metal-evaporated tape, particularly for digital video recording. Doubtless, however, the introduction of this new type of tape will engender new problems to be overcome before it matures as the next standard recording medium.

2.5.3.2 Demagnetization of the recorded video signal.

The video signal recorded on tape can be adversely demagnetized as the result of a number of factors: magnetic contamination, head-tip magnetization, mechanical stress, and thermal stress.

Magnetic contamination. Many components of the machine which contact the tape can be contaminated by magnetized material from the tape; this contamination can erase the signal that has been recorded on that part of the tape. In order to eliminate demagnetization by this type of magnetic contamination, the tape binder should be made resistant to heavy contact. In addition, the tape transport mechanism should be sufficiently smooth to avoid scratching the tape; in particular, the video head tip must be finished to as smooth a surface as possible.

Magnetization of the video head tip. A magnetized head tip can erase a recorded signal during playback. To prevent such erasure, the coercivity of the video head core should be as low as possible. Fig. 2.54 shows the demagnetization of a recorded tape by magnetic fields caused by either magnetic contamination or by magnetization of the video head tip.

Mechanical stress. Mechanical stress on the tape can result from a number of causes, such as pinch-roller pressure and contact with either the video head or small-radius tape guides. Depending upon the magnetic properties of the tape, this mechanical stress can cause magnetostrictive demagnetization. The demagnetization caused by repeated passes of the tape will be accelerated if the acicularity (the ratio between the long axis and the short axis) of the magnetic particles is low. Different acicularities are obtained by different types of magnetic pigment; the highest acicularity is obtained in chromium dioxide, the lowest is obtained by granular-shaped (cubic) cobalt-doped ferric oxide. The gamma ferric oxide widely used at present in video tape has an acicularity approaching that of chromium dioxide.

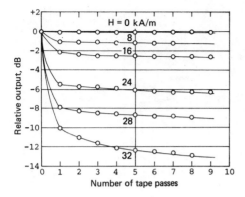

Figure 2.54 Demagnetization of a signal (λ = 5 μm) recorded on Co-γ-Fe$_2$O$_3$ tape after repeatedly applied magnetic fields of different strengths.

If various tapes on which are recorded signals of different wavelength are played, the effect of demagnetization on each pass is greater at the shorter wavelengths. This is true even for tapes with magnetic particles of high acicularity, such as chromium dioxide, run under identical mechanical pressure (Fig. 2.55).

The hardness of the pinch-roller material (when used) is also a crucial factor affecting mechanical demagnetization and should not exceed a figure of 80. As the size of the modern video cassette has become smaller, the tape guides and other projections over which the tape must pass become so small in radius that the surface presented to the tape approximates a sharp edge. The sharper the edges of these projections, the greater the mechanical stress on the tape as it passes over them. Figure 2.56 shows this phenomenon. The mechanical stress on the tape will also

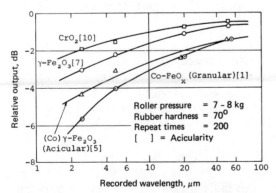

Figure 2.55 Losses in output produced by mechanical stress as a function of wavelength and particle acicularity (shown in brackets). The stress consisted of 200 passes through a pinch roller (pressure = 7 to 8 kg, hardness of rubber = 70).

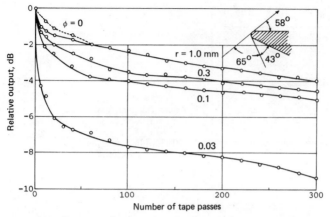

Figure 2.56 Demagnetization of a signal ($\lambda = 2.5$ μm) recorded on particulate Co–γ-Fe$_2$O$_3$ tape, after several passes under $40g$ of tension in contact (on the magnetic coating side) with sharp edges of different radii r.

be greater as the angle of the tape pass becomes greater over these sharp-edged projections (Woodward, 1982). Similarly, the tension of the tape as it is run over these projections directly affects the mechanical stress. These three factors, the radius, the wrap angle, and the tension, determine the amount of mechanical compression to which the binder system of the tape is subjected.

If the contact is with the tape backing, the demagnetization will be drastically reduced; thus an extremely sharp edge of 0.03-mm radius will produce a comparatively low demagnetization. This is approximately equal to that which would be produced by the much gentler surface of a 0.3-mm radius projection in contact with the magnetic pigment side of the tape.

Thermal stress. Heavy, prolonged contact between the video tape and either the head tip or the rotary-head drum portion will produce heat by friction, demagnetizing the signal recorded on the tape. In theory, the rate of demagnetization by thermal stress can be determined in the main by the temperature dependencies of crystal anisotropy and magnetization, and second, by the Curie temperature (T_c) of the magnetic pigment (see Table 2.6). The magnetic properties of nonacicular Co-doped ferric oxide are strongly affected by both mechanical and thermal stresses. This has proved to be a fundamental defect in Co-doped ferric oxide and caused it to be abandoned as a material for video tape. Chromium dioxide tape, which has a very low T_c (120°C), is not subject to large demagnetization by thermal stress since the coercivity is relatively constant below T_c.

2.5.4   Head-tape interface problems

2.5.4.1   Dropouts. A dropout is the disappearance, temporary or otherwise, of the recorded signal on playback. Dropouts may be caused by physical peeling of the recorded portion, by spacing loss caused by asperities on the tape surface, or by debris on the head or tape (Geddes, 1965). The dropout rate increases when the recorded wavelength is very short. For broadcast use, a typical maximum acceptable dropout rate is 5 dropouts per minute; for consumer use a typical maximum rate is 15 dropouts per minute. In general, the dropout rate will decrease during the first 100 passes of the tape, as the video head polishes the tape, removing particles of dust and other debris from the tape surface. The dropout rate will then remain fairly steady during the next 100 passes, increasing gradually after approximately 200 passes. Tape life, as calculated by the maximum acceptable rate of dropouts, is approximately 300 passes when the tape transport mechanism is in good working order; this value may, however, be greatly affected by poor environmental conditions. In the absence of adverse environmental conditions, the dropout rate may be minimized by several means, including careful control of the tape production process, use of a conductive backcoating, use of a durable binder formulation, and cleanly slit tape edges. The video head surface should be both smooth and durable. Soft metal material will be damaged by tape friction; conversely, a rough head surface will damage the tape surface. Finally, greater head-tip projection is desired to ensure minimum dropouts (this is particularly important for quadruplex recorders).

All of the above factors need to be taken into consideration, but each of these measures by itself can have negative influences on other areas of the recording process. Thus the design of the video tape and recorder mechanism must be based on a series of compromises carefully chosen to optimize all aspects of the video recording process.

2.5.4.2   Head clogging. *Head clogging* refers to the accumulation of debris or contamination from the tape coating on the surface of the video head. Head clogging interferes with both the recording of a signal and the playback of a recorded signal in the short-wavelength range. This occurs as the result of the large spacing loss created by the clogging and is an extremely serious problem in video recording. Head clogging may occur under several conditions. Contamination of the tape surface may cause a clog and is most frequently encountered in the form of fingerprints or dust on the tape. The use of a cassette or cartridge tape container is essential for the prevention of head clogging. Heavy friction between the head and the tape may also cause a clog. Friction is a particularly serious problem when both metal tapes and metal heads are used. Also, the binder formulation and the surface treatment of metal tape is crucial. Environmental conditions are important, the major problems being extremes of low or high humidity and dust in the air. Professional machines are normally used in a controlled environment, but most consumer equipment must be operated in ambient temperature and humidity.

Finally, the composition and formulation of the binder are the most important factors in the prevention of head clogging. The binder must meet two conflicting conditions: on the one hand, it must act as a lubricant; on the other, it must serve as an abrasive to clean the contaminated head surface as well as to polish the head surface appropriately to keep it in good condition.

2.5.4.3 Brown stain. When alloy magnetic materials (such as Sendust) or amorphous materials are used for the tip portion of the video head, another type of head-to-tape interface problem can occur; this is particularly evident in conjunction with the use of metal-powder tapes.

The friction between the tape and the head can affect the surface of the video head tip, turning it into a brown-colored coating. This "brown stain" is an extremely thin nonmagnetic coating caused by a chemical reaction that occurs as the result of the friction. This stain can be the cause of a substantial spacing loss.

The moisture in the air tends to act as a natural lubricant, and when the relative humidity is low (e.g., less than 10 percent), the formation of the brown coating is accelerated. Although a solution to this problem has yet to be discovered, a tape binder with abrasive properties can be used to remove the brown stain once it has formed. The use of metal-evaporated tape, which has no binder, will necessitate a different type of solution which does not rely on the use of a tape binder. We can anticipate that the special treatment of tape surfaces will constitute an important part of the future technology of video recording.

2.6 Duplication of Video Tape

2.6.1 Development of video tape duplication

Originally, the video tape recorder was used primarily for time-shift purposes, both for broadcast and home use. Only later did video recording come to be used extensively for the playback of professionally recorded (prerecorded) programs.

In the production of prerecorded audio cassette tapes, the dubbing speed may be as high as 64 times normal playback speed. An increase of this extent translates the audio frequencies up to approximately 1.3 MHz, which may be effectively recorded on specially designed high-speed master and slave tape recorders. At this speed, if the master tape recorder is connected to 10 slave tape recorders, the tape duplication efficiency will be 640 times that of simple copying from machine to machine in real time.

The typical consumer-use video tape recorder, on the other hand, operates at high frequencies of almost 7 MHz, and it is very difficult, if not impossible, to increase the head-to-tape speed for duplication purposes. Consequently, video tape duplication is carried out at present by the rather primitive means of connecting a series of slave recorders to the master video recorder and recording a program played back from the master recorder in real time.

There is, however, an attractive alternative in the contact duplication of magnetic tape by either anhysteretic or thermal magnetization mechanisms. In the contact duplication method, a copy can be made directly from the magnetically recorded master tape to blank slave tapes, in much the same way that contact prints can be made from photographic film. The negative film is analogous to the master tape, which is recorded in a mirror-image pattern; the photographic paper is analogous to the slave tape on which the magnetically recorded signal is transferred; the light corresponds either to the magnetic transfer field in the anhysteretic transfer method or to heat in the thermal transfer method.

The original concept of contact duplication by the anhysteretic method is rather old (Mueller-Ernesti, 1941), and was applied quite early to audio tape duplication (Herr, 1949; Camras and Herr, 1949). The audio signal on tape is, however, distributed over a very wide range of wavelengths from a few micrometers up to several thousand micrometers. The contact duplication efficiency of the lower frequency (longer wavelength) signals is lower than that of the higher frequency (shorter wavelength) signals, as shown in Fig. 2.57 (Sugaya, 1973). As a consequence, contact duplication of audio signals was actually applied only for rather specialized institutional equipment. The FM video recorded signal is, however, concentrated mainly in the short-wavelength range. Thus, the major problem to be faced in contact duplication of a frequency-modulated video signal centers around the increase in the transfer efficiency of very short-wavelength signals (Sugaya et al., 1969). Later, the development of chromium dioxide video tape, which has a very low Curie temperature (125°C), made it possible to consider the use of a thermal transfer method of contact duplication (Hendershot, 1970). Reports of this possibility stimulated further development of the contact duplication process for video tape (Van den Berg, 1969; Ginsburg, 1970). At the same time, theoretical investigations and analyses of contact duplication were begun (Hokkyo and Ito, 1970; Kobayashi and Sugaya, 1971a, 1971b; Tjaden and Rijckaert, 1971; Mallinson et al., 1971).

In spite of the concerted efforts of numerous researchers into both the

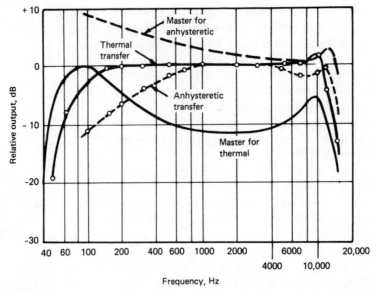

Figure 2.57 Comparison between thermal and anhysteretic transfer methods at audio frequencies (relatively long wavelengths). The tape speed is 19 cm/s.

theory and feasibility of contact duplication, its practical implementation has been delayed. This delay is attributed partly to technical difficulties, such as the slow development and availability of suitable master tape. Perhaps more important is the fact that direct machine-to-machine duplication has actually proved to be sufficient to handle the demand for tape copying. This has become particularly true as mass production techniques have resulted in a sharp decrease in the price of recorders, along with improved reliability; these can then be economically used in quantity as slave machines in copying. The rapid growth of the prerecorded video tape business may, however, eventually render direct machine-to-machine video copying obsolete and spur further efforts into the development of the contact duplication process.

2.6.2 Head-to-head duplication method

In this method, both the video and the audio signals from the master recorder are sent, via a distribution amplifier, to a series of recorders operating in real time. At large duplication houses, several hundred such slave recorders may be connected in parallel in order to gain duplication efficiency. The decreased price of video cassette recorders, accompanied with comparable increases in both reliability and head life, has enabled the head-to-head method to become the most effective and inexpensive way of duplicating video tapes. This is in spite of the necessity for replacement by hand of the video cassettes in the slave recorders and for the mainte-nance of the extremely large number of slave recorders.

The head-to-head method provides the following advantages:

1. *A high degree of flexibility.*   Any desired quantity of duplication can be done efficiently by simply changing the number of slaves used in the duplication process, and inexpensive consumer recorders can be used with only slight modification.

2. *Excellent picture and sound quality.*   Modern mass-produced recorders are sufficiently reliable and of high enough quality to give good results in both sound and picture quality.

There are, however, some disadvantages to the head-to-head duplication:

1. *Manual cassette changing.*   This is a laborious process, particularly in the duplication of shorter tapes, such as music and instructional pro-grams, which have an average playing time of about 30 min.

2. *Labor-intensive quality assurance.*   Since video recordings are still subject to such problems as head-clogging during the recording pro-cess, duplicated tapes must be thoroughly checked by the human eye. Unfortunately, it has proven impossible thus far to develop a practical automated procedure for quality control.

As a solution to some of these problems, extremely long video tapes (for example, 5 km, or some 20 times the length of the 250-m VHS cassette tape) on flangeless open reels ("pancakes") are used on the slave recorders instead of the standard cassettes. The long tape on these open reels is run continuously without stopping, thus recording the material from a series of several master programs back-to-back. Either a pair of master record-ers is used to switch the source-program cassettes back and forth contin-uously, or an endless tape in a special bin is used as a continuous program source. This sequential duplicating procedure using a long, open-reel tape can greatly decrease the necessity for changing tapes. Quality-inspection time is also shortened, since most of the problems occur predictably dur-ing the starting and stopping of the tape and not while it is continuously running. This duplication method is most effective when the recorded program is a short one, such as a 30-min music program. Once a series of these programs are duplicated sequentially on the long tape, they can be cut to length and loaded into cassettes by an automatic tape-loading machine, which can be operated independently, off-line from the dupli-cation process. The next improvement anticipated in the head-to-head duplication process is the doubling or tripling of the speed of both the head drum and the tape, provided the cost can be kept commercially feasible.

Although the head-to-head duplication method is quite primitive in concept, it has proved to be the most practical method of video tape duplication thus far; and with the modifications discussed here, it is still the main method for commercial video tape duplication.

2.6.3   Contact duplication methods

2.6.3.1   The anhysteretic process.   In this process, the video signal, along with the audio and control signals, is recorded on a special master tape with approximately three times the coercivity of the slave tape. These signals are recorded by a specially designed master recorder in a format that results in a mirror-image recording of the original signals. The magnetically coated surfaces of both the master and the slave tapes are then placed in direct contact with each other while an alternating magnetic transfer field of carefully adjusted intensity is applied. This instantaneously causes the magnetic field of the master tape to be printed anhysteretically onto the slave tape. Two types of anhysteretic contact duplication have been developed: the parallel-running method and the bifilar-winding method (Fig. 2.58).

The parallel-running method, in principle, can be applied to the contin-

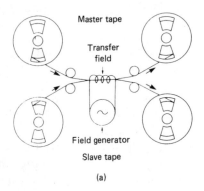

(a)

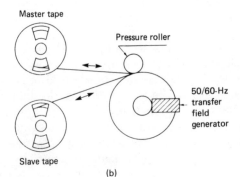

(b)

Figure 2.58 Principle of anhysteretic tape transfer. (*a*) Tape running in contact; (*b*) bifilar winding.

uous (sequential) duplication process described above, as well as to a type of double or triple duplication process which makes use of a single master tape and two or three duplication stations (Crum and Town, 1971). The greatest problem encountered in this method has been the decrease of the duplicated output of short-wavelength signals as the result of slippage between the master and slave tapes during the duplication or transfer process. A compressed-air drum technique was developed to reduce this problem (Kihara et al., 1983).

An alternative is to use a bifilar-winding method to eliminate the slippage problem entirely. In this method, the master and slave tapes are first wound together on a separate reel hub, using a pressure roller to squeeze out any air which is trapped between the two tapes; then the magnetic transfer field is applied. Tape slippage is zero, eliminating any error from this source, and the quality of the output of the transfer-recorded signal, as well as the reliability, is higher than that of the parallel-running method. After this transfer-recording procedure, both the transferred slave tape and the master tape are rewound back onto their respective reels. The drawbacks to the bifilar-winding method arise mainly from this winding and rewinding process. The necessity for rewinding limits the length of both the master and slave tapes. Also, audio print-through of the longer wavelengths occurs as the result of the tapes being tightly wound together during the magnetic transfer process. Separate audio recording is needed to avoid this print-through. This type of video duplication in its present state of development is not suitable for mass duplication; however, its flexibility should make it well-suited to medium-quantity duplication needs.

2.6.3.2 Characteristics of the master tape.

In audio tape recording with a high-frequency bias field, the magnetic signal field decreases at the same rate as the bias field in the recording process. In the anhysteretic transfer process, on the other hand, the signal field which is supplied from the master tape surface remains constant, and only the bias, or transfer, field decreases. This is known as the *ideal anhysteretic magnetization process* (see Chapter 2 of Volume I). The anhysteretic susceptibility of the slave tape increases in proportion to the intensity of the transfer field, as shown in Fig. 2.59. The transfer field for contact duplication should, therefore, be at least 1.5 times higher than that of the slave-tape coercivity; at the same time, however, the bias field will tend to erase the recorded signal on the master tape, as shown in Fig. 2.60. Under optimum conditions, however, the master-tape signal should not be erased by more than 1 dB at the first duplication, and further erasure should be negligible. Thus, the slave tape's output signal produced by contact duplication can be maintained at a constant level even after thousands of duplications.

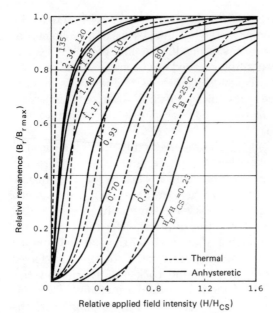

Figure 2.59 Remanent magnetization versus applied field curves for various values of transfer field H_B or transfer temperature T_B using CrO_2 tape. The anhysteretic curves are computer-simulated. H_{cs} is the coercivity of the slave tape.

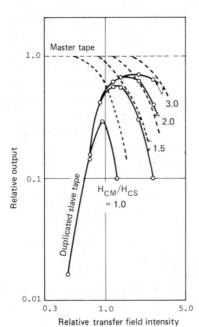

Figure 2.60 Recorded master-tape output and duplicated slave-tape output for different transfer fields (H_{cm} and H_{cs} are the coercivities of the master and slave tapes) *(Sugaya, 1973).*

According to both theoretical and experimental results, the coercivity of the master tape should be at least 2.5 times higher than that of the slave tape. The coercivity will, however, be limited by other factors, such as the availability of suitable magnetic materials for the master tape and video head saturation. The coercivity of the master tape required for VHS or Beta tape duplication is, for example, at least 120 kA/m (1500 Oe). The magnetic material used for such a master tape was originally high-coercivity, cobalt-modified iron oxide, but this has begun to be phased out in favor of metal particles with both a higher coercivity of 154 kA/m (1930 Oe) and a higher magnetic remanence of 250 kA/m (emu/cm^3) (Chubachi and Tamagawa, 1984).

In order to obtain a higher transfer efficiency for signals of long wavelengths, the remanent magnetization and the coating thickness of both the master tape and the slave tape should be as great as possible. For short-wavelength signals, the surface roughness of both tapes should be as small as possible. The signal deterioration at short wavelengths, in dB, in general follows the spacing-loss formula 54.6 d/λ, where d, in this case, is the sum of the surface roughness of both tapes. In practice, these conditions for master tapes are not easily satisfied, and problems still exist stemming from head clogging, head wearing, and dropouts.

2.6.3.3 Contact duplication by the thermal process. At present, chromium dioxide is the only available material suitable for use as a slave tape in the thermal duplication process. The thermal characteristics of CrO_2 tape are shown in Fig. 2.61. In order to minimize the demagnetization effect during the cooling process, H_c should recover faster than B_r does. The apparent

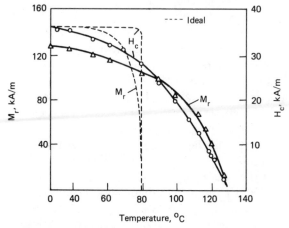

Figure 2.61 Temperature dependence of remanent magnetization M, and coercivity H_c of CrO_2. Dashed lines, ideal thermal characteristics.

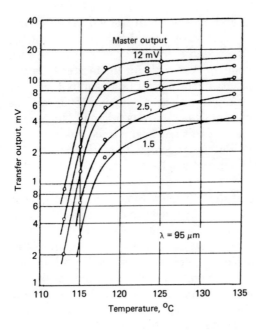

Figure 2.62 Temperature dependence of the thermal transfer of CrO_2 slave tape from γ-Fe_2O_3 master tapes having different outputs.

susceptibility of CrO_2 tape achieved by thermal transfer is higher than that achieved by anhysteretic means (Fig. 2.59); thus the recording efficiency of thermal transfer is higher than that of the anhysteretic method.

In the thermal duplication process, the characteristics of the slave tape are more important than those of the master tape. In principle, any video tape can be used for the master tape in this process provided it has a high Curie temperature, although a higher magnetic field from the recorded signal will yield better results, as shown in Fig. 2.62. In practice, high-coercivity tape, such as $(Co)FeO_x$ or metal tape, is used for the master tape in thermal duplication. In carrying out a duplication, a CrO_2 slave tape is heated up to the Curie temperature and then cooled by placing the master tape in contact with the surface of the heated slave tape. In this cooling process, the magnetic moment acquired by the slave tape will be fixed in accordance with the magnetic pattern on the master tape. Cooling can be very rapid; therefore, the slippage between the master tape and the slave tape is not as critical as it is in the anhysteretic transfer process.

The main problem encountered in the thermal duplication process is the thermal deformation of the tape-backing material (polyester film has 2 to 3 percent shrinkage at 150°C). Significant deformation spoils the tape compatibility, particularly for such narrow-track recording systems as VHS and Beta. It appears, however, that these problems can be solved by the application of a high-power laser beam (Cole et al., 1984). In this procedure, the laser beam is focused onto the CrO_2 coating from behind, through the backing of the slave tape, for a period of 2 ms; this brief exposure is insufficient to cause any deformation of the polyester film, but it is enough to heat the CrO_2 coating to the Curie temperature at a tape speed of 1.65 m/s, which is nearly 50 times VHS or Beta standard-play tape speed. The maximum tape speed in this duplication process is steadily increasing along with developments in high-powered lasers. The laser-beam thermal-duplication process makes it possible to duplicate video signals at very high speeds. It is anticipated that this process may be successfully applied, for example, to the mass duplication of music video tape.

2.6.3.4 The double-transfer method.

One of the least satisfactory features of contact duplication is the necessity for a specially designed, mirror-image master recorder. The double transfer method was proposed as a solution to this (Nelson, 1970; Dickens and Jordan, 1970).

As shown in Fig. 2.63, the first step entails the transfer, by thermal means, of the original (master) tape, which has the same tape format as the slave tape, onto a CrO_2-surfaced transfer drum. Then this transferred signal, which now exists as a mirror image on the surface of the transfer drum, is duplicated, either by anhysteretic or thermal means. In this manner, the original tape signal can be transferred onto the slave tape without the necessity of either a special master tape or a mirror-imaging master recorder. If the thermal-transfer method is used in the second step, the slave tape should be CrO_2 tape; if the anhysteretic method is used, any tape can be used for the slave tape, provided the coercivity on the transfer drum is at least 2.5 times higher than that of the slave tape.

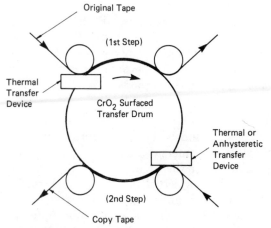

Figure 2.63 Principle of the double-transfer method.

2.6.4 The future of video tape duplication

In the modern "information era," people are becoming more and more dependent on information transmitted by television. Both cable television networks and direct broadcast by satellite have become extremely important means of information transmission, particularly in the United States. It is expected, however, that the development of inexpensive video recording equipment and software rental will bring about drastic changes in this situation.

Information broadcast over the air waves is limited in terms of both transmission time and the number of programs. Recorded video software, on the other hand, permits the viewer to see whatever he or she wants at any time, and the number of programs is virtually unlimited. It is expected, therefore, that prerecorded video tape media will soon come to occupy a significant share of the information market. Tape duplication in the future will become a highly specialized enterprise, with the choice of methodology depending on the application or purpose of the duplicated tape, more or less as follows:

1. *Mass duplication of long programs (e.g., movies).* The head-to-head method, using many slave recorders, will continue to be the main approach. Tape systems will eventually shift from tape-cassette format to the use of pancake tape.

2. *Mass duplication of short programs (e.g., music).* A similar trend is envisioned; the shift from cassettes to pancake tape, however, will be accelerated. Thermal transfer will be used where quick delivery is necessary.

3. *Duplication of medium-sized lots.* The head-to-head approach will continue to be used for the time being; contact duplication may be used more extensively in the future.

4. *Duplication of many kinds of small-sized lots.* Both the head-to-head method and the automatic contact-duplication method will be used. If the double-transfer machine can be commercially developed, then the copying of video tape will become as common an everyday process as xerography and other plain-paper, dry-copying methods.

It can be anticipated that should extra-long-play (ELP) video software begin to be commercially successful, and if the price drops sufficiently (e.g., by contact duplication), then consumers will find it advantageous to purchase video software outright rather than rent it. This phenomenon would have the effect of increasing the actual number of prerecorded video tapes produced and would result in an increase in the extent of the mass production techniques involved in the tape-duplication industry. Accompanying this trend would be the usual benefits accruing from the introduction of wide-scale mass production—increases in quality and standardization, accompanied by even further decreases in price.

2.7   Future Prospects

Continued increases in recording density will soon make feasible such new possibilities as the video disk recorder, wideband signal recording (using high-definition television recorders and digital video recorders), and advanced recording processes (for example, perpendicular magnetic recording and nonmechanically scanned recording).

2.7.1   Video disk recording

The video tape recording formats which have been discussed so far in this text have all used longitudinally wound tape. These formats are volumetrically efficient in terms of the amount of information that can be stored; they do have the drawback, however, of a comparatively long access time for the retrieval of data recorded on the tape. For quick retrieval in television broadcasts, limited use is being made at present of video disks to record selected portions of the video signals in a program. The shortened access time afforded by this procedure is used, particularly in sports and news broadcasts, for instant replay, for slow motion, and for inserts of still pictures as an overlay on the main video picture. Figure 2.64 shows the principle behind this type of slow-motion recording. In order to be able to record and erase continuously without interruption, two plated rigid disks (with a total of four recording surfaces) and four heads are used alternatively in the record-erase-move process shown in Fig. 2.64. A complete single field of the video signal is recorded in a concentric circle on the disk; this requires a very high recording density. To reproduce a recording in slow motion, the video head makes multiple passes over the same concentric track. Conventional helical-scan recorders can also be used for slow-motion and still-picture playback, but the paths of the head when recording and then playing back in slow motion are not identical, as shown in Fig. 2.30. Hence, with helical-scan recorders, it is difficult to achieve either precise random access of the desired portion of the program or slow-motion and still pictures of perfect quality.

More advanced slow-motion video recorders use a digital signal, with the result that the single video field can be broken up and recorded on several tracks instead of on one concentric track, as is done in analog recording. This type of recording allows a lower recording density, thus permitting the use of a conventional coated rigid disk which is today less expensive than the plated disk.

In order to decrease the cost of disk recording media even further, flexible magnetic material can be used for video disk recording. This can con-

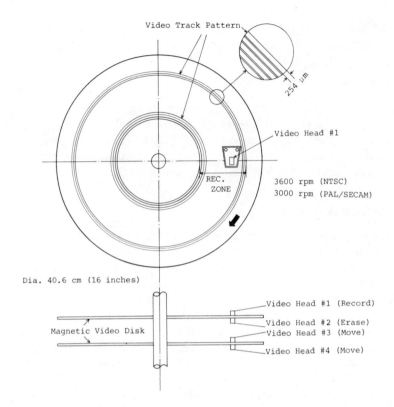

Figure 2.64   Principle of slow-motion video disk recorder using hard disk.

sist of unoriented coated video tape material in the web form, provided by the manufacturer before slitting. In one embodiment, a flexible disk approximately 30 cm in diameter is used; this is kept taut by clamping it around the perimeter in a fashion similar to the membrane of a drum. The resulting flexibility of the magnetic sheet in this mounting format yields excellent head-to-medium contact but this disk is more expensive to manufacture than the more conventional, center-supported flexible disk (Kihara and Odagiri, 1975).

The center-supported type of magnetic disk first came into use in video recording as a comparatively inexpensive means for single-frame recording and for use in slow-motion video recorders. To keep both the size (approximately 20 cm) and the cost of the disk to a minimum, the recording time is limited to 10 s (yielding 600 field pictures); this relatively small memory capacity limits its use to such areas as the medical field (x-ray recording) and professional sports (on-the-spot analysis of the movements of athletes).

As more advanced types of recording media (such as metal powder) become available for use in video disks, it will prove feasible to utilize extremely small disks of only a few centimeters in diameter to record still images. Devices using this format are known as video (or electronic) still cameras. Figure 2.65 shows the design of a 47-mm flexible video disk with sufficient storage capacity to record 25 frames (50 fields). This type of miniature flexible disk, used in conjunction with a charge-coupled-device (CCD) miniature television camera, can form the basis of an electronic still camera comparable in size and weight to a conventional 35-mm single-lens reflex camera (Kihara et al., 1982). The picture quality of the electronic still camera is inferior to that produced by the conventional camera; but unlike conventional film, the video disk can be used repeatedly, and the recorded picture can be viewed enlarged on a television screen and can be transmitted over a telephone line by a simple encoding.

The resolution of a still picture needs to be higher than that of a moving picture. Taking the specifications of the 8-mm video format ($f_{max} = 5.4$ MHz) as a standard for moving video pictures, the recordable frequency in an electronic still camera should therefore be higher than 5.4 MHz. The head drum diameter of 8-mm video is 20 mm for single-head helical-scan recording or 40 mm for the two-head type. It has been found that to

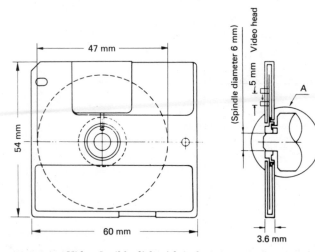

Figure 2.65 Video flexible disk with jacket.

achieve the same minimum recordable wavelength in the two-dimensional format of the video floppy, a disk diameter of 47 mm (slightly less than 2 in) is the minimum size. A video disk of this size can therefore be used to record a frequency of 7.5 MHz with a single video head. This is a sufficient amount higher than the 5.4 MHz of the 8-mm video format for use in the video recording of still pictures. In order to achieve improved color quality, the color signal in this format is recorded sequentially as a frequency-modulated color-component signal (R−Y, B−Y). The proposed frequency allocation is shown in Fig. 2.66.

It is expected that many more applications of the video flexible disk will be found in the future. This will be the result of advances in both hardware and supporting software. Advances in hardware will consist of technological improvements and decreases in prices of materials and components used in the flexible disk format. Advances in software will include the development of new businesses, such as the production of new types of color photo albums, together with facilities for displaying customers' photos on television screens. Also, the development of new methods of taking and processing pictures will be possible, such as the ability to take many pictures rapidly of the same scene using the same disk and then

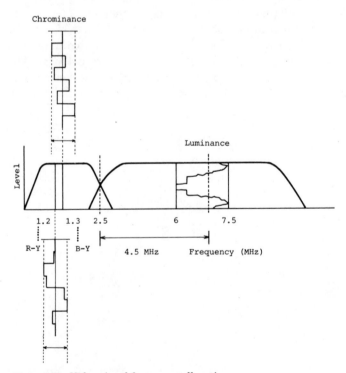

Figure 2.66 Video signal frequency allocation.

select the best picture by projecting them on a television screen for inspection.

2.7.2 Recorders for high-definition television

The majority of the television systems throughout the world at the present time use the NTSC, PAL, or SECAM systems. In the future, however, we can look forward to the development of a new format—high-definition television. One type of high-definition television is shown in Table 2.1. This system requires a bandwidth of 20 to 30 MHz, which is almost five times wider than that needed for conventional television. In order to record such a wide frequency range (from 50 MHz up to approximately 70 MHz for FM recording), a single-crystal Mn-Zn ferrite head could be used (Abe, 1981). As a first step in an attempt to achieve the necessary rotation speed (double or triple that of conventional video recorders), standard type B and type C recorders were modified to increase the head-to-tape relative speed. The second step in the development of high-definition baseband-signal video tape recording for broadcast use will be the establishment of broadcast standards for the worldwide exchange of programs among television studios.

Satellite transmission is limited by international agreement to a base bandwidth of 8 MHz; this is insufficient for high-definition television broadcasts, which need up to four times this bandwidth. To solve this problem and to provide for the satellite broadcasting of high-definition signals, the MUSE (multiple sub-nyquist encoding) system was developed. This compresses the frequency bandwidth to one-quarter of its original width. This system makes it possible to transmit the high-definition signal through the conventional satellite bandwidth.

The third step will be the development of consumer-type recorders for high-definition television signals. Consumer-use recorders must be able to record both satellite broadcast signals and prerecorded programs. The tape format must be both reasonably priced and in a convenient cassette form. This will necessitate even greater advances in the recording density to be able to produce high-quality recordings over such a wide frequency band while maintaining a compact cassette size comparable to that of present-day cassettes.

2.7.3 Digital-process analog recorders

In the audio field, recording technology is rapidly shifting from analog to digital recording in the consumer area, and the optically recorded compact disk is becoming an increasingly popular audio recording medium. The same trend toward an emphasis on digital technology can be expected in the video field as well. However, as we have seen, the video signal bandwidth is more than 100 times wider than that of the audio signal. From the point of view of recording efficiency, digital recording needs several times the area of tape that is required by conventional analog recording. Although the rate of tape consumption is still decreasing every year, it is unlikely that digital video recording techniques will become feasible for consumer use. In the near future, digital techniques will probably be limited to studio use, which must have a minimum of deterioration in picture quality after multiple generations of duplication for editing.

On the other hand, very large digital circuitry can now be integrated into a small silicon chip, and digital processing has been introduced successfully in the consumer color television set. In the same way, modulation and demodulation for video recording are processed by digital means in conjunction with many very precisely tuned digital filters. Thus, conventional video circuitry can be integrated into two chips, an analog-to-digital and digital-to-analog converter, and a digital-process video circuit (Mehrgardt, 1985). This is referred to as the *digital-process analog recording method* (Fig. 2.67). With this method, a conventional video tape format, such as VHS, can be used with complete compatibility. Furthermore, any television standard such as NTSC, PAL, or SECAM can be operated with the same circuitry simply by changing the software. Many other new functions, such as noise reduction using field-picture relationships, enhancement of a deteriorated picture, and compensation for dropouts and crosstalk, can be introduced in conjunction with digital memory devices. In summary, the digital-process analog recording method is a very attractive approach for the next generation of analog consumer video recorders.

2.7.4 Advanced recording processes

The longitudinal format is beginning to reach its practical limits in terms of recording loss through self-demagnetization. The minimum recordable wavelength in commercially available video recorders is 0.57 μm at present (for 8-mm video PAL); extrapolating from the historical trend line (Fig. 2.7), a minimum wavelength of less than 0.4 μm can be anticipated. This

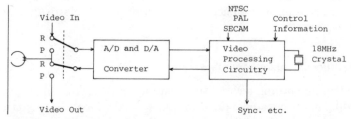

Figure 2.67 Digital signal processing for a video tape recorder using two LSI circuits.

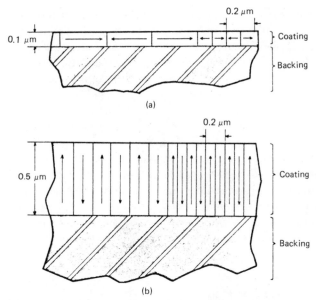

Figure 2.68 Typical example of very short-wavelength signal recording. (a) Longitudinal recording on thin coating; (b) perpendicular recording on thicker coating.

has already been achieved in the research laboratory and will become commercially available in approximately 1990. At such very short wavelengths, three main losses affect video recording. First, the gap loss at short wavelengths will demand very narrow gaps, on the order of 0.1 μm. This can be achieved by special lapping techniques to eliminate the worked layer on the head core material; however, a gap depth shallow enough to achieve a sufficiently high output signal can be created only at a substantial sacrifice of head life (Sugaya, 1985). Next, spacing losses will emphasize the need to solve head-to-tape interface problems as tape surfaces become even smoother. Finally, self-demagnetization losses will increase as the recorded wavelength becomes shorter. Two methods have been developed to minimize self-demagnetization loss, as shown in Fig. 2.68: (1) the use of magnetic tape with extremely thin magnetic coatings (e.g., metal-evaporated tape with a coating thickness on the order of only 0.1 μm) can substantially reduce self-demagnetization loss (Mallinson, 1985); (2) self-demagnetization loss can, in principle, be eliminated by using the perpendicular vector component in the recording process, instead of the conventional longitudinal vector component (Iwasaki, 1980, 1984). Perpendicular recording may be accomplished by the use of a specially designed recording medium, such as Co-Cr film—which favors recording perpendicular to the tape surface—used in conjunction with a perpendicular head (Iwasaki et al., 1980). This process will permit the recording of signals with wavelengths even shorter than 0.4 μm. The technology underlying perpendicular recording (see Chap. 2 of Volume I) yields other benefits applicable to the improvement of the recording density, particularly when video information is recorded digitally.

One traditional method of decreasing tape consumption is to decrease the track to very narrow widths. A magnetoresistive head can be used to play back extremely narrow tracks (2 μm) with sufficient output (Kanai et al., 1975). Another possibility for narrow-track recording is to use magnetooptical recording, which is described in detail in Chap. 6. Magnetooptical recording technology will be effective for video disks as well as for tape recording, provided high-power lasers, semiconductor devices, and commercially viable, high-sensitivity (high signal-to-noise ratio) magnetooptical recording media are successfully developed for this purpose. The combination and mutual reinforcement of the applications of these various advanced recording techniques will yield even higher recording densities in the future.

2.7.5 Nonmechanically scanned recording

The video recording techniques discussed so far in this chapter have been limited mainly to rotating-head formats. In video recording technology, it is possible that the rotating-head mechanism will eventually be supplanted by a nonmechanically scanned device. Schemes for such nonmechanically scanned recording formats have been proposed (see, for example, Camras, 1963b; Peters, 1964), but so far these have not yet been made commercially practical. In this section, we review some nonmechanically scanned magnetic tape recording methods and discuss their future prospects.

2.7.5.1 Magnetically scanned recording. Several schemes have been proposed to modulate the permeability of a single wide-track recording gap so that only a small zone of the total head has high enough permeability to allow recording to take place. This high-permeability zone is then scanned along the gap to produce transverse recording on a slowly moving tape. One approach to achieving control of the permeability is to saturate a portion of the head in one direction and another portion in the opposite direction (Peters, 1964). If these two portions each correspond to one-half the head width, there will be an unsaturated zone between them that can, in theory, be used for recording from a coil wound around the complete head. The position of this small recording zone can then be moved by increasing the width of one saturated portion with respect to the other. Difficulties with this type of scanning head lie in the requirement that certain portions of the head core be saturated without inadvertently producing a field from that portion of the head. Another magnetic reluctance scanning head, which does not rely on balanced magnetic circuits, is used to drive a sonic pulse along the direction of the head gap. By choice of a magnetic core material with the correct magnetostriction, the narrow sonic pulse can, in principle, produce a narrow zone of high permeability. This ingenious scheme suffers from the lack of control of the speed of the sonic pulse. For most practical schemes, the pulse travels too quickly and only a portion of a television line is recorded each scan. While the examples cited have exhibited problems in development, the advantage of avoiding high-speed head-to-medium interfaces is still desirable, and efforts to produce a workable magnetic scanning head are still worthwhile.

2.7.5.2 Thermomagnetic direct-image recording. The forms of video recording described in this chapter have been concerned with the recording and reproduction of a television signal, in which a visual image is encoded as a series of dots and scanning lines. It could be argued, however, that the ultimate objective of video recording should be to record images directly from the image source onto the tape. If an image could be recorded directly, as a photograph is, without the need for any special intervening encoding, new possibilities for video recording would be opened up (Waring, 1971). The advantage of direct-image recording is that it would be compatible with any type of television system, whereas conventional types of mechanical-scanning recorders need to be set for a specified rotating speed of the head drum as well as a specified tape speed, according to the frame rate. Direct-image recording would, therefore, be an attractive format for prerecorded video information. A single-format tape could be used throughout the world without the need for adjustment to the various differing types of television systems. A successful direct-image recording scheme would have to compete with conventional video recording in terms of tape consumption.

One possibility of direct-image recording is by thermomagnetic means using a chromium dioxide medium. As discussed previously (Section 2.6.3), chromium dioxide has a relatively low Curie temperature (120° C) near which the apparent susceptibility becomes very high, and magnetization of the tape becomes possible in a weak magnetic field. Figure 2.69a shows how this property can be used to record directly from a half-tone photographic image. A chromium dioxide tape is used which is magnetized in a direction opposite to that of the bias field. The tape is preheated by a heated plate and subjected to a weak bias field which is not strong enough to change the polarity and thus cannot erase the premagnetized flux. Then, when the preheated coating is exposed to a flash from a powerful light source through the half-tone film, the light will further heat up the chromium dioxide through the transparent part of the film to the Curie temperature and switch the direction of the magnetization of the tape to that of the bias field. Thus, a black-and-white image can be recorded in this way as a magnetized pattern. A possible technique for playing back this magnetically recorded image pattern is shown in Fig. 2.69a. A thin magnetic film, such as Permalloy evaporated on a glass substrate, is placed in proximity to the recorded tape. The magnetization of this film corresponds to the recorded pattern on the tape, and reproduction can be effected using a polarized-light beam and detecting the magnetization by the Kerr effect. When a color image is required, the chrominance signal can be recorded separately using compression.

2.7.5.3 Thermomagnetic laser-beam recording.

Another possibility for thermomagnetic recording uses a high-power laser, as shown diagrammatically in Fig. 2.69b. The laser beam heats up a portion of the tape to the Curie temperature so that a video signal can be easily recorded at the spot where the laser beam is focused (Nomura and Yokoyama, 1979).

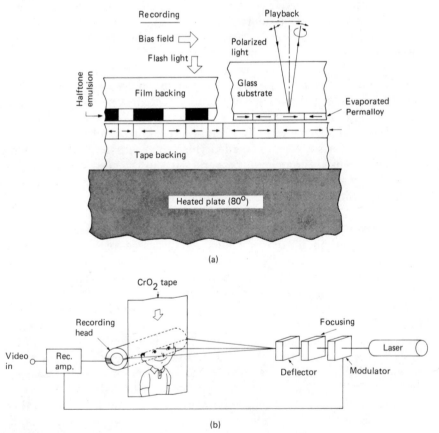

Figure 2.69 Principle of thermomagnetic recording. (a) Half-tone film, and (b) thermomagnetic recording with magnetic head.

To record a complete image directly on the tape by means of this technique, the laser beam must be deflected back and forth transversely on the tape, as indicated in the figure.

2.8 Digital Video Tape Recording

2.8.1 Video recording by digital means

All of the various video tape recorders which have been discussed thus far in this chapter have been analog video recorders. Conventional analog machines record the video signal by means of frequency modulation, and the analog information is converted to differences in distance between the zero-crossing points of the FM signal (Fig. 2.70a).

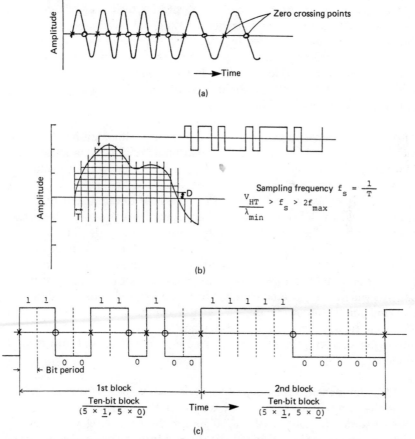

Figure 2.70 Typical example of FM signal and digital signal. (a) FM signal; (b) analog-to-digital conversion by PCM; (c) 8-10 group coded signal.

The most commonly used coding method in digital video recording is pulse-code modulation (PCM). In this coding method, the distance between the zero-crossing points is a discrete value which is an integer times the bit period (Fig. 2.70c). To encode the signal into a digital format, the analog signal is sliced into segments of width T and amplitude D (Fig. 2.70b), and the discrete values are coded into a digital binary format. These processes are known as *sampling, quantizing,* and *coding.* In an example of quantizing, a value of 2^8, or 256 (8 bits), is just sufficient for the binary-coded video information in terms of picture quality. This type of coding is effective in reducing the influences of waveform distortion and time-base instability, phenomena intrinsic to magnetic recording.

Other errors which occur can be compensated for by various error-correction techniques. For a number of reasons, digital recording is the ideal recording method for use in magnetic recording, but it does have two disadvantages. First, very high density recording technology is required because of the extremely wide frequency range involved. Second, a great number of circuits is required for digital encoding and decoding, as well as for error correction and concealment. As continuing advances and developments in magnetic recording and semiconductor technology yield solutions to these problems, digital recording will become increasingly important in the field of video recording.

2.8.1.1 Digital technology advantages. Digital technology has been the focus of intensive investigation in video broadcasting. It is anticipated that the application of digital technology will provide workable solutions to many of the problems now associated with the standard analog machines. It is also expected that the successful development of digital video recording will bring dramatic changes in future program production systems and in the operations of broadcasting stations. The primary expectation of these investigations is the prospect of high picture quality combined with the capability of multiple program generation, free from the deterioration of picture quality inherent in analog systems. Additional benefits which are expected from the introduction of digital recording systems include the following: improvement in overall operation as the result of automatic adjustment and self-diagnostic functions, the capability of improving copied picture quality, greater reliability, and future cost reductions.

2.8.1.2 High-density recording. One serious drawback of digital recording is the necessity to record a gross bit-rate PCM signal in excess of 86 Mb/s in order to contain the total information for a television signal. High-density recording technology is thus essential to the development of

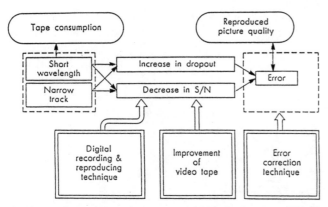

Figure 2.71 Main points of consideration for high-density digital video recording.

a digital recorder with a rate of tape consumption competitive with that of analog recorders. Figure 2.71 shows factors that affect the relationship between picture quality and rate of tape consumption in digital video recording at present levels of development. In general, a decrease in tape consumption resulting from the application of high-density recording techniques brings about a concomitant increase in the error rate; this directly affects the quality of the reproduced picture. As a consequence, it is essential to develop digital magnetic recording techniques for reliable readout, to devise a system of error correction matched to the television signal, and to provide improved video tapes and heads (Yokoyama et al., 1979a).

2.8.1.3 Basic structure of the digital video tape recorder. Figure 2.72 shows an example of the structure of a digital tape recorder. The input video signal is first pulse-code-modulated by an analog-to-digital converter (ADC). Then the PCM signal is fed to a time-base-compressing memory to facilitate the composition of the error-correction code and to adjust the timing of the PCM signal in conjunction with the rotary-head system. It is necessary to record a maximum bit rate of approximately 100 Mb/s (including parity bits for error correction). Thus, the signal may be recorded in several channels in order to reduce the recorded bit rate. Both the parity bit necessary for error correction and concealment and the synchronizing bit are added and then separated into the necessary channels for magnetic recording (see Fig. 2.72). Following this, the PCM signal is converted into a recording signal by the modulator. A recording equalizer may also be used.

In the reproduction system, the output signal is extracted by a process

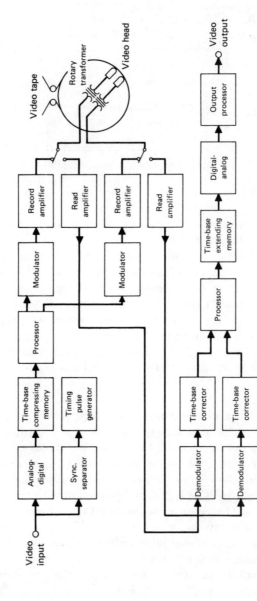

Figure 2.72 Block diagram of experimental digital video tape recorder.

opposite to that used in the recording system. The demodulator in Fig. 2.72 contains an equalizer for reproduction (Katayama et al., 1975), an automatic threshold controller (Katayama et al., 1976), a clock regenerator, and other essential devices for digital reproduction. Where horizontal and vertical blanking signals and synchronizing signals are added by the digital method, the jitter of the reproduced signal is absorbed by either the time-base corrector or the time-base extending memory.

2.8.2 Development of the digital video recorder

Investigation of the possibility of digital video recording began in the 1970s, and various experimental recorders were devised as a result of this research. Early studies in this area have reported that digital technology offered many advantages, including high picture quality, high reliability, and adjustment-free operation (Jones, 1973; Davidoff, 1975; Bellis, 1976).

These early digital recorders had the disadvantage, however, of an undesirably high rate of tape consumption due to the very wide bandwidth of the PCM-encoded video signals. Consequently, research soon focused on the reduction of tape consumption in PCM recording. Eventually, in 1979, a digital recorder with tape consumption comparable to that of conventional analog machines was developed (Yokoyama et al., 1979b, 1980). Experimental recordings made on this recorder demonstrated the commercial feasibility of such equipment.

International standardization of the digital recording format has been making steady progress under the aegis of the Comité Consultatif International des Radio Communications (CCIR).

2.8.3 Review of digital video recording technology

2.8.3.1 Composite versus component digital coding systems. Numerous visual experiments have been carried out on the picture quality resulting from different signal-encoding systems for television broadcast use. It has been found, as the result of many experimental and theoretical studies, that the optimum sampling frequency should, in general, be three times the color subcarrier frequency f_{sc} and that quantizing should be eight bits per sample for both NTSC and PAL systems (Godall, 1951; Ishida, 1970; Goldberg, 1973). This leads to 86 Mb/s for NTSC and 105 Mb/s for PAL. These coding conditions provide only marginal luminance bandwidths. It has also been argued that in such signal-processing procedures as editing, picture-quality adjustment, montage, and mixing, the processing is facilitated when the sampling points are located on a vertical line in the tele-

vision image. Taking this into account, it has thus proven more practical to provide a sampling frequency of four times the color subcarrier frequency; there is also a body of opinion which favors a quantizing of 9 bits per sample when signal processing is taken into consideration. This leads to about 129 Mb/s for NTSC.

Until recently, the $3f_{sc}$ and 8-bit composite coding system was used in almost all of the experimental digital video recorders because minimization of the recordable bit rate was the primary design limitation. Conversion between NTSC, PAL, and SECAM systems is of vital concern. To facilitate such conversion, a component coding system in which the luminance signal and the two color-difference signals are individually coded was proposed, and common sampling frequencies were standardized by the CCIR in 1981. This component coding is based on a sampling frequency of 13.5 MHz for the luminance signal and 6.75 MHz for each color-difference signal. Quantization is at 8 bits per sample for a total bit rate of 216 Mb/s. With these bit rates, the recording of a component signal became conceivable, and by 1984 both broadcasters and manufacturers had selected this approach. The final figures agreed on are given in the Appendix, Table 2.13. As far as possible, the standard is a worldwide one with, for example, equal samples per line on both 50- and 60-Hz systems. Since the drum rotation rate is unchanged between 50 and 60 Hz, a 60-Hz field is recorded on 10 tracks and a 50-Hz field on 12. The standard intentionally does not specify the recorder parameters such as drum size, drum speed, or the number of heads. Several manufacturers are taking different approaches, dividing the signal into 2, 3, or 4 parallel channels. It is not yet clear where the optimum lies.

2.8.3.2 Digital recording versus analog recording. It has been well established in magnetic recording and reproduction theory that both the high- and low-frequency ends of the spectrum have low sensitivity. Thus, to record wide-band television signals, it is necessary to modulate the signal and to increase the relative head-to-tape speed. It is for this reason that the analog video recorders use low-index FM recording. Table 2.7 compares the basic ($3f_{sc}$ and 8 bits per sample) specifications of digital video recorders with those of standard analog recorders in use at the present time. Because the relationship between tape consumption (recording density) and picture quality is a crucial one, it is necessary to explore in detail the important differences between the two systems.

It has been experimentally determined that the signal-to-noise ratio of FM video recording is nearly proportional to the relative head-to-tape speed and to the square root of the track width. If the usable area of the tape is designated as T, then

$$(S/N)_{FM} \propto T^n$$

TABLE 2.7   Comparison of Digital and Analog Recording Techniques for Broadcast Use

	Current analog recording for broadcast use	Digital recording
Modulation method	FM: 　　Carrier: White level, 10 MHz 　　　　　　　Black level, 8 MHz	PCM (e.g., NRZI): 　　Sampling frequency, $3f_{sc}$†
	Frequency deviation, 2 MHz	Quantizing, 8 bit/sample
	Video emphasis, 8 dB	Data rate, 86 Mb/s
Recording bandwidth required	1–15 MHz	0.1–86 MHz
Carrier-to-noise ratio required	Approx. 50 dB	Approx. 30 dB
Picture quality deterioration occurring in tape-head system	Carrier-to-noise ratio: Tape noise limited	Depend on coding system; no deterioration
	Waveform distortion: Main cause is recording characteristics of sideband spectrum	
	Banding occurs	No Banding
	Dropout occurs	Error correction possible to considerable extent

†NOTE: f_{sc} = color subcarrier frequency.

where n is between 0.5 and 1.0. In contrast to this, the signal-to-noise ratio of pulse-code-modulation video recording improves by as much as 6 dB for each integral increase in the bit number, hence

$$(S/N)_{\mathrm{PCM}} \propto 2^{T/T_0}$$

where T_0 is the tape area necessary to record one bit of information. Figure 2.73 compares these results; line A gives values for a standard broadcast analog recorder; B_1 and B_2 are corresponding values for a representative experimental digital recorder. It can be seen that when the signal-to-noise ratio surpasses the points P_1 and P_2, digital video recording has a lower tape consumption (Yokoyama and Nakagawa, 1978) and, as digital recording technology continues to develop, we can anticipate an even smaller tape consumption.

2.8.3.3   Causes of code error.   The fundamental parameter for the evaluation of digital recording is the code error rate. According to evaluations which have been performed on the relationship between code error rate

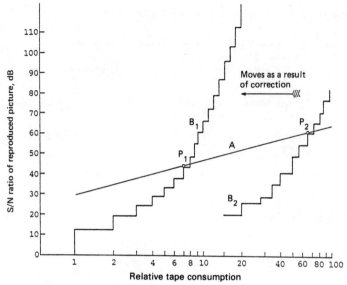

Figure 2.73 Comparison of tape consumption using PCM and FM video recording.

and picture quality, it has been determined that a value not exceeding 10^{-7} is necessary for broadcast quality pictures; this is the threshold level at which the error is no longer visible (Geddes, 1965; Ishida, 1971).

The causes of code errors in digital magnetic recording can be classified, as shown in Fig. 2.74, into discontinuous noise, continuous noise, and crosstalk. Discontinuous noise occurs when the tape is separated from the head because of a defect in the magnetic layer of the tape, scratches on the surface of the tape, or the adherence of dust to the tape. In some cases, recording or reproducing becomes impossible. The phenomenon in which the reproduced signal becomes practically defective is called a *dropout*. Continuous noise, which has a Gaussian distribution of amplitude, consists of tape noise, playback amplifier noise, and head noise (thermal agitation noise generated by the resistive component of head impedance).

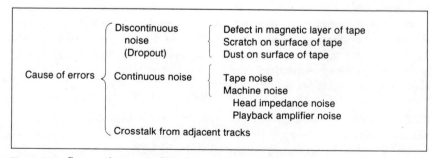

Figure 2.74 Causes of errors in digital magnetic recorders.

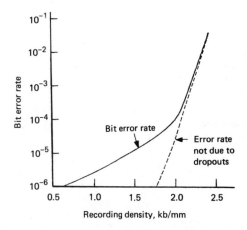

Figure 2.75 An example of error rate versus recording density.

2.8.3.4 Recording density and error rate. There is a close relationship between the recording density and the error rate in digital magnetic recording, with the error rate increasing in proportion to the recording density; this is shown in Fig. 2.75. When the recording density is comparatively low, dropouts are the predominant causes of error because the signal-to-continuous-noise ratio is sufficiently high at lower recording densities. As the recording density increases, however, the signal-to-continuous-noise ratio reaches its threshold value, which is 20 dB (p–p/rms) for the broadcast-quality error rate of 10^{-7}. Recording becomes impractical at higher recording densities because the errors rapidly become excessive. To a certain extent, code errors can be corrected by using an error-correction code. But to compensate for errors related to continuous noise, the recording density must be increased by adding redundant bits for error correction; if the increase in error rate is large as the result of these added redundant bits, the correction effect will be lost. With consideration given to this inherent increase in error rate, the practical limit for recording density is considered to be the point where the threshold value of the signal-to-continuous-noise ratio is reached. The more general relationships between error rate and recording density are shown in Fig. 2.76 in terms of the various sources of error (Yokoyama and Nakagawa, 1978).

2.8.3.5 Signal-to-noise ratio. It is necessary to record short wavelengths on extremely narrow tracks in pulse-code-modulation video recording; consequently, any deterioration in the signal-to-noise ratio due to a drop in the output signal poses a serious problem. The signal-to-equipment-noise ratio has a direct relationship to the output signal. Therefore equipment noise has a greater effect on the signal-to-noise ratio than does tape noise and is thus the main limitation on the threshold recording density. Figure 2.77 shows the theoretical value of the error rate relative to random

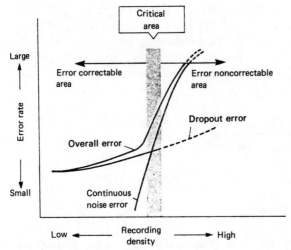

Figure 2.76 General relationship between error rate and recording density.

noise. Note that to reduce the error rate to the required value of 10^{-7}, it is necessary to realize a minimum signal-to-noise ratio of 20 dB (p–p/rms) relative to equipment noise. In actual practice, an additional marginal allowance of approximately 10 dB is required to allow for such additional factors as pulse interference, clock jitter, output level fluctuations, and tape life; thus, as shown in Table 2.7, approximately 30 dB of carrier-to-noise ratio is necessary.

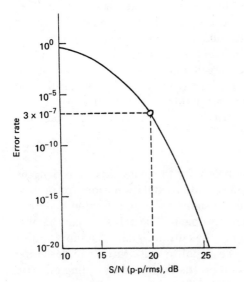

Figure 2.77 Theoretical value of error rate as a function of signal-to-random-noise ratio.

2.8.3.6 Rotary-head versus stationary-head recorders. There are two types of recorders suitable for the recording of high-bit-rate, pulse-code-modulated signals: first, a wideband, rotary-head system (either helical or transverse scan) using relatively few recording tracks; second, a narrow-band, multitrack system with stationary heads. Table 2.8 shows a comparison of the specifications of these two types of recording methods.

With both these types of drive mechanisms, it is necessary to improve the linear density (the density in the head-scanning direction) as well as the track density in order to achieve a high overall recording density. The

TABLE 2.8   Comparison of Various Types of Digital Video Recorders

	Rotary head		Stationary head
	Transverse scan	Helical scan	Multichannel
Typical number of channels	1	2–3	20–40
Head:			
Bandwidth	50 MHz	20 MHz	1–2 MHz
Life	Comparatively short (300–400 h)		Comparatively long (1000 h)
Coupling between head and circuit	Wideband rotary transformer or low-noise slip-ring coupler	Wideband rotary transformer or low-noise slip-ring coupler	Direct coupling
Tape:			
Reduction in total thickness	Possible	Somewhat difficult	Difficult
Packing density:			
Linear density	High	High	High
Track density	Very high	High	Moderately high
Mechanics:			
Timing jitter	Small	Moderate	Large
Track retrace	Easy	Somewhat difficult	Difficult
Track splicing	Possible	Difficult	Difficult
Electric circuitry:			
Number of high-speed circuits	Very large	Large	Small
Number of low-speed circuits	Small	Small	Small
Memory capacity for error correction	Large	Large	Moderate
Total number of circuits	High	Moderately high	Moderate

tracking accuracy of the stationary-head format is insufficient; this is due both to the meandering of the tape and to the crosstalk between adjacent head cores at higher track densities. As discussed earlier, the track density of a rotary-head recorder can be made much higher than that of a stationary-head recorder; the rotary-head approach has the further advantage of such special-effects capabilities as slow-speed, still, and high-speed search modes (Yokoyama and Habutsu, 1976).

2.8.4   High-density recording technology

2.8.4.1   Digital magnetic recording technology.   Because of the necessity for wideband recording, nonsaturation pulse-recording techniques are better suited to digital video recording technology than are the saturated recording methods used in computer applications. This is shown in Fig. 2.78, where the experimental results for both the output level e_0 and the pulse width p_{50} of the output signal are plotted as a function of the recording current. The optimum recording current should be selected in the nonsaturation recording range, as indicated in the figure, in order to obtain a shorter reproduced pulse width and a higher output level. In nonsaturation recording, waveform distortion due to intersymbol interference occurs easily because the condition of tape-to-head contact during recording is critical. As shown in Fig. 2.79, however, equalization of the reproduced signal is effective in reducing error occurrence.

2.8.4.2   Coding of the digital signal.   A magnetic recording and reproducing system has basically a bandpass characteristic—low sensitivity at very low- and high-frequency regions. It is also nonlinear and has problems

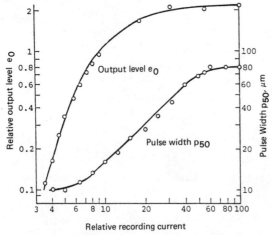

Figure 2.78   Characteristics of pulse recording.

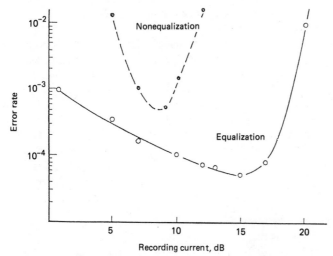

Figure 2.79 Effect of equalization on error rate. Tape velocity = 11.1 m/s; recording density = 13730 bits per inch; H_c = 24 kA/m.

such as amplitude instability and time-base error. Thus, in both analog and digital recording, the recording signal should be modified to recognize these system limitations. In frequency-modulation video recording, the necessary frequency components are distributed in the higher frequency range; thus high-frequency recording (i.e., short-wavelength recording) is a major problem to be solved while the output signals of lower frequencies are less problematic.

In digital recording, the simple channel-coding schemes can give rise to dc, or very low frequency components, which represent a problem. Several types of channel codes have been proposed to eliminate both the dc and the very low frequency components from the signal recorded on the tape, and to record digital information magnetically with maximum efficiency while using the conventional inductive type of video head. These include:

1. The Miller square code (M^2) (Mallison and Miller, 1977)

2. The 8–10 group code (Baldwin, 1978; Morizono et al., 1980)

3. The scramble-interleaved NRZI code (S-I-NRZI) (Kobayashi and Tang, 1970; Yokoyama and Nakagawa, 1981)

4. The trilevel code with partial response processing (Eto et al., 1981)

5. The NRZ code with adapted spectral energy (NRZ-ASE) (Heitmann, 1984)

6. The scrambled NRZ (S-NRZ)

The Miller square code is a derivative of the Miller code which is widely

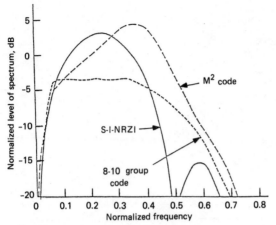

Figure 2.80 Power spectrum density.

used in computer disk drives (usually under the acronym MFM code) and it has no dc component. However, all Miller codes must be sampled twice per bit cell to decode them unambiguously. This is a serious problem with the very high bit rates found in digital video recording.

The 8–10 group code takes an 8-bit word and, using a look-up table, replaces it with a 10-bit word (Fig. 2.70c). Since there are four times as many 10-bit combinations as there are 8-bit combinations, only those with no dc component need be used. However, the bit rate has been increased 25 percent so that a higher-density recording is needed.

The S-I-NRZI code employs decoding using partial-response processing. Partial-response decoding is attractive because its spectrum goes to zero both at dc and at half the Nyquist rate. The narrow spectrum distribution (Fig. 2.80) reduces the effect of low-frequency cutoff. Although this detection is 6 dB more sensitive to noise due to trilevel decoding, the overall signal-to-noise ratio at the detection point is almost equal to that of the NRZ code when decoded by integrated detection (Nakagawa et al., 1980). Since low-frequency components of crosstalk signals from the adjacent tracks are eliminated at detection, this decoding method can improve track density.

The S-NRZ code takes the input data stream and multiplies it by a long pseudorandom stream. Thus, where a fixed video level produces long, repetitive patterns, the randomizer breaks them up. In theory there is always some video pattern which produces the inverse of the random generator and so reintroduces the low-frequency component; in practice this occurs so infrequently that it has as yet not been visually detected. S-NRZ code with integrated detection has the advantage of phase margin and high signal-to-noise ratio, but it is prone to be affected by crosstalk from the adjacent tracks and is inadequate for high-track-density recording. The S-NRZ code with integrated detection or the S-I-NRZI code with partial-response detection have their own merits and demerits. The S-NRZ code has been selected in the proposed CCIR standard (see Appendix, Table 2.13). However, experiments on the other codes continue, and they may yet find practical applications.

2.8.4.3 Optimization of linear and track densities.   As we have seen, large tape consumption is the major problem to be overcome in the development of a successful digital video recording format. In order to decrease the tape consumption, both the linear density and the track density must be maximized. The maximum linear density (i.e., the minimum recordable wavelength) is directly related to both the head output and the signal-to-noise ratio. The track density, on the other hand, is limited to a certain extent by the signal output, the crosstalk from adjacent tracks, and tracking error.

Once the basic parameters for the fundamental components of the recording format (i.e., the recording medium, the head, the bandwidth, the channel coding, signal-to-noise ratios, crosstalk, tracking error, etc.) have been established, the optimum combination of linear and track density (i.e., maximum areal density) can be obtained as shown in Fig. 2.81. Line A shows the simple relationship between bit interval (half the wavelength) and track width, provided the signal-to-tape-noise ratio is simply proportional to the bit interval and the square root of the track width. In practice, background noise, crosstalk, and tracking error must be taken

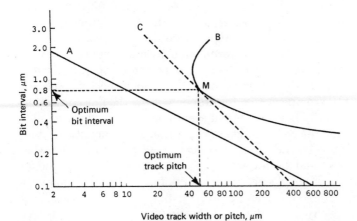

Figure 2.81 Optimum combination of bit interval and track width. Line *A* is for the signal-to-tape-noise ratio constant at 30 dB (p–p/rms); curve *B* represents a tracking error of 15 μm; line *C* is a plot of constant areal density.

into consideration, and this relationship is expressed by curve B. Since the crosstalk increases at longer bit intervals, the track pitch also increases above the minimum value of about 40 μm. Line C is a constant areal density plot. The point M, where curve B is tangent to line C, gives the optimum recordable bit interval and optimum track pitch (Nakagawa and Yokoyama, 1981).

2.8.4.4 Error correction and concealment technology. To achieve practical packing densities, digital video recorders may have raw error rates as poor as 10^{-4}; these give visually unacceptable pictures. There are two possible methods for dealing with errors: error correction or error concealment. The latter is a method of compensating for code errors visually by means of the correlation of television signals. One example of error concealment is the dropout canceler used in analog video recorders. In either method, error detection is imperative. We shall consider three methods of error detection:

1. Formulation of error-detection codes.

2. Detection of errors by means of the regularity of the reproduced signal and channel coding. For example, the S-I-NRZI code or M^2 code have certain laws, and violations of these give limited error detection. There are no illegal patterns in the S-NRZ code.

3. Detection of erroneous portions by abnormalities in the reproduced signal (Croll, 1978).

Studies of the general theory of the use of codes for the detection or correction of errors are well advanced, and are described in Chap. 5. The following points should be noted regarding the formulation of error-detecting and correcting codes in digital recorders. For convenience of program editing, codes should be formulated with either field or frame resolution. It is necessary to take into account the comparatively long burst errors which are due to dropouts. Codes should be formulated in such a way that the compensation for dropouts using the line co-relationship (i.e., the similarity between two adjacent horizontal lines) is possible; this makes possible the combination of correction and concealment functions. In order to keep the recorder compact and easy to handle when recording, codes should be formulated so as to facilitate reproduction without correction. Finally, since the amount of digital information to be recorded is very large and must be processed at very high speeds, such areas as the simplicity of the design of the circuit structure, the capacity of the memory, and the power consumption should be taken into consideration.

Error-correction codes which have been proposed thus far include the following:

1. Hamming (63, 56) code (Yokoyama et al., 1977)

2. Concatenated code—formulated on erasure correction (Yokoyama et al., 1979*a*)

3. Single-erasure correction code (Hashimoto and Eguchi, 1980)

4. Fire code (Eto et al., 1981)

5. Reed-Solomon code (Reed and Solomon, 1960)

In the 4:2:2 digital video recording standard proposed in CCIR, video data from 50 contiguous lines are divided and recorded onto the four separate areas on the tape called *sectors*. In each sector a product code is composed. Considering the video data words to be arranged in a rectangular array with a row dimension of 600 bytes and a column dimension of 30 bytes, 2 check bytes are first added to each column to produce a total of 600 outer-code blocks of 32 bytes each. Each of the 32 rows is then divided into 10 blocks of 60 bytes each, to which 4 check bytes are added to form inner-code blocks of 64 bytes each. Using the Reed-Solomon system, these check bytes allow correction of very high error rates and, since the stream is truly corrected, the output is identical to the input. The limitation of a Reed-Solomon system is that it cannot handle long error bursts. Since dropouts produce bursts, the bit stream is interleaved before recording, so that adjacent input samples are spaced apart on tape. The inverse process in playback then breaks up a dropout into a number of 1- or 2-byte errors which the decoder can correct. However, in any practical system, the error-correction system will occasionally overload. For example, a head may clog. The final protection is then error concealment. The distribution of samples between heads is such that samples around a missing one (both on the same line and on adjacent lines) come from other heads. When the decoder reports an overload, an interpolation is made using these adjacent samples. Note that concealment is only an approximation which can be defeated by some critical patterns; on the other hand, correction is mathematically exact. Nevertheless, concealment can be visually so successful that one permanently clogged head out of four can produce a barely detectable degradation.

2.8.5 Performance of digital video recorders

Various experimental digital recorders have been developed taking the various points described earlier into consideration. The performances of these experimental digital recorders for broadcast use have the following advantages over analog recorders:

1. *High picture quality.* The signal-to-noise ratio of the reproduced video signal is theoretically in excess of 55 dB (p–p/rms), while that of an analog recorder is approximately 48 dB.

2. *Multiple generation of programs with no degradation.* The cumulative degradation of the signal-to-noise ratio in an analog machine is theoretically 3 dB at the second generation, 4.5 dB at the third generation, and 6 dB at the fourth generation.

3. *Freedom from dropouts.* One example of an experimental digital recorder gave a raw block error rate on the order of 10^{-4} and the remaining errors after correction were almost zero.

In addition, the digital video recorder has many possible advantages associated with automatic adjustment, self-diagnostic functions, greater reliability, and cost reduction. It is expected to be considered for the next-generation machines for broadcast use, and standardization is being discussed in the international organizations such as EBU and SMPTE.

Appendix

Important historical developments and technical data are summarized in Tables 2.9 to 2.13 and Figs. 2.82 to 2.84.

TABLE 2.9 Milestones of Video Tape Recorder Developments (Mainly NTSC)

Year	Country	New product	Number of video heads	Tape width	Lin. tape speed (cm/s)	Max. rec. time	Recording mode
1953	U.S.A.	Fixed-head color VTR (RCA)	F 4	$\frac{1}{2}$ in	914	$\frac{1}{3}$ h	D + FM
1954	U.S.A.	Fixed-head multitrack VTR (Bing Crosby Enterprises and GE)	F 10	$\frac{1}{2}$ in	254	$\frac{1}{4}$ h	D
1956	U.S.A.	†The first rotary-head broadcasting QUAD VTR (Ampex)	R 4	2 in	38.1	1 h	FM
1956	U.S.A.	Fixed-head color home VTR (RCA)	F 2	$\frac{1}{4}$ in	305	$\frac{1}{4}$ h	D + FM
1957	U.S.A.	†Broadcasting color QUAD VTR (RCA, Ampex)	R 4	2 in	38.1	1 h	FM
1958	Japan	Broadcasting VTR trial model (NHK, Matsushita)	R 8	2 in	38.1	1 h	FM
1958	U.K.	Broadcasting VTR trial model—VERA—(BBC)	F 2	$\frac{1}{2}$ in	508	$\frac{1}{4}$ h	D + FM
1959	Japan	The first 1-head helical-scan broadcasting (Toshiba)	R 1	2 in	38.1	1 h	FM
1960	Japan	The first 2-head helical-scan broadcasting (JVC)	R 2	2 in	38.1	1 h	FM
1961	Japan	†Helical-scan industrial VTR (Matsushita)	R 2	1 in	27.9	1 h	FM
1961	U.S.A.	†α-lap helical-scan industrial VTR (Ampex)	R 1	2 in	12.7	3 h	FM
1962	Japan	†Ω-lap helical-scan industrial VTR (Sony)	R 1.5	2 in	14.4	1 h	FM
1963	U.K.	Simplified fixed-head VTR (Telcan)	F 1	$\frac{1}{4}$ in	305	$\frac{1}{6}$ h	D
1964	Holland	Ω-lap helical-scan industrial VTR (Philips)	R 1	1 in	19.1	$\frac{3}{4}$ h	FM
1964	U.S.A.	†HiBand color QUAD VTR (Ampex)	R 4	2 in	38.1	2 h	FM
1964	Japan	†Field skip VTR (Sony)	R 2	$\frac{1}{2}$ in	19.1	1 h	FM, FS
1965	Japan	†Small-size VTR (Matsushita)	R 2	$\frac{1}{2}$ in	25.4	$\frac{2}{3}$ h	FM
1965	Japan	Small-size VTR (Ikegami)	R 2	$\frac{2}{5}$ in	30.5	$\frac{3}{4}$ h	FM
1966	Japan	RGB sequential camera and VTR (Matsushita)	R 2	$\frac{1}{2}$ in	6.4	$\frac{1}{3}$ h	FM
1966	Japan	Magnetic-sheet spiral recorder "VIDEO MAT" (Sony)	1	10 ϕ in	—	30 s	FM
1966	U.S.A.	†The first segment sequential VTR (Westel)	R 2	1 in	25.4	1 h	FM
1966	U.S.A.	Newell sys. fixed-head VTR	F 1	$\frac{1}{4}$ in	305	1 h	FM
1966	U.S.A.	†α-lap small VTR (3M)	R 1	$\frac{1}{2}$ in	19.1	1 h	D
1967	Japan	Magnetic-sheet concentric recorder "VSR" (Matsushita)	1	10 ϕ in	—	20 s	FM
1967	U.S.A.	†α-lap helical-scan industrial VTR (IVC)	R 1	1 in	17.5	1 h	FM

TABLE 2.9 Milestones of Video Tape Recorder Developments (Mainly NTSC) (*Continued*)

Year	Country	New product	Number of video heads	Tape width	Lin. tape speed (cm/s)	Max. rec. time	Recording mode
1968	Japan	†The first azimuth recording VTR (Matsushita)	R 2	$\frac{1}{2}$ in	12.7	1.5 h	FM, AZ
1969	Japan	†The first video-tape contact printer (VTP) (Matsushita)	—	$\frac{1}{2}$ in	—	—	AT
1969	Japan	†CP-504, unified standards for small VTR (EIA-J)	R 2	$\frac{1}{4}$ in	19.1	1 h	FM
1969	Japan	Cassette VTR (sometimes called magazine VTR) (Matsushita)	R 2	CS $\frac{1}{2}$ in	19.1	$\frac{1}{2}$ h	FM
1969	Japan	Cassette VTR (Sony)	R 2	CS $\frac{3}{4}$ in	9.5	1 h	FM
1970	U.S.A.	†VTR mainly for recorded-tape playback (CTI)	R 3	CS $\frac{1}{2}$ in	9.6	2 h	FM, FS
1970	Japan	Broadcasting video-tapes contact printer (VTP) (Matsushita)	—	2 in	—	—	AT
1970	Holland	†Tandem cassette home VTR (Philips)	R 2	CS $\frac{1}{2}$ in	14.2	$\frac{2}{3}$ h	FM
1970	U.S.A.	†Segment sequential broadcasting VTR (IVC)	R 2	2 in	20.3	1–2 h	FM
1971	Japan	†U-Format VTR (Sony, Matsushita, JVC)	R 2	CS $\frac{3}{4}$ in	9.5	1 h	FM
1971	Japan	†Color VTP by anhysteretic transfer (Matsushita)	—	$\frac{1}{2}$ in	—	—	AT
1971	U.S.A.	Color VTP by thermal transfer (CVS)	—	$\frac{1}{4}, \frac{3}{4}$ in	—	—	TT
1971	Japan	†CP-507, unified standards for small color VTR (EIA-J)	R 2	$\frac{1}{2}$ in	19.1	1 h	FM
1971	Japan	Small cartridge VTR (Matsushita)	R 2	CT $\frac{1}{4}$ in	19.1	$\frac{1}{2}$ h	FM
1971	U.S.A.	†Super HiBand QUAD VTR (Ampex)	R 4	2 in	38	1–2 h	FM
1972	Japan	†CP-508, unified standards for small cartridge VTR (EIA-J)	R 2	CT $\frac{1}{4}$ in	19.1	$\frac{1}{2}$ h	FM
1972	Japan	Standard-alignment tape recorder for EIA-J Type I (Matsushita)	R 2	$\frac{1}{2}$ in	19.1	2 h	FM
1972	Japan	Color still-picture cassette player (Matsushita)	R 3	3.81 mm	4.8	$\frac{1}{2} \times 2$ h	FM
1973	Japan	†Small broadcasting QUAD VTR (Asaka)	R 4	1 in	25.4	1 h	FM
1973	U.S.A.	Thermal/anhysteretic transfer (3M)	—	$\frac{1}{4}, \frac{3}{4}, 1$ in	—	—	TT, AT
1974	Japan	†Cartridge color video tape printer (Matsushita)	—	$\frac{1}{2}$ in	—	—	AT
1974	Japan	Magnetic-sheet recorder "MAVICA" (Sony)	R 2	6.5×8.5 in	—	$\frac{1}{6}$ h	Phase mod.
1974	Japan	†Cassette VTR V-Cord (Sanyo, Toshiba)	R 2	CS $\frac{1}{2}$ in	13.4	$\frac{1}{2}$ h	FM

Year	Country	System	Head	Medium	(value)	Time	Modulation
1975	Japan	Magnetic-sheet recorder "Panaslow" (Matsushita)	R 2	8 in ϕ	—	20 s	FM
1975	Japan	†Azimuth recording color VTR "Betamax" (Sony)	R 2	CS $\frac{1}{2}$ in	4.0	1 h	FM, AZ
1976	Japan	Long-play VTR V-Cord by FS (Sanyo, Toshiba)	R 3	CS $\frac{1}{2}$ in	6.7	1 h	FM, FS
1976	Japan	†1-head tandem cassette VTR (Matsushita)	R 1	CS $\frac{1}{2}$ in	5.2	1.5 h	FM
1976	Japan	†Azimuth recording color VTR "VHS" (JVC)	R 2	CS $\frac{1}{2}$ in	3.3	2 h	FM, AZ
1976	Germany	†Type B segment sequential-broadcast VTR (Bosch)	R 2	1 in	24.5	3 h	FM
1977	Japan	†2-h Beta-format VTR with overlap recording (Sony)	R 2	CS $\frac{1}{2}$ in	2.0	2 h	FM, AZ
1977	Japan	†4-h VHS-format VTR with overlap recording (Matsushita)	R 2	CS $\frac{1}{2}$ in	1.7	4 h	FM, AZ
1978	Japan, U.S.A.	†Type C VTR (Sony, Ampex)	R 1.5	1 in	24.4	3 h	FM
1979	Germany	Newell-type fixed-head VTR (BASF)	F 1	CT $\frac{1}{4}$ in	305	3 h	FM
1979	Japan	†6-h VHS format VTR (VHS Group)	R 2	CS $\frac{1}{2}$ in	1.1	6 h	FM, AZ
1980	Holland	†2×4-h V-2000 VTR for PAL (Philips)	R 2	CS $\frac{1}{2}$ in	2.4	8 h	FM, AZ
1980	Japan	Video movie (Sony)	R 2	CS 8 mm	2.0	$\frac{1}{3}$ h	FM, AZ
1980	Japan	MAG VIDEO (Hitachi)	R 2	CS $\frac{1}{4}$ in	1.6	2 h	FM, AZ
1981	Japan	Microvideo (Matsushita)	R 2	CS 7 mm	1.4	2 h	FM, AZ
1981	Japan, U.S.A.	†Type-M ENG VTR using 2-track with VHS cassette (Matsushita/RCA)	R 4	CS $\frac{1}{2}$ in	6.8	$\frac{1}{3}$ h	FM AZ I/Q
1981	Japan	†Type-L ENG VTR using 2-track with Beta cassette (Sony)	R 4	CS $\frac{1}{2}$ in	11.9	$\frac{1}{3}$ h	FM, Y, R−Y/ B−Y Timeplex
1981	Japan	Endless-tape fixed-head VTR (Toshiba)	F 1	CS $\frac{1}{2}$ in	610	2 h	FM
1981	Japan	Electronic still camera "Mavica" (Sony)	F 1	CT 50 mm ϕ	—	50 fields	FM Y&C
1984	Japan	†8-mm-video format (8 mm video conference)	R 2	CS 8 mm	1.4/0.7	2 h/4 h	FM, AZ
1984	Germany	†Quartercam ENG VTR using 2-track (Bosch)	R 4	CS $\frac{1}{4}$ in	12	$\frac{1}{3}$ h	FM, AZ Timeplex
1985	Japan	†M-II Broadcast VTR using 2-track (Matsushita)	R 4	CS $\frac{1}{2}$ in	6.8	1.5 h	Y FM AZ (R−Y/ B−Y) Timeplex

†Put on the market. NOTE: Only the new systems are listed.

ABBREVIATIONS: R = Rotary-head type; F = Fixed-head type; CS = Cassette (2-reel); CT = Cartridge (1-reel); ϕ = Disk diameter; FM = Freq. modulation; D = Direct; FS = Field skip; AZ = Azimuth; AT = Anhysteretic transfer; TT = Thermal transfer.

TABLE 2.10 Characteristics of Typical Broadcast Video Tape Recorders (NTSC)†

	2 in, QUAD	1 in, type C	1 in, type B	¾ in, U format	½ in, type M	½ in, type L	½ in, M II‡
				Tape width, name			
Recording time, min	120 (14 in reel)	90 (10.5 in reel)	90 (10.5 in reel)	60	20	20 (90)*	90 (20)*
Tape/speed, mm/s	381	244	245	95.3	204.5	118.6	67.7
Cassette size, mm	—	—	—	221W× 140D×32H	188W×104D×25H	156W×96D×25H (254W×145D× 25H)	188W×106D×25H (130W×87D×25H)
Relative tape speed, m/s	39.6	25.59	24.08	10.26	5.6	6.9	7.1
Head drum diameter, mm	52.5	134.6	50.33	110	62.0	74.487	76.0
Drum, rotation, r/s	240	60	150	30	30	30	30
Tape lap angle, degrees	90	150	180	180	180	180	180+30
Video track pitch, µm	250+125	130+52	(160+42)×5	85+52	175+21+65+21**	86+1+73+1**	44+2.25+36+2.25**
Azimuth, degrees	0	0	0	0	0	±15	±15
Format (one field)	4-head transverse (segment)	1.5-head helical (nonsegment)	2-head helical (segment)	2-head helical (nonsegment)	4-head helical (nonsegment)	4-head helical (nonsegment)	4-head helical (nonsegment)
Signal channel	1	1	1	1	2 (Y+C)	2 (Y+C)	2 (Y+C)
Color recording method	Direct FM	Direct FM	Direct FM	Color-under	Y: FM C: FM (I,Q)	Y: FM C: TIMEPLEX (R−Y,B−Y)	Y: FM C: TIMEPLEX (R−Y,B−Y)
Y FM deviation, MHz	9–12.5	7–10	7–10	3.8–5.4 (5–6.6)	4.3–5.9	4.4–6.4 (5.7–7.7)	5.6–7.7

†See Fig. 2.82 for head and track formats.
‡MP tape.
*Different cassette size.
**Y+G+C+G.

136

TABLE 2.11   Characteristics of Typical Home Video Tape Recorders (NTSC)†

	VHS			Beta			8-mm video	
	SP	LP	ELP	β-I	β-II	β-III	SP	LP
Recording time, min	120	240	360	60	120	180	120	240
Tape speed, mm/s	33.3	16.7	11.1	40	20	13.3	14.3	7.2
Track pitch, μm	58	29	19	58	29	19	20	10
H-Alignment difference (α_H)	1.5H	0.75H	0.5H	1.5H	0.75H	0.5H	1.0H	0.5H
Head drum diameter, mm	62			75			40	
Luminance FM carrier, MHz	3.4–4.4			4.0–5.2‡ 3.6–4.8 (4.4–5.6)§			4.2–5.4	
Color-under carrier, MHz	0.63			0.69			0.74	
Cassette dimensions, mm	25H×188W×104D			25H×156W×96D			15H×95W×62.5D	
Tape length, m	240			150			108	
Tape thickness, μm	20			20			10	
Relative tape speed, m/s	5.8			7.0			3.8	
Tracking method	Control signal			Control signal			4 Freq. pilot signals	
Azimuth, degrees	±6			±7			±10	
Track angle (running)	5°58′99″			5°01′42″			4°54′13.2″	
Fixed-head audio channels	1 or 2			1 or 2			1	
Audio track width, mm	1.00 or 0.35×2			1.05 or 0.35×2			0.6	
Rotary-head audio channels	FM 2			FM 2			FM 1	
	(Double component)			(Freq. division)			(Freq. division) PCM 2 (8-bit with noise reduction)	

†See Figure 2.83 for head and track formats.
‡Audio FM model.
§HiBand model.

TABLE 2.12 Characteristics of Typical Home Video Tape Recorders (PAL/SECAM)†

	VHS	Beta	V-2000	8-mm video
Recording time, min	180/360	130	360 (both ways)	90/180
Tape speed, mm/s	23.4/11.7	18.7	24.4	20/10
Cassette size, mm	25H×188W×104D	25H×156W×96D	26H×183W×62D	15H×95W×105D
Head drum diameter, mm	62	75	65	40
Luminance FM carrier, MHz	3.8–4.8	3.8–5.2	3.3–4.7	4.2–5.4
Color-under carrier, MHz	0.63	0.69	0.63	0.74
SECAM color-recording method	Color-under	Color-under	Color-under	PAL/SECAM Trans Coder
Tape length, m	258	150	260	108
Tape thickness, μm	20	20	15	10
Track pitch, μm	49/24.5	33	23	34/17
H-alignment difference (α_H)	1.5H/0.75H	1.5H±0.5H	1.5H	2H/1H
Relative tape speed, m/s	4.8	5.8	5.0	3.1
Tracking method	Control signal	Control signal	4 freq. pilot	4 freq. pilot
Azimuth, degrees	±6	±7	±15	±10
Track angle (running)	5°57′50.3″	5°00′58″	2°63′50″	4°54′58.8″
Fixed-head audio channels	1 or 2	1 or 2	1 or 2	1
Audio track width, mm	1.00 or 0.35×2	1.05 or 0.35×2	0.65 or 0.25×2	...
Rotary-head audio channels	FM 2 (double component)	FM 2 (double component)		FM 1 (freq. division) PCM 2 (8-bit with noise reduction)

†See Fig. 2.84 for head and track formats.

TABLE 2.13   Proposed Digital Video Tape Recorder Specifications (SMPTE/EBU)

	525/60 System	625/50 System
Coded signals	Y, R − Y, B − Y	
Sampling frequency:		
Y	13.5 MHz	
R − Y, B − Y	6.75 MHz	
Coding form	8-bit linear quantization	
Number of samples per total line:		
Y	858	864
R − Y, B − Y	429	432
Number of samples per active line:		
Y	720	
R − Y, B − Y	360	
Number of recorded lines per field	250	300
Number of tracks per field	10	12
Cassette type	D1-S (11 min), D1-M (34 min), D1-L (94 min)	
Tape material	Improved metal oxide	
Tape width	19.01 mm	
Tape thickness	16 μm (13 μm for long-play)	
Linear tape speed	286.6 mm/s	286.9 mm/s
Helical-track total length	170/1.001 mm	170 mm
Track width	40 μm	
Channel coding	Scrambled NRZ	
Error protection system	Reed-Solomon product code	
Inner code	(64,60) Reed-Solomon code	
Outer code	(32,30) Reed-Solomon code	

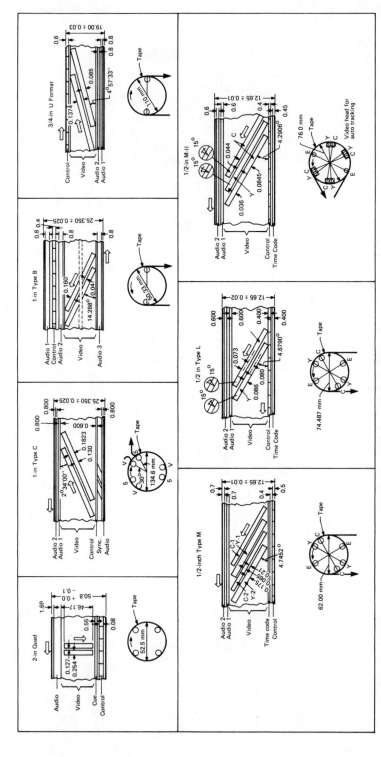

Figure 2.82 Head and track formats of typical broadcast video tape recorders (NTSC).

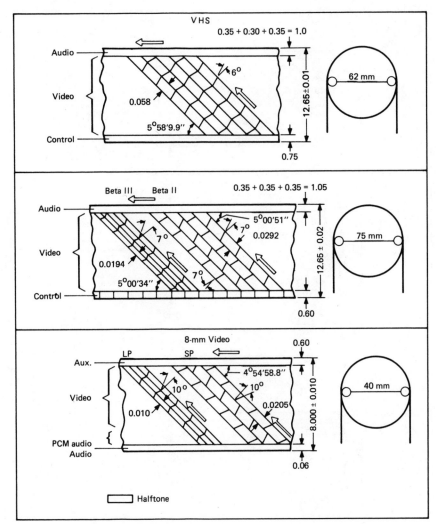

Figure 2.83 Head and track formats of typical home video tape recorders (NTSC).

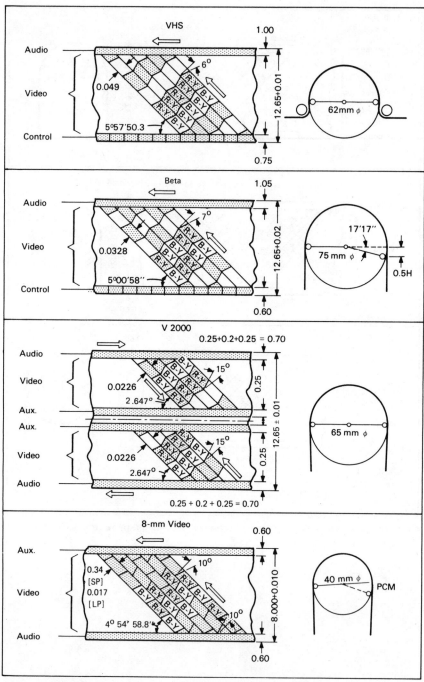

Figure 2.84 Head and track formats of typical home video tape recorders (PAL/SECAM).

References

Abe, H., "Magnetic Recording of a High-Definition Television Signal," *J. SMPTE*, **90**, 192 (1981).

Abramgon, A., "A Short History of Television Recording: Part II, " *J. SMPTE*, **82**, 188 (1973).

Alden, A. E., "The Development of National Standardization of the One-Inch Helical Video Tape Recording Systems," *J. SMPTE*, **86**, 952 (1977).

Amari, S., "Magnetic Recording and/or Reproducing Apparatus with Chrominance Crosstalk Elimination," U.S. Patents 4,007,482 and 4,007,484, 1977.

Anderson, C. E., "The Modulation System of the Ampex Video Tape Recorder," *J. SMPTE*, **66**, 132 (1957).

Anderson, C. E., "The Problems of Splicing and Editing Color Video Magnetic Tape, " *IEEE Trans. Broadcast.*, **BC-15**, 59, (1969).

Arimura, I., and K. Sadashige, "A Broadcast-Quality Video/Audio Recording System with VHS Cassette and Head Scanning System," *J. SMPTE*, **92**, 1186 (1983).

Arimura, I., and H. Taniguchi, "A Color VTR System Using Lower-Frequency-Converted Chrominance Signal Recording" (Japanese), *Nat. Tech, Rep.*, **19**, 205 (1973).

Axon, P. E., "Electronic Recording Apparatus," *J. Telev. Soc., U.K.*, 399 (Nov., 1958).

Baldwin, J. L. E., "Digital Television Recording with Low Tape Consumption," *Int. Broadcasting Convention Record*,133 (1978).

Bellis, F., "An Experimental Digital Television Recorder," BBC Res. Dept., Report BBC RD 1976/77, 1976.

Benson, K. B., "Video-Tape Recording Interchangeability Requirements," *J. SMPTE*, **69**, 861 (1960).

Bertram, H. N., and J. C. Mallinson, "A Theory of Eddy Current Limited Heads," *IEEE Trans. Magn.*, **MAG-12**, 713 (1976).

Camras, M., "Magnetic Transducer Head," U.S. Patent 3,079,470, 1963*a*.

Camras, M., "Experiments with Electron Scanning for Magnetic Recording and Playback of Video," *IEEE Trans. Audio*, **11**, 93, (1963*b*).

Camras, M., and R. Herr, "Duplicating Magnetic Tape by Contact Printing," *Electronics*, **22**, 78 (1949).

Chynoweth, W. R., "Ferrite Heads for Recording in the Megacycle Range," *Tele-Tech. and Elec. Ind.*, 82 (Aug, 1955).

Chubachi, R., and N. Tamagawa "Characteristics and Applications of Metal Tape," *IEEE Trans. Magn.*, **MAG-20**, 45 (1984).

Cole, G. R., L. C. Bancroft, M. P. Chovinard, and J. W. McCloud, "Thermomagnetic Duplication of Chromium Dioxide Video Tape," *IEEE Trans. Magn.*, **MAG-20**, 19 (1984).

Coleman, C. H., "A New Technique for Time-Base Stabilization of Video Recorders," *IEEE Trans. Broadcast.*, **BC-17**, 29 (1971).

Croll, M. G., "A Digital Television Error-Protection Scheme Based on Waveform Estimates," BBC Res. Dept., Report BBC RD, 1978/79, 1978.

Crum, C. W., and H. W. Town, "Recent Progress in Video Tape Duplication," *J. SMPTE*, **80**, 179, (1971).

Darnell, F. J., "Magnetization Process in Small Particles of CrO_2," *J. Appl. Phys.*, **32**, 1269 (1961).

Davidoff, F., "Digital Video Recording for Television Broadcasting," *J. SMPTE*, **84**, 552 (1975).

Dickens, J. E., and L. K. Jordan, "Thermoremanent Duplication of Magnetic Tape," *108th SMPTE Conf.*, No. 18 (1970).

Dieter, P., "Method for Transmission and/or Recording of Color Television Signals," German Patent P2629706.3, 1978.

Dolby, R. M., "Rotary-Head Switching in the Ampex Video Tape Recorder," *J. SMPTE*, **66**, 134 (1957).

Dolby, R. M., "An Audio Noise Reduction System," *J. Audio Eng. Soc.*, **15**, 4 (1967).

de Lange, H. K. A., "A Video Tape Recorder for Nonprofessional Use," *Phil. Tech. Rev.*, **26**, 186 (1965).

Duinker, S., "Durable High-Resolution Ferrite Transducer Heads Employing Bonding Glass Spacers," *Philips Res. Rep.*, **15**, 342 (1960).

Eto, Y., S. Mita, Y. Hirano, and T. Kawamura, "Experimental Digital VTR with Trilevel Recording and Fire Code Error Correction," *J. SMPTE, 90, 611 (1981).*

Fisher, E. P. L., "Telcan," *Int. Telev. Tech. Rev.*, **4**, 238 (1963).

Fujimori, H., M. Kikuchi, Y. Obi, and T. Masumoto, "High Permeability Properties of Amorphous Co-Fe Base Alloy" (Japanese) *J. Jpn. Inst. Metals*, **41**, 111 (1977).

Fujio, T., "A Study of High-Definition TV System in the Future," *IEEE Trans. Broadcast.*, **BC-24**, 92 (1978).

Fujita, M., "Color Video Signal Recording and Reproducing System," U.S. Patent 3,715,468, 1973.

Fujiwara, T., M. Issiki, Y. Koike, and T. Oguchi, "Recording Performances of Ba–Ferrite Coated Perpendicular Magnetic Tapes," *IEEE Trans. Magn.*, **MAG-18**, 1200 (1982).

Fujiwara, Y., T. Eguchi, and K. Ike, "Tape Selection and Mechanical Considerations for 4:2:2 DVTR," *J. SMPTE*, **93**, 818 (1984).

Geddes, W. K. E., "Dropouts in Video Tape Recording," BBC Monograph No. 57, June, 1965, p. 5.

Ginsburg, C. P., "A New Magnetic Video Recording System," *J. SMPTE*, **65**, 302 (1956).

Ginsburg, C. P., "Contact Duplication of Quadruplex Video Tapes," *J. SMPTE*, **79**, 43 (1970).

Godall, W., "Television by Pulse Code Modulation," *Bell Syst. Tech. J.*, **30**, 33 (1951).

Goldberg, A., "PCM Encoded NTSC Color Television Subjective Test," *J. SMPTE*, **82**, 649 (1973).

Harris, A., "Time-Base Errors and Their Correction in Magnetic Television Recorders," *J. SMPTE*, **70**, 489, (1961).

Hashimoto, Y., and T. Eguchi, "One-Inch Digital VTR" (Japanese), Tech. Group on Magnetic Recording, IECE Jpn. Tech. Rep. MR79-42, 1980.

Hayashi, K., "Research and Development on High-Definition Television in Japan," *J. SMPTE*, **90**, 3 (1981).

Heitmann, J., "Digital Video Recording: New Result in Channel Coding and Error Protection," *J. SMPTE*, **93**, 140 (1984).

Hendershot, W. B., "Thermal Contact Duplication of Video Tape," *Proc. Int. Broadcast. Conf. (London)*, 204 (1970).

Herr, R., "Duplication of Magnetic Tape Recording by Contact Printing," *Tele-Tech.*, **8**, 28 (1949).

Hirota, A., "Recording and Reproducing Method of Color Video Signal," Japanese Patent 56-9073, 1981.

Hirota, K., M. Sugimura, and E. Hirota, "Hot Pressed Ferrites for Magnetic Recording Heads," *Ind. Eng. Chem. Prod. Res. Dev.*, **23**, 323 (1984).

Hitotsumachi, S., "VHS HiFi VTR" (Japanese), *J. Inst. Telev. Eng. Jpn.*, **37**, 1009 (1983).

Hokkyo, J., and N. Ito, "Theoretical Analysis of the Process of Contact Printing of Magnetic Recording," *Proc. Third Hungarian Conf. on Mag. Rec. (Budapest)*, 2 (1970).

IEC Standard, Publication 766, *Helical-Scan Video-Recording Cartridge and Reel-to-Reel System (EIA-J Type I) Using 12.70 mm Magnetic Tape*, 1983.

Iijima, T., and H. Hanafusa, "Plasma Polymerized Protective Film for Magnetic Recording Thin Film Tape" (Japanese), *J. IECE Jpn.*, **J67-C**, 88 (1984).

Iijima, S., K. Fujii, H. Sugaya, and M. Kano, "Pleasing the Consumer While Playing to Win," *IEEE Spectrum*, **14**, 45 (1977).

Ishida, J., "Television Signal Codes" (Japanese), *NHK Tech. Rep.*, **13**, 169 (1970).

Ishida, J., "Design Basis of PCM Television Transmission System," (Japanese) *Trans. IECE Japan*, **J54-A**, 589 (1971).

Iwasaki, S., "Perpendicular Magnetic Recording," *IEEE Trans. Magn.*, **MAG-16**, 71 (1980).

Iwasaki, S., "Perpendicular Magnetic Recording—Evolution and Future," *IEEE Trans. Magn.*, **MAG-20**, 657 (1984).

Iwasaki, S., K. Ouchi, and N. Honda, "Studies of the Perpendicular Magnetization Mode in Co-Cr Sputtered Films," *IEEE Trans. Magn.*, **MAG-16**, 1111 (1980).

Jacobs, I. S., "Magnetic Materials and Applications—A Quarter Century Overview," *J. Appl. Phys.*, **50**, 7294 (1979).

Johnson, W. R., "Reproducing Color Television Chrominance Signals," U.S. Patent 2,921,976, 1960.

Jones, A. H., "Digital Television Recording: A Review of Current Developments," BBC Res. Dept. Rep., BBC RD 29, 1973.

Kanai, K., F. Kobayashi, and H. Sugaya, "Super-Narrow Track MR Head," *IEEE Trans. Magn.*, **MAG-11**, 1212 (1975).

Katayama, H., K. Yokoyama, and S. Nakagawa, "An Equalizing Method of Recording Pulse Waveforms for NRZ Recording" (Japanese), *Seventh Annual Conf. Rec. Magn. Soc. Jpn.*, **PA-14**, 25 (1975).

Katayama, H., K. Yokoyama, and S. Nakagawa, "A High Speed ATC for NRZI Recording" (Japanese), *IECE Japan, Nat. Conv. Rec.*, 199 (1976).

Kawamata, T., Y. Mizoh, H. Ushigome, and H. Hagiwara, "Abrasivity Effect of Magnetic Recording Tape Rubbing Noise" (Japanese), *J. IECE Jpn.*, **J67-C**, 62 (1984a).

Kawamata, T., K. Inoue, and M. Kittaka, "Study for Wear Resistance and Abrasivity of Magnetic Recording Tape" (Japanese), *J. IECE Jpn.*, **J67-C**, 227 (1984b).

Kawasaki, M., and S. Higuchi, "Alloy Powders for Magnetic Recording Tape," *IEEE Trans. Magn,*, **MAG-8**, 430 (1972).

Kazama, K., and H. Itoh, "Automatic Storage and Retrieval of Videotaped Programs," *J. SMPTE*, **88**, 221 (1979).

Kihara, N., "Magnetic Recording and Reproducing System," U.S. Patent 3, 188, 385, 1965; Japanese Patent 450,256 (Appl. June 21, 1961).

Kihara, N., "Compact Type VTR" (Japanese), *J. Inst. Telev. Eng. Jpn*, **17**, 667 (1963).

Kihara, N., and Y. Odagiri, "Slow-Motion Magnetic Sheet Videocorder and its Applications," *J. SMPTE*, **84**, 789 (1975).

Kihara, N., F. Kohno, and Y. Ishigaki, "Development of a New System of Cassette Type Consumer VTR," *IEEE Trans. Consumer Electron*, **22**, 1 (1976).

Kihara, N., K. Nakamura, E. Saito, and M. Kambara, "The Electronic Still Camera: A New Conception in Photography," *IEEE Int. Conf. Consumer Electron*, 325 (June, 1982).

Kihara, N., Y. Odagiri, and T. Sato, "High-Speed Video Replication System Using Contact Printing," *IEEE Int. Conf. Consumer Electron.*, 72 (1983).

Kimura, T., N. Kobayashi, H. Fujiwara, Y. Shiroishi, M. Kudo, and T. Iimura, "Rubbing Noise of Mn-Zn Ferrite Single Crystal for Magnetic Heads," *Summaries Int. Conf. Ferrite*, 138 (1980).

Kobayashi, F., and H. Sugaya, "Computer Simulation of Contact Printing Process," *IEEE Trans. Magn.*, **MAG-7**, 528 (1971a).

Kobayashi, F., and H. Sugaya, "Theoretical Analysis of Contact Printing on Magnetic Tape," *IEEE Trans. Magn.*, **MAG-7**, 244 (1971b).

Kobayashi, H., and D. T. Tang, "Application of Partial Response Channel Coding to Magnetic Recording Systems," *IBM J. Res. Dev.*, **14**, 368 (1970).

Kono, T., "Beta HiFi VTR" (Japanese), *J. Inst. Telev. Eng. Jpn.*, **37**, 1014 (1983).

Kornei, O., "Magnetic Head Has Megacycle Range," *Electronics*, 172 (Nov. 1956).

Maesawa, Y., M. Takao, H. Hibino, M. Odagiri, and K. Shinohara, "Metal Thin Film Video Tape by Vacuum Deposition," *Fourth Int. Conf. Video Data Processing*, 54 (1982).

Makino, Y., K. Aso, S. Uedaira, S. Eto, M. Hiyakawa, K. Hotai, and Y. Ochiai, "Amorphous Alloys for Magnetic Head," *Proc. Third Int. Conf. on Ferrites*, 699 (1980).

Mallinson, J. C., "The Next Decade in Magnetic Recording," *IEEE Trans. Magn,*, **MAG-21**, 1217 (1985).

Mallinson, J. C., and J. Miller, "Optimal Codes for Digital Magnetic Recording," *Radio Electron. Eng.*, **47**, 172 (1977).

Mallinson, J. C., H. N. Bertram, and C. W. Steele, "A Theory of Contact Printing," *IEEE Trans. Magn.*, **MAG-7**, 524 (1971).

Marzocchi, L., "Electromagnetic Sound Recording," U.S. Patent 2,245, 286, 1941.

Masterson, E. E., "Magnetic Recording of High Frequency Signals," U.S. Patent 2,773, 120, 1956.

Matsumoto, H., and T. Yamamoto, "New Alloy 'Sendust' and Magnetic and Electric Properties of Fe-Si-Al Base Alloy," *J. Jpn. Inst. Metals,* **1,** 127 (1937).

Matsuura, K., K. Oyamada, and T. Yazaki, "Amorphous Video Head for High Coercive Tape," *IEEE Trans. Magn.,* **MAG-19,** 1623 (1983).

Mehrgardt, S., "Digital Processing in Video Tape Recorders," *IEEE Int. Conf. Consumer Electron. Dig. Tech. Papers,* **IX.** 132 (1985).

Mizushima, M., "Mn-Zn Single Crystal Ferrite as a Video Head Material," *IEEE Trans. Magn.,* **MAG-7,** 342 (1971).

Mohri, K., Y. Yumde, M. Umemura, Y. Noro, and S. Watatani, "A New Concept of a Handy Video Recording Camera," *IEEE G-CE, Spring Conf.,* **CE-27,** 3 (1981).

Morio, M., Y. Matsumoto, Y. Machinda, Y. Kubota, and N. Kihara, "Development of an Extremely Small Video Tape Recorder," *IEEE Trans. Consumer Electron.,* **CE-27,** 331 (1981).

Morizono, M., H. Yoshida, and Y. Hashimoto, "Digital Video Recording—Some Experiments and Future Consideration," *J. SMPTE,* **89,** 658 (1980).

Mueller-Ernesti, R., German Patent 910,602, 1941.

Mullin, J. T., "Video Magnetic Tape Recorder," *Tele-Tech. and Electron. Ind.* **13,** 77 (1954).

Nagai, K., Japanese Patent 318,961, 1961.

Nakagawa, S., "A Constitution of Concatenated Code for Digital VTR" (Japanese), ITE Jpn., Tech. Report VR 32-3, (1978).

Nakagawa, S., and K. Yokoyama, "A Design Method of Linear Density and Track Density for Maximizing Area Density in Digital Videotape Recorder" (Japanese), *Trans. IECE Jpn.,* **J 64-C,** 386 (1981).

Nakagawa, S., K. Yokoyama, and H.Katayama, "A Study on Detection Methods of NRZ Recording," *IEEE Trans. Magn.,* **MAG-16,** 104 (1980).

Nakano, K., H. Moriwaki, T. Takahashi, K. Akagiri, and M. Morio, "New 8-Bit PCM Audio Recording Technique Using an Extension of the Video Track," *IEEE Intl. Conf. Consumer Electron.,* 241 (1982).

Nelson, A. M., "Double Transfer Curie Point and Magnetic Bias Tape Copy System," U.S. Patent 3,496,304, 1970.

Nomura, T., and K. Yokoyama, "Thermomagnetic Video Recording," *IEEE Trans. Magn.,* **MAG-15,** 1932 (1979).

Numakura, T., "Color Video Signal Magnetic Recording Equipment," Japanese Patent 49-44535, 1974.

Okamura, S., "Magnetic Recording Processing Equipment," Japanese Utility Patent S39-23924, 1964.

Olson, H. F., W. D. Houghton, A. R. Morgan, J. Zenel, M. Artzt, J. G. Woodward, and J. T. Fisher, "A System for Recording and Reproducing TV Signals," *RCA Rev.,* **15,** 3 (1954).

Olson, H. F., W. D. Houghton, A. R. Morgan, M. Artzt, J. A. Zenel, and J. G. Woodward, "A Magnetic Tape System for Recording and Reproducing Standard FCC Color Television Signals," *RCA Rev.,* **17,** 330 (1956).

Peters, C. J., "A Magnetically Scanned Magnetic Tape Transducer," *IEEE Trans. Elec. Computers,* 196 (April, 1964).

Pritchard, D. H., and J. J. Gibson, "Worldwide Color Television Standards," *J. SMPTE,* **89,** 111 (1980).

Potgiesser, J. A. L., and J. Koorneef, "Wear of Magnetic Heads," *Proc. Conf. Video and Data Rec.,* 203 (1973).

Reed, I. S., and G. Solomon, "Polynomial Codes over Certain Finite Fields," *J. Soc. Ind. Appl. Math.,* **8,** 300 (1960).

Reeves, A. H., French Patents 833,929, 1937; 837,921, 1937, and 852,183, 1938. U.S. Patent 2,272,070, 1942.

Reimers, U. H., W. H. Zappen, and H. F. Zettl, "KCF I—Using the Leading Edge of Technology in a Small Broadcast Camera," *J. SMPTE,* **94,** 573 (1985).

Ryan, D. M., "Mechanical Design Considerations for Helical-Scan Video Tape Recorders," *J. SMPTE,* **87,** 767 (1978).

Sadashige, K., "An Overview of Longitudinal Video Recording Technology," *J. SMPTE,* **89,** 501 (1980).

Sanderson, H. J., "Method of Controlling the Position of a Write or Read Head and a Device for Carrying out the Method," U.S. Patent 4,297,733, 1981.

Sato, S., K. Takeuchi, and M. Yoshida, "Recording Video Camera in the Beta Format," *IEEE Trans. Consumer Electron.*, **CE-29**, 365 (1983).

Sawazaki, N., "Magnetic Recording Apparatus," Japanese Patent S34-171, 1955.

Sawazaki, N., M. Yagi, M. Iwasaki, G. Inada, and T. Tamaoki, "A New Video-Tape Recording System," *J. SMPTE*, **69**, 868 (1960).

Sawazaki, N., H. Tsukamoto, M. Imamura, and T. Fujiwara, "Endless Tape Fixed Head VTR," *IEEE Trans. Magn.*, **MAG-15**, 1564 (1979).

Schüller, E., "Magnetische Aufzeichnung und Wiedergabe von Fernsehbildern," German Patent 927,999, 1954.

Sekimoto, K., M. Matsui, and I. Obata, "M-II Format VTR," *Sixth Conf. Video, Audio & Data Recording*, 121 (1986).

Shannon, C. E., "A Mathematical Theory of Communication," *Bell Syst. Tech. J.*, **27**, 379 (1948).

Shibaya, H. and I. Fukuda, "The Effect of the B_s of Recording Head Cores on the Magnetization of High Coercivity Media," *IEEE Trans. Magn.*, **MAG-13**, 1005 (1977).

Shimada, Y., "Cobalt Amorphous Magnetic Materials Produced by Sputtering," *J. Jpn. Inst. Metals*, **22**, 11 (1983).

Shinohara, K., H. Yoshida, M. Odagiri, and A. Tomago "Columnar Structure and Some Properties of Metal-Evaporated Tape," *IEEE Trans. Magn.*, **Mag-20**, 824 (1984).

Shiraishi, Y., and A. Hirota, "Magnetic Recording at Video Cassette Recorder for Home Use," *IEEE Trans. Magn.*, **MAG-14**, 318 (1978).

Snyder, R. H., "Ampex's New Video Tape Recorder," *Tele-Tech. and Electron. Ind.*, 15, 72 (1956).

Sugaya, H., "Some Problems of Metallic Magnetic Material at High Frequency" (Japanese), *J. Inst. Telev. Eng. Jpn.*, **18**, 722 (1964).

Sugaya, H., "Newly Developed Hot-Pressed Ferrite Head," *IEEE Trans. Magn.*, **MAG-4**, 295 (1968).

Sugaya, H., "Magnetic Tapes for Contact Duplication by Anhysteretic and Thermal Transfer Methods," *AIP Conf. Proc.*, **10**, 1086 (1973).

Sugaya, H., "Recent Advances in Video Tape Recording," *IEEE Trans. Magn.*, **MAG-14**, 632 (1978).

Sugaya, H., "Home Video Tape Recording and Its Future Prospects," *IERE Proc.*, **54**, 75 (1982).

Sugaya, H., "Video Tape Recorder and Its Future Prospects," *J. Appl. Magn.*, **8**, 305 (1984).

Sugaya, H., "Mechatronics and the Development of the Video Tape Recorder," *ASLE Special Pub.*, **SP-19**, 64 (1985).

Sugaya, H., "The Videotape Recorder: Its Evolution and the Present State of the Art of VTR Technology," *J. SMPTE*, **95**, 301, (March, 1986).

Sugaya, H., and F. Kobayashi, "Theoretical Analysis of Contact Printing on Magnetic Tape," *IEEE Trans. Magn.*, **MAG-7**, 244 (1971a).

Sugaya, H., and F. Kobayashi, "Computer Simulation of Contact Printing," *IEEE Trans. Magn.*, **MAG-7**, 528 (1971b).

Sugaya, H., and A. Tomago, "Metal Evaporated Tape," *Symp. Magn. Media and Mfg. Methods, Hawaii* **C-2**, 1 (1983).

Sugaya, H., F. Kobayashi, and M. Ono, "Magnetic Tape Duplication by Contact Printing at Short Wavelengths," *IEEE Trans. Magn.*, **MAG-5**, 437 (1969).

Sugaya, H., M. Deguchi, H. Taniguchi, and T. Yonezawa, "Standard Alignment Tape Recorder for EIA-J Type I Video Tape Recorder," *J. SMPTE*, 83, 901 (1974).

Sugaya, H., F. Kobayashi, and M. Ono, Japanese Utility Patent S54-6346, 1979.

Sugimoto, M., "Crystal Growth of Manganese Zinc Ferrite," *J. Appl. Phys. (Jpn.)*, **5**, 557 (1966).

Suzuki, K., E. Kimura, and K. Yokoyama, "Magnetic Recording Apparatus," Japanese Patent 480,366 (1979).

Swoboda, T. J., P. Arthur, Jr., N. L. Cox, J. N. Ingraham, A. L. Oppegard, and S. Sadler, "Synthesis and Properties of Ferromagnetic Chromium Oxide," *J. Appl. Phys.*, **32**, 374 (1961).

Tajiri, H., S. Tanaka, I. Sato, M. Yagi, and N. Sawazaki, "Color Video Tape Recorder for Home Use," *J. SMPTE*, **77,** 727, (July, 1968).

Takahashi, K., K. Ihara, S. Muraoka, E. Sawai, and N. Kaminaka, "A High Performance Video Head with Co Based Alloy Laminated Films," *Dig. Intermag. Conf.* 1987, **EB-08** (1987).

Takama, E., and M. Ito, "New Mn-Zn Ferrite Fabricated by Hot Isostatic Pressing," *IEEE Trans. Magn.*, **MAG-15,** 1958 (1979).

Tanimura, H., Y. Fujiwara, and T. E. Mechrens, "A Second Generation Type-C One-Inch VTR," *J. SMPTE*, **92,** 1274 (1983).

Tjaden, D. L. A., and A. M. A. Rijckaert, "Theory of Anhysteretic Contact Duplication," *IEEE Trans. Magn.*, **MAG-7,** 532 (1971).

Tomita, Y., "Two-Head Color Video Tape Recorder" (Japanese) *J. Inst. Telev. Eng. (Jpn)*, **15,** 22 (1961).

Torii, M., "New Process to Make Huge Spinel Single Crystal Ferrite," *IEEE Trans. Magn.*, **MAG-15,** 873 (1979).

Torii, M., U, Kihara, and I. Maeda, "On the Rubbing Noise of Mn-Zn Ferrite Single Crystal," *Proc. Int. Conf. Ferrites*, 01AB2-4, 137 (1980).

Umeki, S., S. Saitoh, and Y. Imaoka, "A New High Coercive Magnetic Particle for Recording Tape," *IEEE Trans. Magn.*, **MAG-10,** 655 (1974).

Van Den Berg, R., "The Design of A Machine for High-Speed Duplication of Video Records," *105th Tech. Conf.*, *SMPTE*, **78,** 709 (1969).

Waring, P. K., "CrO_2-Based Thermomagnetic Information Storage and Retrieval System," *J. Appl. Phys.*, **42,** 1763 (1971).

Woodward, J. G., "Stress Demagnetization in Videotapes," *IEEE Trans. Magn.*, **MAG-18,** 1812 (1982).

Yasuda, I., Y. Yoshisato, Y. Kawai, K. Koyama, and T. Yazaki, "Ultra-High-Density Recording with Sendust Video Head and High Coercive Tape," *IEEE Trans. Magn.*, **MAG-17,** 3114 (1981).

Yokoyama, K., "Basic Studies of Video Tape Recorder Design," (Japanese) *NHK Technical Journal*, **21,** No. 4 (1969).

Yokoyama, K., and H. Habutsu, "PCM-VTR and Editing" (Japanese), *J. Inst. Tel. Engr. Japan*, **32,** 843 (1976).

Yokoyama, K., and S. Nakagawa, "Trends of Research on Digital VTR," (Japanese), *J. Inst. Tel. Engr. Japan*, **32,** 819 (1978).

Yokoyama, K., and S. Nakagawa, "An Experimental Channel Coding for Digital Videotape Recorder," *12th Intl. Tel. Symp. Record*, 251, (1981).

Yokoyama, K., S. Nakagawa, H. Shibaya, and H. Katayama, "PCM Video Recording Using a Rotating Magnetic Sheet," NHK Labs. Note, No. 221, 1977.

Yokoyama, K., S. Nakagawa, and H. Katayama, "Experimental PCM-VTR," NHK Labs. Note, No. 236, 1979*a*.

Yokoyama, K., S. Nakagawa, and H. Katayama, "Trial Production of Experimental PCM VTR" (Japanese), *IECE Jpn. Tech. Rep.*, MR 79-8 (1979*b*).

Yokoyama, K., S. Nakagawa, and H. Katayama, "An Experimental Digital Videotape Recorder," *J. SMPTE*, **89,** 173 (1980).

Yoshida, H., and T. Eguchi, "Digital Video Recording Based on the Proposed Format from Sony," *J. SMPTE*, **92,** 562 (1983).

Zahn, H., "The BCN System for Magnetic Recording of Television Programs," *J. SMPTE*, **88,** 823 (1979).

Chapter 3

Audio Recording

Eric D. Daniel[†]

Woodside, California

John R. Watkinson[‡]

*Ampex Great Britain Limited, Reading,
Berkshire, England*

3.1 Audio Recording Principles

3.1.1 Characteristics of audio signals

In the recording and reproduction of audible sounds the final arbiter of quality is the human ear. It is therefore fitting that the subject of audio recording be introduced by a brief review of some of the more important aspects of psychoacoustics.

The human ear responds to sounds covering a range of frequencies extending from 20 Hz to 20 kHz. Relative to the signals discussed in other chapters, the highest frequency is modest, but the bandwidth, measured in octaves (10), is large. The range of intensities that can be detected is

---

† Sections on recording with ac bias.

‡ Sections on digital audio recording.

also large, the ratio between the loudest (threshold of feeling) and the softest (threshold of hearing) intensities being about one billion to one, or 120 dB. The smallest detectable sound is in fact comparable to the thermal energy of the molecules of air.

Different sounds of interest do not occupy the same space in the intensity-frequency plane and may not require equal treatment to give satisfactory electroacoustic reproduction. The spaces occupied by speech and music are compared with the boundaries of hearing in the highly simplified diagram of Fig. 3.1. In the case of speech, the main components related to intelligibility are centered around 2 kHz, but most of the energy lies below 1 kHz. To reproduce every nuance of a speaker's voice requires a large bandwidth and a high signal-to-noise ratio. Merely to reproduce intelligible speech (e.g., a telephone) demands much less, and to provide maximum intelligibility with minimum power (e.g., a hearing aid) calls for a carefully restricted bandwidth and compression of the dynamic range.

Musical sounds approach all reaches of the audible spectrum, and the dynamic range between the loudest and softest passages of a symphony orchestra can be 70 dB or more. The power is not evenly distributed over the spectrum but tends to drop off at the low and at the high frequencies to an extent that is highly dependent on the particular musical instruments and composition (Sivian et al., 1931; McKnight, 1959). There is really no average power spectrum but only a set of individual spectra that may exhibit the low-frequency power of a pipe organ at one extreme and the substantial high-frequency energy of metal percussion at the other.

A listener possesses definite opinions about which sounds are wanted

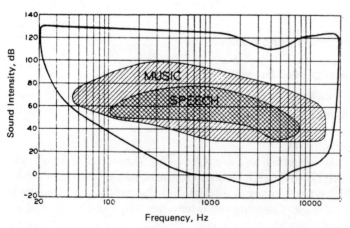

Figure 3.1 Simple representation of the frequency and intensity ranges of music and speech. The solid line indicates the boundaries of normal hearing *(Olsen, 1957)*.

or unwanted and can detect one in the presence of the other with a high degree of discrimination. On the one hand, it is possible to recognize speech in the presence of a higher level of noise. On the other hand, the enjoyment of listening to music may be marred by the presence of a level of noise which, although low enough to be masked by the louder passages, is clearly evident in quiet intervals. A critical listener is also disturbed by the presence of small components of sound introduced by nonlinear distortion of the signal amplitude. Here components that are in a harmonic relationship to the signal are less offensive than nonharmonic components, such as those introduced by intermodulation of the contents of a complex signal. Generally, intermodulation and harmonic distortion components are related, and it is common to specify nonlinear distortion in terms of the more easily measured harmonic components produced when a single-frequency signal is recorded.

The human ear is less sensitive to distortions of signal phase within a channel, but phase linearity has to be maintained within certain limits, otherwise the subjective impression of transients is impaired (Preis, 1983). Also, phase is critically important with respect to the directional properties of normal hearing. The fact that a sound arrives at each ear at a different time (and with a different intensity) allows the source to be located in space. More generally, the listener is aware, not only of the sound emanating straight from a source, but of secondary sounds arriving from other directions after reflections, and multiple reflections, from the boundaries of the room.

Finally, the quality of musical reproduction is degraded by the introduction of unnatural amplitude modulation caused by variations in tape sensitivity or frequency modulation caused by variations in tape speed. Subjectively, the sidebands of amplitude modulation can be perceived as noise components around the signal which detract from its clarity. Frequency modulation can be particularly disturbing, especially when listening to a musical instrument that should have little or no vibrato, such as a clarinet or a piano.

Ideally, a high-performance audio recorder must be capable of reproducing sound quality that is indistinguishable from the live performance. From the foregoing, the requirements for approaching this ideal fall into three categories. First, the recorder must match the frequency response of the ear and the dynamic range of a wide variety of music. Second, all unwanted signals, such as additive noise, modulation noise, print-through, and distortion components, must be inaudible or masked by wanted signals. Third, at least two (stereophonic) channels must be provided and their individual and mutual phase integrity maintained so as to simulate binaural effects and the associated sensations of direction, depth, and reverberation.

3.1.2 Historical development

In 1897, Valdemar Poulsen applied for a patent using the properties of a permanent magnet, in the form of a steel wire, as a means of recording. His work achieved considerable acclaim. Working models of his "telegraphone" were exhibited and numerous, mostly short-lived, companies were set up to exploit his concept. Unfortunately, successful commercial realization could not occur until electronic amplification was developed in the 1920s. The Marconi-Stille equipment produced in the early thirties used steel tape and heads with staggered poles on either side of the tape, and was among the first magnetic recorders to achieve commercial success. Early work in the United States focused on using various compositions of stainless-steel wire as the magnetic medium, an approach that was doomed to be of limited performance.

The greatest practical advance in the early years of magnetic recording was the development of the Magnetophone in Germany in the 1930s and 40s. This equipment had all the essential features of a modern audio tape recorder: coated particulate plastic tape; gapped ring heads; separate erase, record, and reproduce heads; and a capstan and pinch-roller drive. Above all, in the higher-performance Magnetophones, ac bias was used in place of dc bias to produce dramatic improvements in linearity and background noise.

The introduction of ac bias was the most significant step forward in audio magnetic recording technology. The concept of using ac bias was not new but had been invented and patented in the United States by Carlson and Carpenter in 1927. However, the concept languished in obscurity for many years, and the Magnetophone recorders were the first in which ac bias was applied in a practical, well-engineered fashion.

The 1950s saw a rapid increase in audio tape recording activities in Europe, the United States, and Japan. Inexpensive consumer recorders were manufactured, and more sophisticated equipment, mirroring the Magnetophone, was built to satisfy the growing interest on the part of the broadcasting organizations and recording studios. Theoretical studies of magnetic recording, previously neglected, were undertaken, and many of the fundamental principles were established (Wallace, 1951; Westmijze, 1953).

Technical innovations were few; progress occurred more by incremental improvements, particularly in the composition and dimensional precision of tapes and heads. Such improvements led to increased audio performance and reduced tape speed and, eventually, to convenient tape packaging such as the "compact cassette" introduced by Philips in 1963. The development of multitrack heads had naturally made magnetic recording a convenient vehicle for implementing stereo recording, an important step towards providing realistic sound reproduction. Studio recorders were subsequently developed with many more than two-channel capability to provide additional flexibility in mixing and to facilitate the production of special effects. Meanwhile, improvements were made in that key index of audio performance, signal-to-noise ratio, not only by means of better media, heads, and electronics but by introducing electronic noise-reduction techniques.

Incremental advances in ac-biased recording will continue to be made. There are, however, alternative routes to achieving higher levels of audio recording performance, for example, the use of modulation schemes to replace the linear recording approach. The use of modulation techniques received scant attention until the seventies. Since then, momentum has increased rapidly, with emphasis on the development of digitally encoded audio recording as a means of reducing or eliminating many of the limitations inherent in ac-biased recording. Digital audio was first applied to professional studio recorders and became available indirectly to the consumer in the form of phonograph disks produced using digitally recorded master tapes. Later, several developments took place in rapid succession. Full digital recordings were commercially produced in the form of factory-recorded, play-only, nonmagnetic optical disks (*Compact Discs* or *CDs*) produced, again, using digitally recorded master tapes. Home video tape recorders were introduced which used FM or digital techniqus to incorporate the audio signal into the video track; they provide a recording capability comparable in quality to that of a CD. Finally, efforts were accelerated to produce dedicated digital audio recorders for home as well as studio use. A detailed discussion of digital audio technology as applied to magnetic recording is given in Sec. 3.3.

3.2   Recording with AC Bias

3.2.1   Principles of design and operation

3.2.1.1   Recording components.   The essential components of an audio tape recorder are shown diagrammatically in Fig. 3.2. During recording, the tape is moved at constant speed successively over the erase head, which removes any previous recording; the record head, where a new program is recorded; and the reproduce head, which provides a delayed reproduction of the program for monitoring purposes.

The signal to be recorded is fed to the record head via an amplifier and equalizer, and the reproduced signal is amplified, integrated, and subjected to more equalization before being sent to the output terminals. Integration is required to compensate for the fact that an inductive reproducing head responds to the derivative of the recorded flux. The primary function of the pre- and postrecording equalizers is to correct for losses

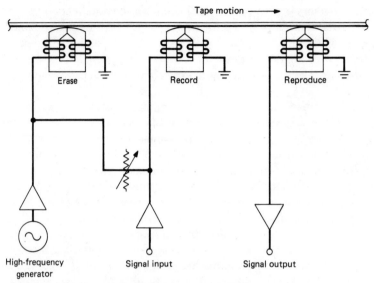

Figure 3.2 Block diagram of the components of an ac-biased audio tape recorder using separate erase, record, and reproduce heads.

that occur in the recording and reproducing processes at short wavelengths, so that the frequency response of the recorder is flat up to the desired highest frequency (shortest wavelength).

An ac-bias current is added to the signal current in the head to linearize the otherwise highly nonlinear, hysteretic recording process. The bias current is considerably greater than the largest signal current. As discussed later, its precise magnitude is critical to all aspects of recording: sensitivity, frequency response, distortion, noise, permanence, and erasability. The frequency of the bias is not critical provided it is several times higher than the highest signal frequency: a frequency of 100 kHz is typical.

The erasing process not only must eradicate previous recordings but must leave the tape in a condition of lowest possible noise. This dictates the use of ac erasure, the head being commonly fed from the same source as the bias. It is essential that the bias and erase-current waveforms be symmetrical. Introducing a small degree of asymmetry is equivalent to adding direct current, or permanently magnetizing a head, either of which can cause a substantial increase in noise.

The heads are of ring construction (Volume I, Chap. 3), with gap lengths suited to their function. The erase-head gap length is made large, often several times the coating thickness, to assist in providing a strong erasing field throughout the whole coating. Sometimes multiple gaps are used for reasons that are discussed later. The record-head gap is typically made comparable to the coating thickness, a compromise between maximizing long- and short-wavelength performances. The reproduce head gap is made small enough to avoid significant gap loss at the shortest wavelength of interest.

3.2.1.2 Linearity of the recording process. The physical mechanism underlying the linearizing effect of ac bias is analogous to a modified form of anhysteresis and has been discussed in detail in Volume I, Chap. 2. It will, however, be useful to review some of the principles of ac-biased recording here in order to support the discussions of audio recording that follow.

Without bias, the relationship between remanent magnetization M_r and recording signal field H follows an S-shaped curve of the type shown dashed in Fig. 3.3. The use of an optimal amount of ac bias, equivalent to the use of a peak bias field approximately equal to the coercivity, promotes a relationship between anhysteretic magnetization M_{ar} and field H of the type shown by the full-line curve. The curve is intially straight, but the linear region is limited, and if the signal field is too high, the curve departs from linearity and distortion occurs. The data from a wide variety

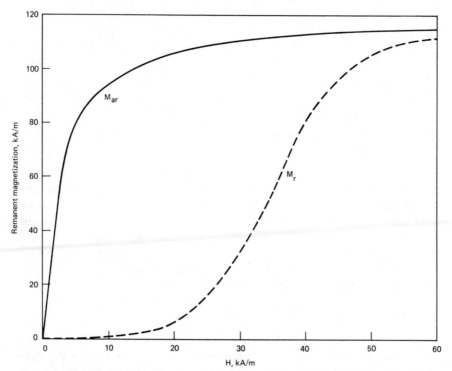

Figure 3.3 Anhysteretic magnetization M_{ar} as a function of signal field H for optimal bias field. The ordinary remanent magnetization curve M_r is shown dashed.

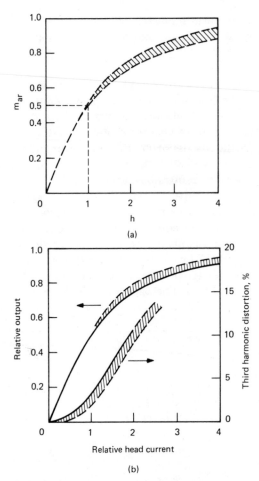

Figure 3.4 (a) The normalized anhysteretic magnetization versus signal field curve. The shaded area contains all the data from a wide variety of longitudinally oriented particulate media, including pure and cobalt-modified iron oxides, chromium dioxide, and metal. (Köster, 1975). (b) Relative output and third-harmonic distortion versus signal current. The shaded areas contain data from the same media as in (a); the full lines represent calculations based on a universal anhysteretic magnetization curve (Köster, 1975).

of longitudinally oriented particulate coatings are practically identical when normalized as shown in Fig. 3.4a, implying that all such tapes have the same distortion characteristics (Köster, 1975). The extent of this distortion can be calculated by expressing the data by the best-fitting, odd-order power series and computing the harmonic content produced when a sinusoidal field is applied. Below a reduced field of $h = 2$, the data can be represented quite accurately by three terms:

$$m_{ar} = 0.5691h - 0.756h^3 + 0.0065h^5 \tag{3.1}$$

Calculations of long-wavelength output and third-harmonic distortion based on assuming a universal anhysteretic curve are shown in Fig. 3.4b. Data on the same wide range of oriented particulate coatings are contained within the shaded area, indicating good agreement between theory and experiment (Köster, 1975). The output corresponding to 3 percent third-harmonic distortion (an upper limit commonly used for audio recording) is about 6 dB below the maximum output, corresponding to a magnetization of $M_{ar} = 0.5M_s$.

At the shorter wavelengths, the departure from linearity takes a different form. The transfer function of the recording process no longer follows the M_{ar} versus H curve but goes through a maximum, then falls rapidly at higher signal currents. Distortion, measured in terms of intermodulation products (harmonics are outside the pass band), does increase at higher currents. But the significant form of distortion is *compression*, a departure from linearity without a concomitant production of distortion components.

3.2.1.3 Bias adjustment.   So far, it has been assumed that the bias field can always be adjusted to the optimal value, but this ignores the fact that the field decreases with distance from the head. In practice the bias field is optimal only for one elementary layer of the coating; layers above and below will be either over- or underbiased. Adjusting the bias current so that the correctly biased layer is at the surface is equivalent to setting the bias for maximum short-wavelength output; the rest of the coating is underbiased and suffers increased nonlinearity of the type illustrated by curve A of Fig. 3.5. Increasing the current so that the correctly biased layer moves to the bottom of the coating is equivalent to setting the bias for minimum distortion at long wavelengths. The surface layers carrying the bulk of the useful short-wavelength recording are overbiased and the short-wavelength output is decreased in accordance with curve C of Fig. 3.5.

Setting the value of bias is therefore a compromise between achieving optimum results at short and long wavelengths. The difficulty of the decision can be gained from Fig. 3.6a, which shows the variation of signal output with bias at a number of different wavelengths, and from Fig. 3.6b, which shows the variation of long-wavelength distortion with bias. Adjusting the bias for optimum short wavelengths leaves the long wavelengths underbiased and with high distortion; but using the bias best suited for long-wavelength performance causes substantial losses in short-wavelength output.

The minimum in the distortion curve can be explained on the following basis (Westmijze, 1953; Fujiwara, 1979). The coefficients of the third- and fifth-order coefficients in the power series representing the underbiased curve B of Fig. 3.5 are of opposite sign. This produces a null in the calculated third harmonic at a certain signal field strength. In like fashion, it is possible to set the bias and signal fields to produce a minimum in the net harmonic content of the whole coating. The minimum is too critical to be used as an operating point; its existence can be more properly

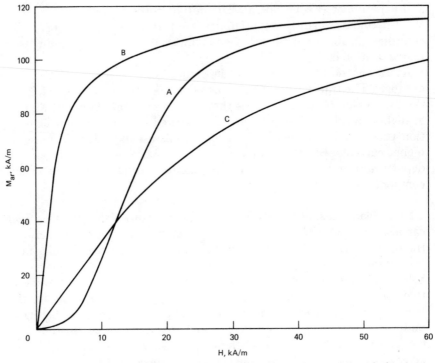

Figure 3.5 Modified anhysteresis curves showing the effects of curve (A) underbiasing, curve (B) optimally biasing, and curve (C) overbiasing.

regarded as an example of the shortcomings of harmonic content in revealing all aspects of nonlinear behavior.

Above the maximum, the decrease in distortion is coupled with a comparable decrease in sensitivity, and the long-wavelength linearity becomes substantially independent of bias. Adjusting the bias on the high side is therefore the safest policy with respect to minimizing distortion. For this reason, the bias in high-speed applications is usually set to a value equal to, or a little greater than, the value that gives maximum long-wavelength sensitivity. In slow-speed applications, where the shortest wavelengths are critically dependent on bias, it may be preferable to adjust the bias to overbias the shortest wavelength by a given amount, for example to produce a 2- or 3-dB loss.

3.2.1.4 High-frequency losses. An audio recorder exhibits appreciable losses at high frequencies. The true frequency-dependent losses are small, such as those associated with losses in the cores of the heads. The significant losses are wavelength-dependent and increase in severity as the tape speed is decreased. For example, at a tape speed of 48 mm/s (1.875 in/s), the total loss at 15 kHz (wavelength of 3.2 μm) can approach 30 dB.

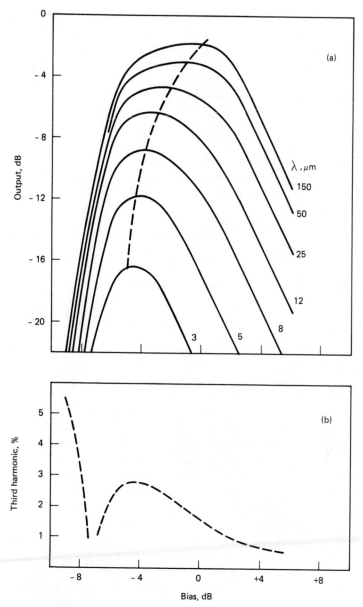

Figure 3.6 Typical curves showing *(a)* reproduced output versus bias current at different wavelengths and *(b)* third-harmonic distortion versus bias at a long wavelength.

Short-wavelength losses arise in many ways, some of which have been discussed in Volume I, Chap. 2. The ones best understood and most easily quantified occur on reproduction and are known as *gap* and *spacing losses*. The gap loss is given by $(\lambda/\pi g_{\text{eff}}) \sin (\pi g_{\text{eff}}/\lambda)$ where the effective gap g_{eff} is some 14 percent longer than the physical gap g. The loss can be controlled to an acceptable value by making the gap sufficiently smaller than the shortest wavelength (the loss is less than 2 dB for $g/\lambda = \frac{1}{3}$), although too short a gap sacrifices head efficiency.

The loss caused by spacing between the medium and the reproducing head increases exponentially with decreasing wavelength, according to $\exp(-2\pi d/\lambda)$ or $54.6d/\lambda$ in dB. Smooth head and tape surfaces and adequate head-tape pressure in the vicinity of the gap are essential to minimize spacing loss.

Even in the ideal case, perfect contact with the head is achieved only for the nearest elementary layer of a medium of finite thickness; the contributions to the output from layers beneath the surface will suffer progressively increasing spacing losses. The net effect, corresponding to integration through the thickness of the medium, is commonly misnamed "thickness loss," and ultimately gives a 6 decibel per octave fall in response for the artificial case of uniform magnetization through the depth of the medium.

Two of the losses that occur during recording have been described earlier in relation to overbiasing and compression. Generally, these and other recording losses are not amenable to the precise analyses possible for reproduction. A major complication is that the field from the head rotates, so that the direction turns toward the perpendicular in the near layers. To a good approximation, however, it can be assumed that the efficiency of the bias field is independent of its direction and depends only on its *resultant* magnitude (Eldridge and Daniel, 1962; Tjaden and Leyten, 1963). Therefore, in the simplest case, recording takes place along a contour of constant resultant bias field approximately equal to the coercivity (Bertram, 1974). An element of medium acquires a magnetization proportional to the signal field at the point where the element crosses the critical bias-field contour. The longitudinal field (and magnetization) along the contour becomes vanishingly small toward the head surface, and this results in a predicted thickness loss that approaches 12 decibel per octave at short wavelengths, as opposed to the 6 decibel per octave deduced for uniform magnetization (or purely perpendicular magnetization). Particulate tapes, well-oriented along their length, give a response in keeping with the 12 decibel per octave prediction (Bertram, 1975).

Further losses in recording sensitivity arise because the particles of a coating do not possess the same critical switching field. Instead, there is a distribution of critical fields, associated with the switching-field distribution defined in Volume I, Chap. 3. Consequently, recording does not take place along a single bias contour but over a zone bounded by contours equal to the highest and lowest particle switching fields. Losses occur when the length of the zone becomes comparable with the wavelength. In the simplest example of a rectangular distribution of critical fields, the zones are as depicted in Fig. 3.7, and it is apparent that the zone length, and hence the loss, increases as bias is increased. Generally, media with narrow switching-field distributions will give superior performance (Köster, 1981).

Demagnetization losses, once thought to be of major significance, are probably negligible down to wavelengths of 2 μm—comparable to the shortest wavelengths used in conventional forms of ac-biased recording (Bertram and Niedermeyer, 1982). This conclusion assumes that intimate contact is maintained between heads and tape so that demagnetizing fields are minimized by the "keeping" action of the high-permeability head core.

3.2.1.5 Equalization. As mentioned briefly above, audio recorders use preequalization before recording and postequalization after reproduction to correct for losses at high frequencies (short wavelengths) and to provide an overall flat frequency response over the working bandwidth. In considering how best to allot the total requirement between pre- and postequalization, there are several factors which favor preequalization:

1. Postequalization increases the reproduced noise, whereas preequalization does not.
2. The power spectrum of much musical program material falls at high frequencies, allowing headroom for preequalization.

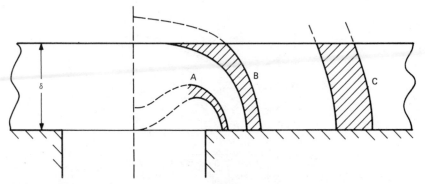

Figure 3.7 Diagrammatic representation of recording zones corresponding to zone *(A)* underbiasing, zone *(B)* optimally biasing for long wavelengths, and zone *(C)* overbiasing. The full lines are contours of constant resultant bias field strength *(Eldridge and Daniel, 1962).*

3. Losses in recording sensitivity can, in principle, be compensated for by preequalization without penalty.

The arguments against preequalization rest mainly on the danger of running into compression, since there is no such thing as a universal power spectrum for program material. Certain programs, or small portions of them, may exceed any nominal limits, and usually the equipment designer has no control over the types of program to be recorded.

In practice, the allotment of equalizations is made on a rather arbitrary basis. The requirement that a program recorded on one machine be playable on any other of the same class makes it essential that the postequalization be standardized. The preequalization is then adjusted to provide a flat response over the working frequency range, using the particular tape-head combination that will be used. Such a procedure cannot produce optimum results under all recording conditions, but it is the basis for the various standards described in subsequent sections.

3.2.1.6 Noise and interference

Types of noise. The fundamental characteristics of noise in magnetic recording are discussed in Volume I, Chap. 5. The purpose here is to expand on those aspects that are specifically pertinent to audio recording.

Spectra of the major types of noise of interest are shown in Fig. 3.8 for a low-speed (48-mm/s) equalized system. The bottom full-line curve shows the noise that results when a tape is reproduced after being erased

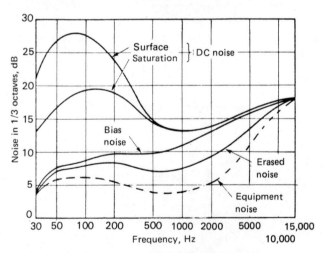

Figure 3.8 Noise spectra of a low-speed (47.6 mm/s or 1.875 in/s audio recorder. The various types of noise are explained in the text *(Daniel, 1972)*.

using a bulk eraser operating at power-supply frequency. The *bulk-erased* noise is associated directly with the particles of the coating (or the micromagnetic structure of a deposited film). When ac bias is applied in the absence of signal, the noise increases by about 3 dB in the mid-band. This *bias noise* is the background noise that is heard in the quiet passages of a recorded program. Also, there are modulation noises that are a function of the recorded signal. The simplest is the *dc noise* that occurs when direct current is fed to the recording head. When the current is sufficient to saturate the tape, the noise increases by about 10 dB at the longer wavelengths. This increase is attributed to nonuniform distribution of particles within the coating. At a lower value of direct current, the long-wavelength noise peaks to a level that may be as much as 20 dB above the bulk-erased noise. The resulting *surface noise* is attributed to variations in head-to-tape spacing caused by asperities on the tape surface.

Head and electronics noise are of concern only at the very shortest wavelengths reached in slow-speed recording. Interference effects are usually not of significance in audio equipment, with the important exception of *print-through,* the transfer of signal between adjacent layers in a stored reel of tape. Print-through is discussed in some detail in Sec. 3.2.1.7.

The audible perception of noise depends strongly upon the frequency content, bearing an inverse relationship to the low-level, equal-loudness contours of the ear (Fletcher-Munson curves). In effect, the sensitivity of the ear to low-level sounds is deficient at low frequencies, peaks around 3 to 4 kHz, and falls off at higher frequencies. For this reason, measurements of audio background noise are often made using a weighting network that purports to take these psychoacoustic effects into account.

Bulk-erased noise. The simplest model of a recording medium is one which contains noninteracting, single-domain, acicular particles. Theoretically, the signal-to-noise power ratio at short wavelengths is then given by

$$\frac{S}{N} = \text{const. } w \frac{p}{\overline{v}} F(\theta) \frac{1}{1 + (\sigma/\overline{v})^2} \frac{1}{k \ dk} \tag{3.2}$$

where w = track width

$\quad\ p$ = volume packing (the volume-fraction of coating occupied by particles)

$\quad\ \overline{v}$ = mean of the particle-size distribution

$\quad\ \sigma$ = standard deviation of the particle-size distribution

$\quad F(\theta)$ = particle-orientation factor

and the noise comes from a narrow band of wave numbers from k to $k + dk$ (Daniel, 1972).

The various factors in Eq. (3.2) have important implications with respect to the design of audio recorders and media:

1. The power ratio is proportional to the track width; halving the track width will cause a 3-dB reduction in signal-to-noise ratio.

2. For best results, the coating should contain the maximum possible content of magnetic material of the smallest possible particle size.

3. The particle orientation factor varies from $\frac{3}{8}$ for random orientation to 1 for perfect particle alignment. There is thus a potential improvement of 4.3 dB to be gained from particle orientation.

4. The signal-to-noise will be impaired by a distribution in particle size. The lognormal type of distribution found in conventional oxide particles, such as that illustrated in Fig. 3.9, gives values of $\sigma/\bar{v}$ of between 1 and 1.5, or the signal-to-noise ratio is 3 to 5 dB lower than it would be with particles of uniform size.

Bias noise. As already mentioned, the noise from a bulk-erased tape increases when it is moved over a recording head energized with ac bias. On poorly designed or maintained equipment, a part of this increase may be caused by an effective direct current component arising from bias waveform asymmetry or a magnetized head. The remaining part is the

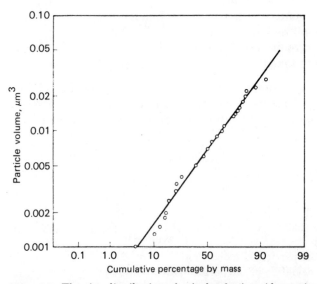

Figure 3.9 The size distribution of acicular ferric oxide particles. The cumulative percentage, by weight, of particles with less than a given volume is plotted on a normal probability scale versus particle volume on a logarithmic scale. The data were obtained from examination of electron photomicrographs and confirmed by sedimentation tests.

true bias noise and has its origin in the reaction of the bias field upon the tape. Bias noise can be explained qualitatively (Ragle and Smaller, 1965) and quantitatively (Daniel, 1972) in terms of a re-recording of noise flux. In effect, the image of the noise that appears in the core of the record head gets anhysteretically recorded back into the tape, for a net doubling (3-dB increase) of noise flux power. The process, which has been confirmed by bulk-erasure experiments, is analogous to that of contact printing (Chap. 2), which has been analyzed in considerable detail (e.g., Tjaden and Rijckaert, 1971). Thus, referring back to Fig. 3.8, the decrease of the bias-noise increment at long wavelengths can be attributed to the inability of a medium of finite thickness to collect all the available flux, while the decrease at short wavelengths can be explained by head-tape spacing. An important conclusion is that the bias-noise increment is essentially independent of the magnetic constituents of a medium.

DC noise. As shown in Fig. 3.8, a direct current of saturation value in the record head causes the noise level of a particulate coating to increase by 10 dB or so at the longer wavelengths. This increase occurs because deficiencies in milling and coating give rise to particle-rich regions (clumps) and particle-poor regions (voids). Such inhomogeneities cause the noise to increase with bulk magnetization because, in simple terms, a magnetized clump (or a void surrounded by magnetized material) acts like a large noise source. The phenomenon of dc noise means that background noise is sensitive to any residual magnetization in the heads and to the presence of even harmonics in the bias current. The latter cause an asymmetry of waveform that is exactly equivalent, as far as the tape is concerned, to adding a direct current to the bias.

The dc-noise behavior of deposited-film media is very different from that of particulate media. Most films show little or no increase of noise above the ac-erased level upon applying a saturating direct current, presumably because the disposition of the noise sources obeys Poisson's statistics.

Surface noise. The maximum dc noise does not occur when tape saturation is achieved but at a somewhat lower value of direct current. As shown by the topmost curve of Fig. 3.8, the additional surface noise is confined to very long wavelengths and is associated with variations in head-to-tape spacing caused by asperities protruding from the tape surface (Eldridge, 1964; Daniel, 1964). The wavelength at which the maximum noise occurs depends primarily upon the stiffness of the tape, which, in turn, depends mainly upon base film thickness. The size and frequency of occurrence of asperities depend upon the quality of particle dispersion and surface treatment associated with the manufacture of the media.

Modulation noise. DC noise is also a measure of the multiplicative modulation noise that occurs when an audio signal is recorded. In the frequency domain, transpositions of the dc-noise spectrum appear as modulation sidebands in relatively close proximity to the signal frequency. In the time domain, dc noise shows up as relatively slow variations in the signal envelope, the extreme excursions of which may be classified as *dropouts*. Either domain can be used as a basis for carrying out specific modulation-noise measurements (Volume I, Chap. 6). Subjectively, the effect of modulation noise is to impair the clarity of the reproduced signal, particularly when the signal is a sustained and relatively pure note (e.g., from a flute).

The modulation noise from particulate media generally decreases slightly with signal frequency. The noise from film media, however, often increases markedly with signal frequency (see Volume I, Chaps. 3 and 5).

3.2.1.7 Print-through

Basic phenomenon. Print-through is a form of interference that is uniquely offensive in audio recording. Therefore it is treated here in some detail. The effect arises when the recorded magnetization M of one layer of a tape stored in a reel creates a field H that magnetizes adjacent layers (Volume I, Chap. 5). In the case of an isolated signal of short duration, the audible effect is a series of "echoes" of logarithmically ascending strength which precedes the signal itself (the *preprints*), followed by a series of echoes of logarithmically descending strength which follows the signal (the *postprints*).

The relative level of the prints increases with the time of storage, exhibiting, in the simplest cases, a logarithmic relationship. The prints also tend to increase logarithmically with the temperature of the storage environment. For storage times of a few hours or more, the printing process is essentially linear and can be written in the form

$$M_p = \frac{\chi_p H_p}{\mu_0} \tag{3.3}$$

where χ_p, the print susceptibility, is an increasing function of time and temperature. As the signal wavelength shortens, the magnitude of H_p increases, reaches a maximum at a wavelength of λ_m, then decreases exponentially as shown in Fig. 3.10. The wavelength for maximum print field, and printed magnetization, is given approximately by

$$\lambda_m = 2\pi t \tag{3.4}$$

where $t = \delta + b$ is the total tape thickness (Daniel and Axon, 1950). Typically, the maximum field can be as high as 2.4 kA/m (30 Oe) and is reached at relatively long wavelengths in the range 75 to 300 μm.

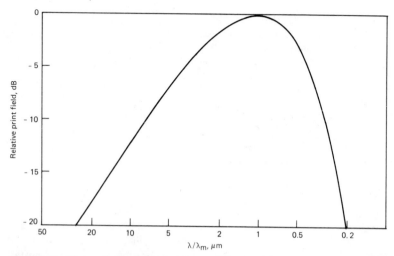

Figure 3.10 Theoretical print-through as a function of the reduced wavelength, λ/λ_m. The wavelength for maximum print is given by $\lambda_m = 2\pi t$, where t is total tape thickness (coating plus base film).

The maximum print-to-signal ratio is given approximately by (Bertram et al., 1980)

$$p_m = \text{const.}\,\frac{\chi_p \delta/b}{1 + \delta/b} \tag{3.5}$$

This equation implies that it is desirable not only to minimize print susceptibility (see below) but also to have as thin a magnetic layer and as thick a base film as possible within the constraints of a given recorder configuration.

Ratio of pre- to postprints. As mentioned earlier, the prints form two series, one that precedes, and another that follows, the signal. The relative magnitudes of these series are generally not the same. Larger prints come from the B layers, in which the coating faces the base film of the recorded signal layer, than from the A layers, lying on the side of the signal layer (see Fig. 3.11). This inequality has its origin in the fact that the print field, and hence the printed magnetization, has both longitudinal and perpendicular components which are 90 degrees out of phase. On one side of the coating, the contributions these components make to the external flux add, while on the other side they subtract. This phenomenon is illustrated in Fig. 3.11. Theoretically, the ratio of the print series should be given by an expression of the type

$$\frac{nP_b}{nP_a} = \frac{|\chi_y - \chi_x|}{|\chi_y + \chi_x|} \tag{3.6}$$

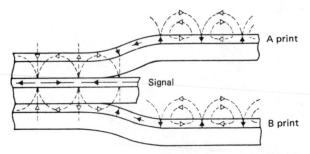

Figure 3.11 Diagrammatic representation of the difference between A and B prints. The components of printed magnetization give rise to external fields which add for B prints but subtract for A prints *(Daniel, 1972)*.

where n indicates the order of the print, and χ_x and χ_y are the effective susceptibilities in the longitudinal and perpendicular directions (Daniel, 1972; Mallinson, 1973). If the coating were isotropic and self-demagnetization effects were negligible, the external flux from the second side would be zero. In practice, unoriented pure oxide tapes can give a print series ratio of 12 dB or more, whereas the same tapes well-oriented give ratios of as low as 3 dB. The important thing is to minimize the worst print-through of the two series, which usually implies maximizing the anisotropy.

Simulated magnetometer measurements indicate that, with many materials, the print susceptibility, as well as the initial reversible susceptibility (which governs the extent of the self-demagnetization), is greater at right angles to the direction of orientation. In cobalt-modified coatings, this situation, in conjunction with demagnetization in the perpendicular direction, can lead to a surprisingly high print-series ratio of about 15 dB, even for well-oriented tapes.

Physical mechanisms. In media made from pure materials, for example gamma ferric oxide, print-through propensity is associated with a tendency for some particles to exhibit superparamagnetic behavior (Volume I, Chaps. 3 and 5). To recapitulate briefly, such behavior occurs when the thermal energy kT, during storage at absolute temperature T, becomes comparable to the anisotropy energy $v\mu_0 M_s H_0$ of a particle. Here v is the particle volume, M_s is the saturation magnetization of the material, and H_0 is the particle switching field. The probability that a particle will switch in a given time is governed by a time constant proportional to

$$\exp{(v\mu_0 M_s H_0/2kT)} \tag{3.7}$$

In the presence of a polarizing field, such as the print field, the particles will tend to be switched in the direction of this field. If all the particles had the same time constant, switching would occur at a simple logarithmic rate with time and with temperature. In fact, the assembly of particles in a coating will have a distribution of time constants governed by the distributions in particle volume, anisotropy, and—most important—interaction field. The rate of switching will therefore be according to a summation of logarithmic rates. These conclusions are in accord with the observed increases of print-through with time and temperature in the case of pure oxide media.

In media made using cobalt-modified oxides, an additional mechanism may exist and is often dominant (Volume I, Chap. 3). This mechanism is associated with the migration of Co^{2+} ions within the crystal lattice and is accelerated by increasing temperature. The print-through from such effects is minimized when the cobalt ions reside primarily at the surface (adsorbed) as opposed to being distributed throughout the body of the particles (doped). A low divalent-iron content (low cation vacancy concentration) is also a key factor minimizing print-through from this cause.

Some examples of print-through versus time for various oxides are shown in Fig. 3.12. The increase of print-through with time is approximately logarithmic for the pure iron oxides. The cobalt-doped oxides show a print-through which rises more rapidly with time and exhibits a strong dependence on divalent iron content (Kishimoto et al., 1979).

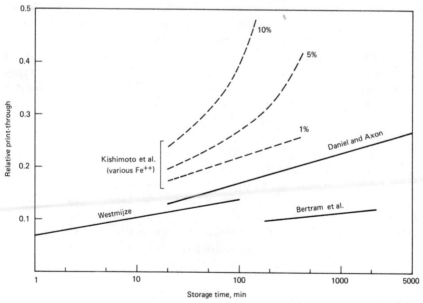

Figure 3.12 Print-through versus storage time. The full-line curves are for pure gamma ferric oxide *(Daniel and Axon, 1950; Westmijze, 1953; Bertram et al., 1980)*, the dashed curves are for cobalt-modified berthollide iron oxides of various amounts of Fe^{2+}, expressed in atom percent *(Kishimoto et al., 1979)*.

3.2.1.8 Erasure. The function of erasure is twofold: first, to reduce any previously recorded signals to an imperceptible level; second, to leave the tape with a noise level that is at least as low as that of the bias noise. The second requirement is easily met using a properly designed and maintained head and well-balanced ac current source. Meeting the first requirement poses some problems in both head and tape design.

The simplest erase head is a ring head, similar to a record head but with a longer gap, often five or more times the thickness of the magnetic medium. When ac bias is used, the ease of erasure is essentially independent of signal level, since it is the bias that dominates the extent to which the medium is magnetized during recording. Erasure does, however, depend on wavelength, in that a short-wavelength signal is inherently easier to erase than a long-wavelength one, in part because the internal demagnetizing field favors erasure. A complication arises when the leakage field from unerased portions of tape is anhysteretically recorded by the decaying field from the erase head. This re-recording effect can cause some major undulations in a partially erased signal spectrum (McKnight, 1963). The use of double-gapped erase heads and a sufficiently high erase field is the practical solution to the re-recording problem (Sawada and Yamada, 1985).

There are also some time effects associated with erasure. In particular, the ease with which a signal can be erased diminishes with the time for which the recorded signal is stored, a phenomenon related to the *viscous remanent magnetization* studied by geophysicists (e.g., Dunlop and Stirling, 1977). The erasure effect is a function of the type of magnetic material used in the media (Volume I, Chap. 3). Pure gamma ferric oxides or chromium dioxides show negligible erasure time effects, but certain types of modified iron oxides can become more difficult to erase after prolonged storage. As in the case of print-through, the worst offenders are cobalt-doped iron oxides with high vacancy (Fe^{2+}) content, and with the Co^{2+} ions uniformly distributed throughout the body of the particles. During storage, an annealing process occurs under the influence of the recorded signal magnetization and its associated internal field. The induced anisotropy created by the annealing process leads to magnetically "hard" regions that conform to the pattern of the recorded magnetization. When the stored signal is bulk erased, the signal may appear to have been removed. Upon subsequently running the tape over a record head energized with bias, the signal may magically return, in the worst case to a level of some 20 dB above the noise.

3.2.2 Open-reel recorders

3.2.2.1 Tape transport. The layout of a simple open-reel audio tape transport is shown in Fig. 3.13. The constant-speed drive usually consists of a hysteresis synchronous motor turning a capstan against which the tape is held with a compliant pinch roller. Speed fluctuations, or flutter, must be kept below the level of perception, which places stringent demands upon the accuracy of all the drive elements. Separate reel motors are usually provided to power the fast-forward and rewind modes and furnish hold-back torque during all operating modes. Tape guiding is provided by a combination of fixed (flanged) and rotating (sometimes crowned) members. Rotating pulleys, fixed or on a sprung arm, also serve to mechanically isolate the reels from the head path and to dampen any tendency of the tape to go into longitudinal oscillation and produce audible squeal.

Professional transports may employ some of the techniques developed for instrumentation recorders (see Chap. 4), such as a closed-loop drive, a differential dual-capstan drive, or a servo-controlled drive. Large diameter capstans that drive the tape without compliant pinch rollers have also been used.

Open-reel audio recorders usually operate at speeds ranging from 95 mm/s (3.75 in/s) on consumer equipment to 760 mm/s (30 in/s) on professional equipment. Corresponding tape widths range from 6.3 mm (0.25 in) to 51 mm (2.0 in).

3.2.2.2 Open-reel heads. The construction of open-reel heads follows the traditional gapped-ring type of design which is amply described in

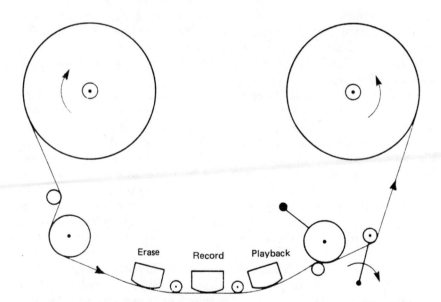

Figure 3.13 Typical configuration of an open-reel recorder, showing the mechanical layout and disposition of heads. The arrows on the supply and take-up reels indicate the direction in which torque is applied.

Volume I, Chap. 4. Record, reproduce, and erase heads differ mainly in the length of their gaps, which are chosen, using the guidelines outlined in an earlier section, to suit the coating thicknesses and shortest wavelengths of interest. Core losses over the audible frequency range can be held to small values using either laminated alloy or ferrite. The relatively low coercivity media generally used in open-reel recording do not place difficult demands upon the saturation characteristics of the core. The choice of core material is therefore governed mainly by such factors as the attainable mechanical and magnetic integrity of the surface and gap edges. Generally, laminated metal cores are preferred; Permalloy is used in the simplest cases, Alfenol or an equivalent is used if head wear is of particular concern. Erase heads with double gaps are often used to avoid the re-recording phenomenon discussed under erasure in Sec. 3.2.1.

It is convenient and often practical for bias and erase currents to have the same frequency, typically about 100 kHz. In professional equipment, erase efficiency is increased and biasing enhanced by making the erase frequency a subharmonic of the bias frequency; frequencies of 80 kHz and 240 kHz are commonly used.

3.2.2.3 Track configuration and equalization. A two-track stereo system is used in the simplest types of quarter-inch professional tape recorders. In consumer equipment, a four-track system is used in which two tracks are used for stereo recording in one direction, and the remaining two tracks for the reverse direction. Studio recorders use wider tapes and many more tracks to accomplish the complex editing, mixing, and special effects used in modern recording studios. The widest tapes are 51 mm (2.0 in) and can accommodate 16 tracks of width 0.18 mm. This can be extended to 24 tracks of width 0.11 mm for a sacrifice of 2.4 dB in signal-to-noise ratio.

Standard postequalizations consist of integration plus linear functions of frequency that are specified in terms of the time constant of the equivalent RC circuit. The more important national and international standards for high-frequency emphasis at various tape speeds are shown in Fig. 3.14. At 15 kHz, the emphasis ranges from 10.7 to 16.5 dB. These levels of postequalization are insufficient to compensate for all the losses in the record and reproduce processes, and preequalization is used to accommodate the balance and achieve an overall flat response. Some of the standards still call for a deemphasis at low frequencies, a dubious practice that had its origin in reducing power-frequency interference, or *hum*.

The standards are implemented on a worldwide basis through the use of standard tapes (McKnight, 1969). Standard tapes are available which not only facilitate the accurate adjustment of frequency response but enable this to be done in terms of the absolute value of the (short-circuit) recorded flux per unit of track width in W/m. Head azimuth alignment is another important function of standard tapes.

3.2.2.4 Open-reel media.

Coated oxide media are universally used for open-reel recording. For consumer use, coatings of 5 to 7 μm are made on a polyester-base film of 25 μm—less for longer playing reels. For professional use, the coating is often increased to 10 μm or more to enhance the maximum output level (MOL) at long wavelengths, and the base film thickness is increased to 38 μm to reduce print-through.

Pure gamma ferric oxide has traditionally been the dominant magnetic material for open-reel media. The low cost and easy manufacturability make γ-Fe_2O_3 attractive for consumer use and also for professional use, since a reel of professional tape can contain a considerable amount of magnetic material. Earlier, it was shown that the key to achieving high signal-to-noise ratio is to develop media having well-oriented, small, uniform particles packed as densely as possible. Over the years, particles of γ-Fe_2O_3 have been made smaller, more acicular, and more easily oriented and dispersed, and the predictable increases in signal-to-noise ratio have

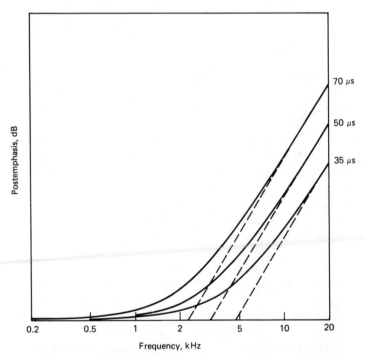

Figure 3.14 One-quarter-inch recorder standards: high-frequency equalization curves corresponding to 35 μs (International Electrotechnical Commission, 38 cm/s), 50 μs (National Association of Broadcasters, 38 and 19 cm/s), and 70 μs (IEC, 19 cm/s).

been realized. Also, better control of particle-size distribution in the raw material and in coating preparation has allowed the average particle size to be reduced without increasing the very small, poorly shaped fraction that is associated with print-through in pure materials.

These advances, in conjunction with progress in coating and surface-treating technology, make it possible to achieve a signal-to-noise ratio of some 60 dB on studio recorders using high quality γ-Fe_2O_3 tapes. This is the unweighted, wide-band ratio, measured relative to the MOL that can be obtained at a low frequency with no more than 3 percent third-harmonic distortion. The minimum signal-to-print ratio after modest storage approaches 60 dB at a rather undesirable frequency ($\lambda_m = 2\pi t$) of about 1.2 kHz, close to the region of maximum audibility. Consumer recorders deliver results some 5 to 10 dB poorer, depending upon the tape speed and the quality of tape and equipment.

3.2.2.5 Noise reduction. In addition to reducing noise at the source, there are ways in which the audibility of noise and some forms of interference, such as print-through, can be reduced by signal processing. Early schemes used signal processing devices on reproduction only and had the disadvantage of modifying the signal as well as reducing noise. Later schemes use the *compandor* concept. This involves precompression and complementary postexpansion of the signal, with the goal of leaving the signal unaltered while reducing noise. In practice, this goal is difficult to achieve, and the first compandor systems produced a variety of unpleasant side effects related to poor input-output tracking, sensitivity to gain errors, transient problems, and audible "breathing" of the signal-modulated noise.

These problems are largely avoided in the professional noise-reduction system illustrated in Figs. 3.15 and 3.16 (Dolby, 1967). In this system, designated "A-type," the signal is operated on by subtracting or adding a small differential component, rather than subjecting the whole signal to the hazards of passage through a variable-gain channel. In other words, as shown in Fig. 3.15, the higher-level signals are transmitted unchanged. Another advantage, shown in the figure, is that the compression and expansion are obtained by using identical differential networks: compression is produced by adding the differential component, expansion by subtracting it. Finally, signal modulation of the noise is made inaudible by splitting the differential component into four frequency bands and relying upon the masking effect of signals of amplitude appreciably higher than the compression thresholds in given portions of the spectrum. The filters employed are 80 Hz low-pass, 80 Hz to 3 kHz band-pass, 3 kHz high-pass, and 9 kHz high-pass. More recent systems (designated *SR* for *Spectral Recording*) use sliding filters controlled by the spectral content of the signal. A noise reduction of 10 dB is delivered, with negligible degradation

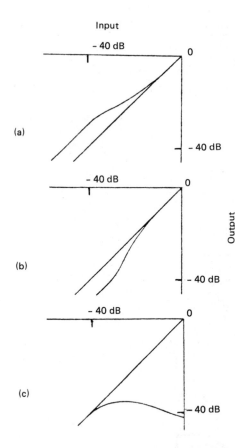

(a)

(b)

(c)

Figure 3.15 Transfer characteristics of a professional noise-reduction system. *(a)* Compression before recording; *(b)* complementary expansion after reproducing; *(c)* the differential component that is added to the signal to give the compression characteristic *(Dolby, 1967).*

of the signal and imperceptible signal-modulated noise effects. This system is the one most widely used in professional recording and has become the standard for the exchange of nondigital master tapes. Other noise-reduction systems have been developed (e.g., DBX, Telcon) but have not achieved the same acceptance as the Dolby system.

In all the noise-reduction systems of the compandor type, the control of signal level on recording and reproduction is critical. This requirement further emphasizes the importance of having the means of absolute calibration, via standard tapes, in audio recording.

3.2.3 Compact-cassette recorders

3.2.3.1 Mechanical aspects. The compact cassette was introduced in the early 1960s as a convenient way of making portable, low-performance tape recorders. The cassette rapidly became accepted as the worldwide standard, and during the 1970s, the performance was progressively improved to the point that cassette recorders became highly competitive with

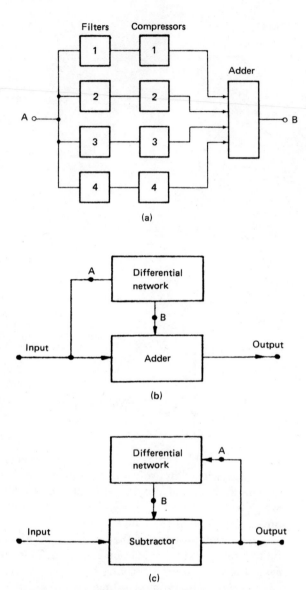

Figure 3.16 Block diagrams of a professional noise-reduction system showing how the same multichannel differential network *(a)* is used to form the record processor *(b)* and the reproduce processor *(c)* by connecting the network in additive or subtractive modes *(Dolby, 1967).*

higher-performance, but less convenient, open reel recording configurations. By 1980, the compact cassette had become the dominant tape format for home use and had displaced the eight-track, endless loop cartridge for automobile use.

The principal features of the compact cassette are shown in Figure 3.17. The cassette contains two flangeless tape reels that are driven by external, splined drive shafts, helped by low-friction liners on either side of the tape packs. Narrow tape, of width 3.8 mm (0.15 in), is fed from one reel to the other via two rotating, flanged rollers, one at each of the two upper corners. The length of tape between the two rollers is accessible at three main points: the two outermost ones give access to the capstan (inserted through the hole in the case) and pinch rollers for either direction of tape travel at the standard speed of 48 mm/s (1.875 in/s). The central opening, which contains a magnetic screen and pressure-pad assembly, provides head access. Mechanical registration with the drive relies upon the engagement of certain keying points and reference planes in the cassette body. Dependence upon plastic parts to provide this critical registration function is the intrinsic shortcoming of the compact cassette.

3.2.3.2 Cassette heads. Originally, the cassette was designed to use a single record-reproduce head inserted through the central opening. Now, most cassette recorders use separate record and reproduce heads. The reproduce head is always positioned centrally to take advantage of the

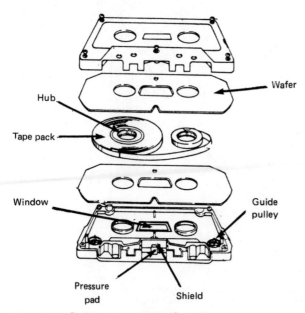

Figure 3.17 Compact-cassette configuration.

pressure pad and screen. The record head either contacts the tape through one of the smaller openings, or is integrated into the reproduce-head structure. Both metal and ferrite (including single-crystal) head cores have been used; but metal cores are required to accommodate high-coercivity iron-particle tapes and, for this reason, most modern drives incorporate metal heads, at least for recording.

3.2.3.3 Track configuration and equalization. Four tracks are used, providing two-way stereo recording. This is similar to four-track open-reel recording, although the disposition of the tracks is different. The original standard postequalization of the compact cassette at high frequencies was equivalent to an RC network of time constant 120 μs. When cassette tapes (initially chromium dioxide) with improved high-frequency response became available, a second standard, corresponding to 70 μs, was established. The high-frequency emphases dictated by these standards are shown in Fig. 3.18. The equalization is normally switched automatically with the bias requirement of the medium, which is described in Sec.

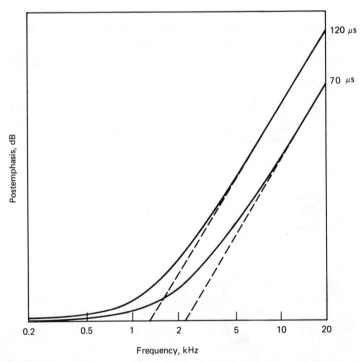

Figure 3.18 Compact-cassette standards: high-frequency equalizations corresponding to 120 μs (type I tape) and 70 μs (type II and IV tapes).

3.2.3.4. As in open-reel recording, standard cassettes are produced that facilitate the interchange of recordings and enable calibration to be carried out in absolute terms.

3.2.3.4 Cassette media. The tapes for compact cassettes come mainly in three sizes, C60, C90, and C120, where the numbers indicate the total two-way playing time in minutes. The tapes differ mainly in the thickness of the base film, which ranges from a nominal 6 to 12 μm. Coating thickness is about 5 μm—rather less on the C120.

To obtain high performance using the narrow tracks and low tape speed of the compact cassette format demands a great deal from the magnetic medium, and improvements in media have been the main instrument for enhancing performance over the years. Every means of increasing signal-to-noise ratio has been implemented: smaller, more uniform particles, improved dispersion and orientation, and higher particle packing density within the coating. This has resulted in cassette media which, in terms of many magnetic properties such as squareness, orientation ratio, and remanence, are superior to all other particulate media.

Cassette media have usually been the first to exploit new magnetic materials offering potential improvements in signal output, particularly at high densities. The innovative approach began with chromium dioxide in 1970 (Naumann and Daniel, 1971). Subsequently, progressively improved versions of cobalt-modified iron oxides were developed, and passivated iron particles were added to the list in the 1980s. Deposited-metal-film media are a later introduction. Media have also been developed in which a thin outer layer of higher coercivity is coated over an inner layer with the result that the bias is more uniform between short and long wavelengths. This configuration is particularly efficacious if the outer layer is metallic (particulate or film).

All these developments have involved significant changes in machine design, particularly in terms of the bias (and erase) fields necessary to match coercivity. The latter has risen from some 28 kA/m (350 Oe) for gamma ferric oxide, to 50 kA/m (650 Oe) for chromium dioxide and cobalt-modified oxides, to 80 kA/m (1000 Oe) for iron particles. All three types of media are in use, and equipment is usually provided with a switch to select between type I (normal bias), type II (high bias), and type IV (metal bias). Dual-coated media (originally designated "type III") now fit into one of the other categories. As a rule, the same switch automatically selects the equalization. The 70-μs equalization is used for chromium dioxide, cobalt-modified iron oxide, and iron-particle media. Compared with the 120-μs, the 70-μs equalization raises the higher frequencies by some 5 dB less, giving a corresponding benefit in high-frequency noise level. This difference puts normal-bias media at an unfair disadvantage with respect to the other media and does not take into account the significant improvements that have been made in gamma ferric oxide over the years.

The unweighted signal-to-noise ratio provided by a high-quality cassette recorder is in the region of 50 dB. The signal-to-print ratio after a few days storage is typically between 40 and 50 dB, depending upon the base film thickness and the type of magnetic material. The minimum ratio occurs at a frequency of about 600 Hz, more favorable from considerations of audibility than for open-reel tapes. Also, the way in which the tape is wound on a cassette (coating outside) means that the postprints are greater than the preprints—a more natural and often a less easily detectable sequence than in open-reel recording.

3.2.3.5 Noise reduction. A major challenge in gaining broad acceptance for the compact cassette has been to improve the signal-to-noise ratio, and Sec. 3.2.3.4 outlined what can be done by working with the media. A complementary approach is to introduce an inexpensive and effective means of electronic noise reduction. A system designed specifically for this purpose was introduced in 1970 (Dolby, 1970). This system, called the "B-type," was derived from the professional A-type described earlier. It uses a single-band approach, but the differential networks contain a variable high-pass filter having a cutoff frequency (around 2 kHz) that moves upward as the level flowing through the differential path increases. The effect is to boost low-level, high-frequency signals by 10 dB on record and attenuate them by the same amount on reproduce, for a net noise reduction of 10 dB. The B-type system has a tendency to cause some compression of high-frequency signals. Recent versions (designated *HX,* for *headroom extension*) are designed to overcome this defect by dynamically reducing the ac bias when the high-frequency energy in the signal is large. Extensive use of the Dolby systems by the major manufacturers has done much to establish the cassette recorder as the dominant form of tape recorder for the home.

3.2.3.6 High-speed duplication. AC-biased magnetic recording can be made to handle frequencies one or two orders of magnitude greater than the highest audio frequency (see Chap. 4). This means that, unlike video programs, audio programs can be duplicated at speeds many times greater than the recording speed. Such high-speed duplication is used for all formats, including open-reel, where large numbers of duplicates are required. But by far the largest application is in duplicating musical programs on cassettes for the mass consumer market.

For duplication onto cassette tapes, the typical procedure is to record a duplicating master at, say, 190 mm/s (7.5 in/s) using the B-type of prerecord noise reduction. The master tape is then reproduced at up to 64 times the recording speed and recorded on a series of slave machines running cassette tape at 64 times the normal speed, or 3 m/s (120 in/s). Convenience dictates that duplication takes place in both the normal or reverse time sequence, so that the master tape need not be rewound. In order to make this possible, the equipment is equalized to have the necessary phase response (see Volume I, Chap. 5 for more details). A bias frequency on the order of 1 MHz is used during recording.

3.3   Digital Audio Recording

3.3.1   Justification

The first sections of this chapter show how the degradation of signals in linear ac-biased recorders can be minimized to provide analog recorders of exceptional quality. The state of the art is, however, such that improvements show diminishing returns because performance is limited by the physics of the process.

The storage of audio waveforms as a series of numbers was proposed decades ago, for reasons which remain valid today. The concept has simply waited for the supporting technology to become available. The attraction of digital audio recording is that nearly all of the degradations encountered in analog recording are not reduced, but eliminated. New types of degradation are caused by the process of expressing the audio waveform as numbers, but these can be made arbitrarily small by design.

A magnetic head cannot know the significance of signals which pass through it, therefore there is no distinction at the head-medium level between an analog recorder and a digital recorder. The distinction is made in the effect of any degradations on the meaning we attribute to the signals. In digital recording, the presence or absence of a flux reversal at a defined time is the only information of interest. Provided that the flux reversal results in a reproduced pulse which is somewhat greater than any noise, the meaning will be unaltered by small changes in the waveform. Such small changes may be caused by nonlinearity, medium inhomogeneity, and head-contact deficiencies, plus crosstalk, interference, or noise.

Large waveform disturbances, due to dropouts or high amplitude noise, may cause flux reversals to be missed, or simulate ones which did not exist in the original signal. This will result in one or more numbers being incorrect. A properly engineered error-correction system will restore these numbers to their original value. Using an error-correction system, the minimum allowable signal-to-noise ratio of the raw digital tape channel can be made very much smaller than that of an analog machine. This permits the use of very narrow tracks, which results in economy of tape use.

Speed variations in the recorder result in numbers appearing at a fluctuating rate. The use of a temporary store enables the numbers to be read out at a constant rate. This approach eliminates wow, flutter, and phase errors between tracks caused by tape weave and azimuth error. Also, if the numbers representing the audio waveform are unchanged by the recorder, the recording process cannot affect the frequency response and the recording may be copied through any number of generations without degradation.

The quality of a digital audio recording is independent of the magnetic recording and reproducing processes provided they are designed to accommodate the necessary bit rate and can provide better than a certain raw error rate. With an error-correction system that can remove all the allowable raw errors, the quality of a digital audio recorder depends only on the precision of the conversions to and from the digital domain. Since quality is central to the subject of audio recording, the conversion process is treated in some detail in this chapter.

The professional user is likely to consider the ability of digital audio to copy many generations just as valuable as the sheer sound quality, whereas the consumer is more likely to appreciate the greater freedom of handling made possible with an error-correction system. The potential to reduce tape consumption will have universal appeal on grounds of both cost and storage space. Digital audio recording relies heavily on computer-related recording technology such as data recording methods, channel codes, and error correction, and the principles of these are comprehensively treated in other chapters. The scope of this chapter will be restricted to specific adaptations to digital audio recording.

3.3.2   The digital audio recorder

A block diagram of a representative digital audio recorder is shown in Fig. 3.19. Each element of the diagram represents a major topic of this chapter. The output of a microphone capsule is an electrical signal in the analog domain, which is continuously variable in time and voltage, and differs fundamentally from the digital domain, in which time increases in fixed steps, and voltages are represented by a range of integers. Conversion in either direction between the analog and digital domains requires separate attention to the time and voltage domains.

Sampling converts a time-continuous analog signal into time-discrete events, but the sample voltage is still infinitely variable. When a dc offset is introduced to make all samples unipolar, the resulting signal is known as a pulse amplitude modulated (PAM) wave. This sample signal can be returned to a time-continuous signal by the process of reconstruction.

To enter the digital domain, each sample undergoes quantizing, in which the infinitely varying sample voltage is represented to finite accuracy by the nearest integer in a fixed range. These integers are then encoded to become binary data. This numerical information is formatted

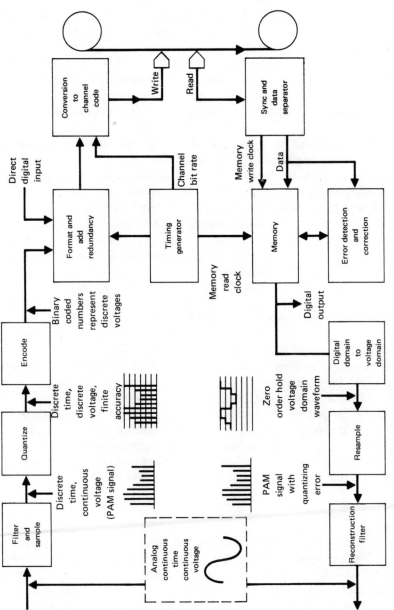

Figure 3.19 The basic elements of a digital audio recorder.

into data blocks, and redundancy is added for the purposes of subsequent error correction. The redundant data blocks are fed in serial form to an encoder, which combines the data with a clock to form a channel code. The channel code is used to reverse the write current in the record head, leaving transitions on the medium. On replay, these transitions generate pulses which can be used to re-create the original channel code. A data separator provides a separate clock and data stream. The data are temporarily stored in a memory which permits the subsequent correction of symbols found to be in error and absorbs any jitter due to speed variation in the medium.

Unfortunately, there is no term which unambiguously describes the process of changing binary data into discrete voltages. Since this process is updated at each sample, the voltage remains steady between samples. Resampling is necessary to return to the pulse amplitude modulated wave and re-create the original signal. The process of reconstruction can be visualized by considering each pulse in the modulated wave individually. The impulse response of a perfect reconstruction filter converts each pulse to a waveform as shown in Fig. 3.20. The waveforms of each pulse add to give the analog output (Betts, 1970).

In the first section of this chapter, the nominal audio bandwidth for high quality was stated as 20 kHz. For reasons which will be explained later, this requires a sampling rate of between 40 and 50 kHz, and the number of bits required per sample is 14 or 16 for most purposes. The required data rate of a digital recorder is equal to the product of the sampling rate and the number of bits per sample or approximately 800 kb/s (100 kB/s) for each audio channel. The necessary storage capacity for 1 hour of such a system is 360 MB. If an allowance is made for the redundant data needed for error correction, the figure rises to about 500 MB/h, and a stereo recorder will require 1 GB of data for each hour of playing time. It has been possible to record these quantities of data at a reasonable cost only with recent high-density storage technology and error-correction strategies.

3.3.3 Conversion

The principles of conversion between the analog and digital domains are used widely in magnetic recorders, but the processes must be performed with unusual precision in digital audio, owing to the acuity of the ear. Since, ideally, these processes alone determine the overall quality of a digital audio recorder, they are described here in corresponding detail.

3.3.3.1 Sampling. The sampling process seeks to represent a continuous process by discrete events. Changes in the input between samples are not recorded, thus the sampling rate will determine the program bandwidth

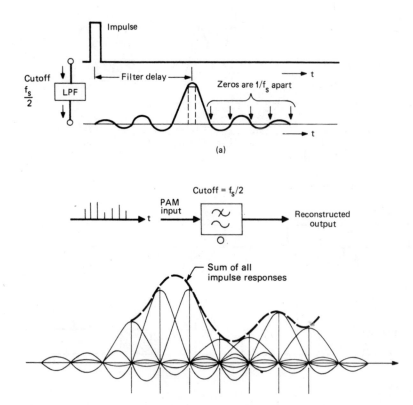

Figure 3.20 Sampling. (a) The impulse response of a low-pass filter which cuts off at $f_s/2$ has zeros at $1/f_s$ spacings as shown in (b). The output is a signal which has the value of each sample instant but with smooth transitions from sample to sample.

which can be used. Figure 3.21a shows a sampling rate which is intuitively adequate to reconstruct the wave. However, in Fig. 3.21b, the same sampling rate is used with a higher-frequency input, and the samples are now grossly misleading. A completely artificial signal has resulted, of which the frequency is the difference between the input and sampling frequencies. This is known as *aliasing*.

The process of sampling essentially multiplies the input signal and sampling waveforms resulting in the spectrum of Fig. 3.22a. The baseband spectrum reappears symmetrically about the sampling frequency, and its integer multiples. In this case, a reconstruction filter which cuts off as shown can recover the original signal. Figure 3.22b shows the case where the baseband is overlapped by sampling-rate sidebands. Aliasing products appear in this area, as the reconstruction filter cannot tell the source of energy from the frequency. The widest possible baseband for the given

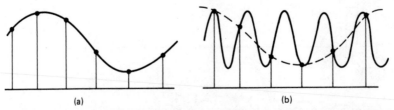

Figure 3.21 Sampling rate. *(a)* The rate is adequate to reconstruct the original signal; *(b)* the rate is inadequate, aliasing occurs, and reconstruction produces an erroneous waveform (dashed).

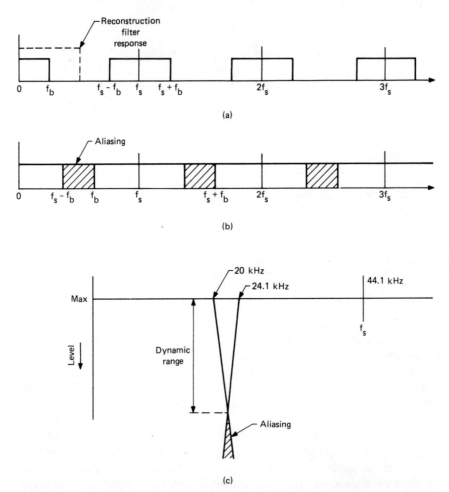

Figure 3.22 *(a)* The spectrum, of highest frequency f_b, repeats about multiples of the sampling rate of fundamental frequency f_s; *(b)* the sampling rate is inadequate, and aliasing takes place; *(c)* the finite slope of the filter causes aliasing, but this is pushed below the dynamic range of the system by raising the sampling rate from the theoretical minimum of 40 kHz to 44.1 kHz.

sampling rate will be seen to be one-half of the sampling rate. This limit was described by Shannon but in audio terminology is usually referred to as the *Nyquist theorem* (Shannon, 1948). As a consequence, it is necessary to sample at twice the highest audio frequency, and two identical filters are required, one prior to sampling, known as an *antialiasing filter,* and one after resampling, known as a *reconstruction filter.* Since all realizable filters have finite slopes, aliasing can always occur, but by raising the sampling rate, as in Fig. 3.22c, aliasing products can be pushed outside the dynamic range of the system.

A characteristic of sharp filters is that they can cause disturbances in phase response at frequencies significantly below the cutoff frequency. For high-quality applications, a phase-equalizing circuit usually precedes the antialiasing filter and follows the reconstruction filter. This equalization results in reduced filter ringing on transients, which permits recording at slighly higher levels with musical signals (Meyer, 1984).

Following filtering, the input signal is sampled. The Nyquist theorem requires an instantaneous sample to be taken of the signal voltage, but practicality requires that the voltage be held constant until the quantizer has operated. For this purpose, a track-hold circuit is almost universally used.

3.3.3.2 Resampling. The sampling process takes instantaneous signal voltages and holds them between samples for the quantizer. Resampling is the opposite process; voltages held between samples are converted to near-instantaneous pulses. The process of holding between samples causes an aperture effect which is analogous to the effect of a finite reproduce-head gap in magnetic recording. As shown in Fig. 3.23a, the response becomes a $(\sin x)/x$ function which is 4 dB down at one-half the sampling rate f_s. The duty cycle of the switch in a resampling circuit is known as the *aperture ratio,* and the effect of various ratios is shown in Fig. 3.23b. The process cannot be taken to extremes since, with a zero aperture, no energy passes the switch. An aperture ratio of 25 percent is a common compromise between frequency response and noise. Since the deviation from flat response is predictable, it can be equalized with precision, and the overall response of the system is usually determined by filter ripple. The output of the resampler is fed directly to the reconstruction filter.

3.3.3.3 Oversampling. Digital audio is an example of trading bandwidth for dynamic range, and oversampling is simply an extension of the principle. In an oversampling converter, the sample rate is several times the Nyquist minimum, and digital filtering is used to convert between the sampling rate of the converter and the Nyquist rate. The principle can be applied to both analog-to-digital converters (ADCs) and digital-to-analog converters (DACs). The information content of an oversampled data

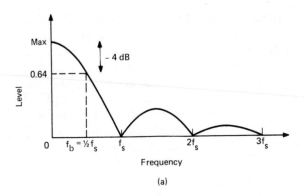

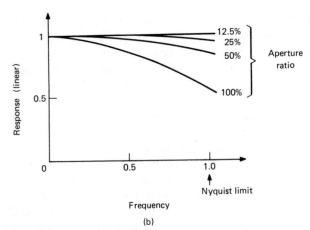

Figure 3.23 *(a)* The sin *x/x* response of an aperture circuit is shown; it is of real interest only up to $f_s/2$; *(b)* the response using various aperture ratios is shown.

stream is proportional to the sampling rate, thus for every doubling of the sampling rate, one fewer bit can be used in the sample word. Thus a 16-bit, 48-kHz data rate can be conveyed with equal dynamic range by a 14-bit channel at 192 kHz.

There are three advantages of oversampling. First, the analog antialiasing filter need have only a relatively gentle roll-off if the sampling rate of the converter is higher. This means that the phase response can be made linear without resorting to phase-compensation circuits, an observation which also holds for the reconstruction filter. In fact, a zero-order hold system can be used instead of resampling such that the (sin *x*)/*x* roll-off becomes part of the reconstruction filter response. Second, the sampling rate conversion in the digital domain can be performed with a filter characteristic which is ideal and unattainable by an analog filter. Third, the required converter accuracy is lower because a shorter word length is required (Van de Plassche and Dijkmans, 1983; Adams, 1985). The disadvantages of oversampling are that a digital filter is required at each converter, and the converters have to run faster. This is not a problem for DACs, which are inherently fast devices, but is more of a problem for ADCs, although the reduced word length requirement helps. For these reasons oversampling DACs are more common, and for consumer equipment the digital filter can be economically realized in large-scale integration (LSI) form.

3.3.3.4 Choice of sampling rate. A high sampling rate eases the design of filters, but the storage requirement is proportional to sampling rate, and excessive sampling rates simply waste the medium. The ability to synchronize with video equipment sometimes determines the sampling rate, since digital audio recording is often carried out using modified video recorders. Suitable sampling rates are obtained by multiplying the frame rate by the number of usable lines in a frame and by the number of audio samples stored per line (usually three). For the NTSC system, the nominal 30-Hz frame rate was reduced by 0.1 percent on the introduction of color broadcasting to prevent subcarrier/sound interference. The resulting sampling rate for three samples per line is $29.97 \times 490 \times 3 = 44.056$ kHz, which is sufficiently in excess of twice the nominal audio bandwidth of 20 kHz.

For the PAL system the comparable rate is $25.00 \times 588 \times 3 = 44.100$ kHz. These two rates are used for audio adaptors which use consumer video recorders for storage. The second rate is the standard sampling rate for the Compact Disc and its mastering equipment.

For professional applications, the ability to run at variable speed without allowing sampling sidebands to pass the reconstruction filter means using a higher sampling rate than otherwise. The current standard for professional applications is 48 kHz since it relates simply to the 32 kHz standard used in transmission links between broadcast studios and FM transmitters in Europe and Japan.

3.3.3.5 Quantizing. Figure 3.24 shows the principle of quantizing. The infinitely varying sample voltages are expressed as integers within a given range. Since binary coding is universally used, the number range will be a power of 2. Audio recording generally uses 14 or 16 bits, which means that every signal will be described by 16K or 64K different numbers (K = 1024). Each binary number thus represents a signal voltage range known as a *quantizing interval (Q)*, where the center of the range is the nominal voltage of that quantizing level. The maximum error will be $\pm 0.5Q$ which

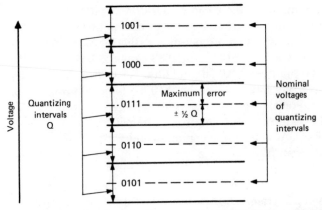

Figure 3.24 Quantizing. All voltages within a particular quantizing interval are assigned the same number corresponding to the voltage of the center of the interval. The maximum quantizing error is ±0.5Q.

is known as the *quantizing error*. The type of quantizing described is the one most commonly used for digital audio recording and is known as *linear quantizing,* even though this seems like a contradiction in terms. *Uniform quantizing* is preferable but not commonly used.

Nonlinear and floating-point pulse-code modulations both suffer from noise modulation which is of little consequence in their main application, telephony, but which precludes their use in high-quality audio. An alternative approach to quantizing is delta modulation, which is practicable only with companded signals and is not in widespread use (Adams, 1984; Betts, 1970).

The reverse of the quantizing process is discussed first, for the simple reason that some analog-to-digital converters use digital-to-analog converters in feedback loops. The basic principle of digital-to-analog conversion is that each binary weighted input produces an output proportional to its weighting, and these are summed together prior to resampling and reconstruction. When several current sources having a binary weighted relationship are switched by the input sample bits, the currents can be summed to produce a voltage output using a virtual-ground amplifier. The problem in such a simple device is arranging for the current sources to have the required accuracy. The accuracy of each current source is a function of the significance of the bit, and the output error due to the most significant bit (MSB) current source should be less than the output change which results from toggling the least significant bit (LSB). Thus in a 14-bit converter, the MSB current source must be accurate to within $100/8000 = 0.012\%$. For a 16-bit converter, the tolerance goes down to $100/32,000 = 0.003\%$.

In order to prevent interaction, the currents must be switched to ground or into the virtual-ground amplifier by changeover switches. The on-resistance of these switches is a source of error, particularly in the MSB, which passes the most current. The solution in monolithic converters is to fabricate switches that have an area proportional to the weighted current, such that the voltage drops of the switches are all the same. The error can then be removed with a suitable offset. The layout of such a device will be dominated by the MSB switch, since it will, by definition, be as big as all the others put together. Since the required accuracy cannot be provided by a conventional ladder network, a different approach is required. Figure 3.25 shows a current source feeding a pair of resistors, each of which can be switched between two outputs. If the switches are driven according to a 50 percent duty cycle, each resistor will be in series with one output for half the time. Thus, the average currents in each output will be identical, because the effective series resistance of each will be $0.5 \ (R_1 + R_2)$. Current averaging is achieved using a pair of ordinary decoupling capacitors. The principle is known as *dynamic element matching* and makes possible an accurate 16-bit converter, needing no

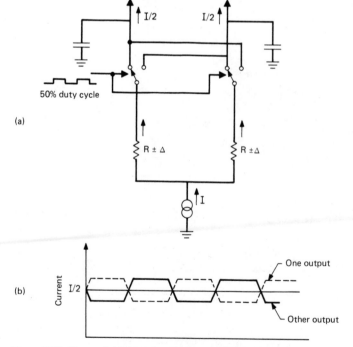

Figure 3.25 Dynamic element matching. *(a)* Each resistor spends half its time in each current path; *(b)* the average current of both paths is identical provided the duty cycle is accurately 50 percent.

trimming (Van de Plassche and Goedhart, 1979). The weighted currents can be obtained by cascading the dynamic elements.

Another alternative to the weighted current source principle is the ramp converter. Figure 3.26a shows a constant current source feeding an integrating capacitor, such that the peak voltage reached by the capacitor is a function of the time for which the current flows. The linearity of the device is a function of timing accuracy and leakage of the capacitor charge, either internally or through a finite load. The simple approach of Fig. 3.26a is of no use without some adaptation since, in a 16-bit, 48-kHz system, the clock would have to be an impracticably high 64×48 kHz $= 3$

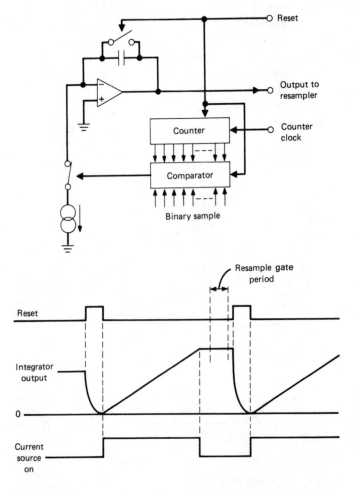

Figure 3.26a A current source drives an integrator for a period proportional to the binary input code, giving a highly linear conversion; this simple circuit is too slow for audio use and requires modification.

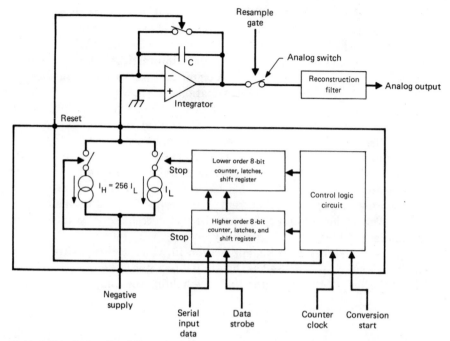

Figure 3.26b A simplified diagram of a practical digital-to-analog converter has two current sources I_H and I_L having a ratio of 256:1; these are controlled by the high and low bytes of the 16-bit input, which enters the chip serially.

GHz. The practical approach, shown in Fig. 3.26b, is to use two current sources with a ratio of 256 to 1, where the larger is timed by the high byte of the sample and the smaller is timed by the low byte. The clock frequency is then reduced by a factor of 256. Any inaccuracy in the 256 to 1 current ratio causes monotonicity errors, but in a monolithic device, accurate tracking is easier to achieve. The output of the integrating capacitor will remain constant once the current sources are turned off, and the requisite pulse-amplitude modulated wave can be produced by an analog FET switch which passes the converted level for the required aperture period. Following this resampling period, the integrator is discharged by a further FET, prior to the next conversion. The conversion count must take place in rather less than one sample period, in order to permit the resampling and discharge portions of the cycle. A clock frequency of about 20 MHz is sufficient for a 16-bit, 48-kHz unit which permits the ramp to be completed in 12.8 μs, leaving 8 μs for resampling and reset.

Analog-to-digital conversion makes use of many of the techniques used in digital-to-analog conversion. The general principle of audio ADCs is that different quantized voltages are compared with the input voltage until the quantized voltage closest to the input is found. The most primitive method of generating different quantized voltages is to connect a counter to a DAC. The resulting staircase voltage is compared with the input and used to stop the clock to the counter when the DAC output has just exceeded the input. This approach is impracticable for audio purposes because it is inherently slow. A significant improvement in speed is obtained by using successive approximations, as shown in Fig. 3.27. Here, each bit is tested in turn, starting with the most significant bit. If the input is greater than half range, the MSB will set and be used as a base to test the next bit, which will set if the input exceeds three-quarters range and so on. The number of decisions is equal to the number of bits in the word, rather than the number of quantizing intervals, which was the case for the example of Fig. 3.26. Analog-to-digital conversion can also be performed using the dual current source type of digital-to-analog converter. The major difference is that the two current sources must work sequentially rather than concurrently.

Figure 3.28 shows a 16-bit application in which the capacitor of the track-hold circuit is also used as the ramp integrator. When the track-hold FET switches off, the capacitor C will be holding the sample voltage, and

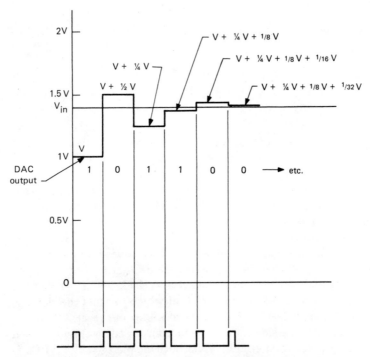

Figure 3.27 A successive approximation method tests each bit in turn, starting with the most significant one. The digital-to-analog converter output is compared with the input. If the DAC output is below the input, the bit is made 1; if it is above the input, the bit is made 0.

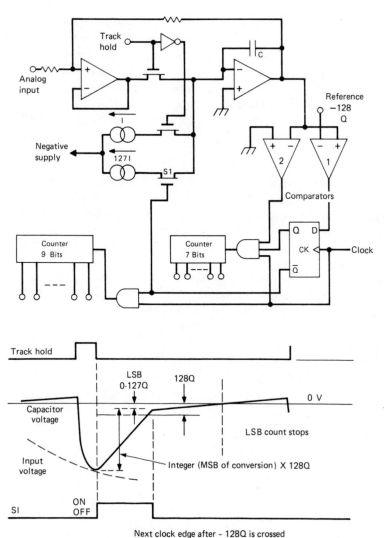

Figure 3.28 A dual-ramp analog-to-digital converter (ADC) using a track-hold capacitor as an integrator. See the text for details.

two currents of ratio 128 to 1 are capable of discharging the capacitor. The smaller current will be used to determine the least significant seven bits ($2^7 = 128$), and the larger current will determine the nine most significant bits. The currents are provided by two sources of ratio 127 to 1 so that, when both run together, the current produced is 128 times the current supplied by the smaller source only. With both current sources enabled, the high-order counter counts up until the capacitor voltage has fallen below the reference of $-128Q$ supplied to comparator 1. The larger current source will be stopped at this time, when the high-order counter will contain a 9-bit accurate conversion, and the remaining voltage on the capacitor can be considered to be the quantizing error due to that conversion. However, only the low-current source is now enabled, which ramps the capacitor down to ground to give a further seven bits of conversion. The track-hold circuit can now operate again for the next cycle. The high-order bits require a maximum count of 512, and the low-order bits 128, a total of 640. Allowing 25 percent of the sample period for the track-hold circuit to operate, a 48-kHz converter would need to be clocked at some 40 MHz.

3.3.3.6 Sources of signal degradation. The areas to which most attention must be paid are the filters, the sample-hold system, and the converters. Figure 3.29 shows the various events during a track-hold sequence and the sources of possible shortcomings. Timing jitter on changing signals results in noise, and the accuracy in timing necessary to avoid significant noise is a fraction of a nanosecond.

An ADC converter may only be as accurate as the DAC it contains, and because of the higher operating speeds, ADCs are generally responsible for more signal degradation than DACs. The two devices have the same transfer function, since only the direction of operation distinguishes them, and the same terminology is used to classify the shortcomings of both. Figure 3.30 shows the transfer functions resulting from the main types of converter errors. These are

1. *Offset error.* A constant appears to have been added to the digital signal. Offset has no effect on sound quality, and DAC offset is of little

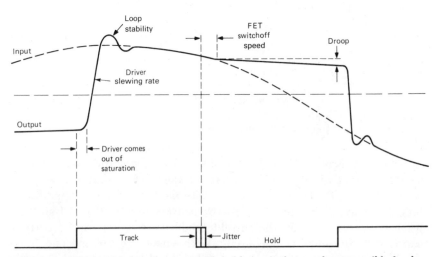

Figure 3.29 Various shortcomings in a track-hold circuit that can be responsible for degradations in audio quality.

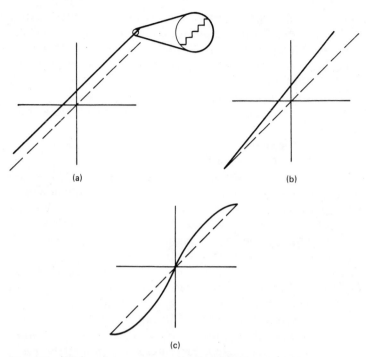

Figure 3.30 The main converter errors (solid line) compared with the perfect transfer function (dotted line). These graphs hold for ADCs and DACs, and the axes are interchangeable: if one is chosen to be analog, the other will be digital. *(a)* Offset error causes the transfer function to miss zero; *(b)* gain error causes wrong transfer function slope and an offset error; *(c)* linearity error causes distortion.

consequence. ADC offset, however, is undesirable as it can cause audible thumps at edit points.

2. *Gain error.* The slope of the transfer function is incorrect. Since converters are referred to one signal limit, gain error causes an offset error. Again there is no audible effect, but severe gain errors will cause asymmetrical clipping.

3. *Linearity error (sometimes called integral linearity).* The transfer function deviates from a straight staircase. This has the same significance as departures from linearity of an analog circuit. Normally, linearity should be better than $0.5Q$ for a quality device. Two subclassifications of linearity are found. The first is differential nonlinearity, which is the amount by which adjacent quantizing intervals differ in size (measured in fractions of Q). The second is nonmonotonicity, which is a special case of differential nonlinearity and can result from major overflows in the binary code. For example, with a code of 01111111 (127), the seven low-order current sources in a DAC will be operational. The next code will be

10000000 (128), where the eighth current source takes over. If this latter is in error on the low side, the analog output for 128 may be less than for 127, which results in nonmonotonicity. If the device has better than $0.5Q$ linearity, it must be monotonic. Absolute accuracy, the difference between actual and ideal output for given input, is not as important as linearity. A good audio converter will have excellent thermal tracking and will maintain linearity with temperature changes, even though the absolute accuracy seems poor.

The degradations of the signal due to shortcomings of components have been described in the major areas of filters and converters, and methods are available to reduce these to acceptable levels. A perfect sampling system causes no degradation whatsoever, but the essence of quantizing is that an approximation is made, which causes a quantizing error. For the purposes of this discussion, the system can be assumed to be perfect, such that any degradation will be due only to quantizing error. For a high-quality digital audio recorder, this is in fact not far from the truth.

Figure 3.31a shows an arbitrary waveform which is superimposed on the pulse amplitude modulated wave which results from quantizing and reconstructing that waveform. Figure 3.31b shows the difference between the DAC resampler output and the original instantaneous voltages of the waveform. This difference can be thought of as an error signal added to the perfect original. The error signal has interesting properties. First, because it is a PAM wave, it has an infinite spectrum, but the reconstruction filter will reduce this to an audio-band signal. Second, the peak-to-peak amplitude of the quantizing error signal cannot exceed Q. Third, the quantizing error can be calculated from the input waveform and the quantizing structure.

For complex signals such as music, the quantization error is random, and adjacent samples are uncorrelated. The quantizing error will thus have a uniform probability density function from $-0.5Q$ to $+0.5Q$ as shown in Fig. 3.32a. The quantizing error is in the form of spikes which are essentially uncorrelated delta functions. Therefore, the autocorrelation function of the error signal is itself a delta function as shown in Fig. 3.32b.

The power spectrum of a signal is the Fourier transform of the autocorrelation function, and the transform of a delta function is a uniform spectrum. True white noise has, however, a Gaussian probability density function, which is the consequence of summing a large number of uniform probabilities. The quantizing error in large complex signals will not sound exactly the same as white noise, but for the large signal case, it is adequate to calculate the equivalent of a signal-to-noise ratio. Many treatments calculate this ratio as $6.02n + 1.76$ dB, where n is the number of bits in the system. A useful rule of thumb is 6 dB per bit. Unfortunately, when signals are not large, and particularly when they are spectrally simple, suc-

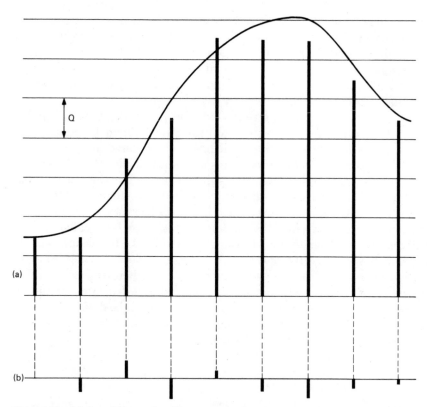

Figure 3.31 *(a)* An arbitrary signal is represented to finite accuracy by pulse-amplitude-modulated (PAM) needles whose peaks are at the center of the quantizing intervals. The errors caused can be thought of as an unwanted signal, shown at *(b)*, added to the original.

cessive samples are not uncorrelated, and the probability density function of the error is no longer uniform. It is no longer acceptable to use the term *noise,* because the quantizing error has become a deterministic function of the input signal, the sample instants, and the quantizing intervals. A deterministic error function has to be described as a distortion of the waveform.

The effect of quantizing distortion depends upon the level. As the level is reduced, the random noiselike behavior gives way to a form of noise modulation where the quantizing error fluctuates with the low-frequency content of the music. The subjective effect is rather "gravelly" and is sometimes called *granulation.* In the case where the sampling rate is a multiple of the signal frequency, harmonic distortion results.

The audibility of quantizing distortion is about 12 dB greater than the equivalent-power white noise, but fortunately with the correct application of dither, this source of degradation can be eliminated.

3.3.3.7 Use of dither. The foregoing has assumed a perfect system in which low-level signals become distorted by quantizing. The correct use of a dither signal at the ADC can convert the distortion into wideband noise and endow the system with a real signal-to-noise ratio. This ratio is not significantly less than the large signal-to-quantizing-error ratio obtained without dither.

A typical dither signal is a wideband noise, with a rectangular probability density function. The peak-to-peak value should be equal to Q, since this is the nature of the quantizing error with large complex signals. Figure 3.33 shows that the waveforms of signals smaller than Q are preserved in the duty cycle of the level switching caused by the dither. It has been shown that harmonic distortion is virtually eliminated by using dither and that the error signal at a low level becomes wideband noise (Vanderkooy and Lipshitz, 1984). The generation of rectangular-probability noise is not inexpensive, since a Galois field (a pseudo-random sequence) generator driving a DAC is necessary (Blesser, 1983). For this reason, Gaussian dither is generally employed, where the rms level is equal to Q. The required noise can be provided by a dedicated circuit, or, at lower cost, the noise from the antialiasing filter can be designed to have the right level. There is no doubt that the correct use of dither is essential to a high-quality audio ADC, although it is sometimes neglected.

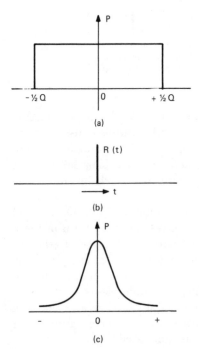

Figure 3.32 *(a)* The amplitude of a quantizing error needle is shown to exist from $-0.5Q$ to $+0.5Q$ with uniform probability. For large, complex signals, the autocorrelation function $R(t)$ has one spike, as shown in *(b)*, giving a uniform spectrum. The nominally white noise in analog circuits follows the Gaussian amplitude distribution shown in *(c)*.

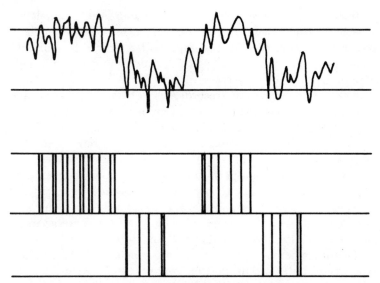

Figure 3.33 The use of wideband dither preserves the waveform of signals smaller than Q (here a sine wave of approximately $0.75Q$ p-p) as duty cycle modulation. The system is thus perfectly linear, as can be tested by time averaging.

3.3.3.8 Recording level and headroom. A digital audio recorder with dither is linear over the entire dynamic range, unlike linear recording machines, where the onset of saturation is a gradual process as the level increases. In AC-biased recorders it is customary to leave signal headroom to avoid running into nonlinearity on large signals. A digital recorder does not, in principle, need any headroom, and the best possible recording will be one in which the largest sample just reaches the limit of the number range. For this reason, current digital level meters indicate "zero dB" at full range, and all operating levels should be below zero. In practice some clipping of transients will take place, but it is observed that such clipping is much harder to detect than in an analog machine, possibly because the time for which the system is nonlinear is much shorter.

3.3.4 Digital audio processing

Once the audio signal has entered the digital domain, all of the traditional processes, such as gain control, filtering, and so on are performed by computation of sample values in logic circuits. The type of binary coding used has not been mentioned so far, because it does not much matter to converters as long as they have the same standard. The same can be said of the tape transport, since it is only storing numbers. It is the convenience of processing which determines the binary coding used in digital audio, just as it influences the choice of uniform quantizing. Audio signals are quantized within a range of binary integers which must represent both positive and negative voltages, but the important reference is midrange. The 2's complement binary system has this property and is in virtually universal use in digital audio processing.

A common requirement in audio recorders is to be able to fade in and fade out. It is not possible to commence or terminate the replay of digital audio simply by starting or stopping the samples passing to the DAC, since the result would be step waveforms producing an objectionable click. It is necessary to fade in and out by multiplying samples by coefficients which ramp up or down. When performing an edit, or when a splice is encountered, it is also necessary to perform a cross fade in the digital domain to avoid a click. The two signals are multiplied, one by a rising set, and one by a falling set of coefficients, and the results are accumulated. This process is also necessary during "punch-in" and "punch-out" on a studio recorder, where a short part of a recording is replaced, perhaps to conceal a wrong note, or the portion containing the wrong note is removed.

The main application of digital filtering to audio recorders is in sampling-rate conversion. This process is necessary, not only to interchange recordings made at different sampling rates, but also for the following purposes:

1. When a digital recorder is played at variable speed, the sampling rate changes proportionately. In order to provide a standard sampling rate for interface to other equipment, a rate converter is necessary to avoid the degradation of the only alternative—a DAC feeding an ADC.

2. When using oversampling ADCs and DACs, sampling rate conversion is performed in the digital domain.

3. In addition to the use of floating-point block coding, edit-point locators use sampling-rate reduction to conserve memory at the expense of band width (Ohtsuki et al., 1981).

In order to change the sampling rate, it is necessary to compute the relationship in time of every output sample to adjacent input samples. Once this has been done, the contributions of each sample from the input to the specific output sample are added together. The process is unchanged whether the rate is being increased or decreased (Lagadec, 1983; Gaskell, 1985).

3.3.5　Digital record-replay

In the interests of moderate tape speed and consumption, stationary-head recorders use efficient run-length-limited channel codes (Doi, 1983b). These invariably require a phase-locked loop in the reproduce circuit which will provide a clock to measure the periods between transitions. Digital audio recorders may be required, in professional applications, to operate at variable speed for the purpose of pitch changing. This puts greater demands on the lock-in range of the data separators and makes them more difficult to design than those in data recording disk and tape drives, which can be optimized for one speed. The reproduced frequencies are proportional to tape speed; therefore, a feed-forward system can be used where the free-running frequency of the phase-locked loop is made proportional to capstan speed. If a digital phase-locked loop is used, this effect can be achieved automatically by clocking the loop from the capstan frequency generator.

The reproduced frequencies in the channel code will be considerably lower than in, for example, rigid-disk drives for data recording, and this permits data separation in the digital domain. The reproduced signal is amplified and then passed to an ADC. The digital sample stream produced is equalized and band-limited in digital filters, and fed to a data separator containing a digital phase-locked loop. If the sampling rate is made a multiple of the loop frequency, which is itself related to the channel bit rate, then the system will track over a wide speed range. Since the bandwidth of a digital filter is proportional to the sampling rate, the reproduced signal will be filtered with the correct bandwidth, whatever the speed, a feature difficult to incorporate in analog filters.

Audio samples must be reproduced at precisely the correct frequency and without jitter. This is the function of the time-base corrector and the capstan servo. The *time-base corrector* is a RAM arranged as a ring memory by address counters which periodically overflow. Samples are written into the memory with jitter, but read out at the reference rate. In order to prevent the corrector overloading, the long-term sample rate from the tape must equal the reference sample rate. This can readily be achieved by controlling the capstan from the address relationship of the time-base corrector. If the capstan runs too fast, the write address will tend to catch up with the read address; the address difference consequently becomes smaller, thereby reducing the drive to the capstan. Figure 3.34 shows a typical system.

3.3.6   Error correction and concealment

The goal of digital audio recording is to make the sound quality independent of the medium, and this mandates the use of an error-correction system. The acuity of the ear is such that single corrupted samples are audible, and this demands a period between miscorrections which can be measured in months (approximately one sample in 10^{10}). While the fundamentals of error correction apply to digital audio, the special requirements of audio recording mean that the implementation is often more complex than in other types of magnetic recording.

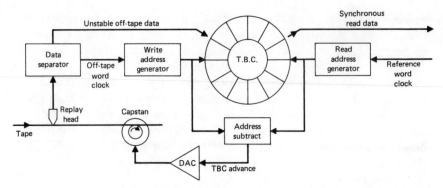

Figure 3.34 The address relationship of the time-base corrector (TBC) is used to control the capstan on reproduction.

In digital recorders associated with computers, the medium can be mapped to prevent recording on defective areas of a disk, or read-after-write can be used to skip over defective parts of a tape. If a readback error occurs, such machines can make several retries before error correction is attempted, and the speed of correction is not of primary importance, as long as correct data are eventually obtained.

In digital audio, the data are required in real time, which makes retries impossible, and only a short time is available for computation during the correction process. The tape is likely to be handled and even spliced, both of which cause severe corruptions of the recording. Real-time operation makes mapping out defects very difficult, and a machine which requires a lengthy formatting process of the blank medium before use would not be acceptable to the industry. Conversely, the nature of audio signals permits interpolation of uncorrectable samples—a process which is of no use in data recorders.

Stationary-head recorders must work at high linear recording density. This means that the shorter run lengths of the channel code will be more prone to random errors. Dropouts caused by debris produce burst errors. Splices cause additional losses because of distortion of the tape and can result in an arbitrary change of clock phase.

3.3.6.1 Interleaving and cross interleaving. Interleaving is an established method of controlling burst errors and is used extensively in digital audio recorders (Watkinson, 1984). Interleaving can take two forms: *block interleaving* and *convolutional interleaving*. In block interleaving, the rearrangement of the sample sequence is done within blocks, which are themselves in the correct sequence. This is commonly used in PCM adaptors where the block structure is related to the frame structure of the pseudovideo, permitting straightforward editing. The interleaving process is endless in convolutional interleaving, where all samples are at constant distances from their original place in the sequence. A block structure may be used after convolutional interleave to prevent the medium having a continuous code, which would make editing difficult.

The process of interleaving can be achieved using a RAM. Incoming samples are written into the RAM at addresses determined by a counter, which overflows periodically, giving the memory a ring structure. Figure 3.35 shows that reading the RAM is done by a sequencer which takes samples out of the memory in the interleave sequence. Thus samples spend different amounts of time in the memory, and the RAM acts as a delay which is a function of sample position in the input sequence.

A reverse process is necessary to provide deinterleaving during reproduction. The delays caused by interleaving are not normally of any consequence, except in multitrack recorders which are used for synchronous recording. The latter is achieved in analog machines by playing back with the record head to eliminate tape-path delay, but interleave delays preclude use of the technique in digital audio recorders. The solution is to have separate reproduce and record heads, separated by the distance tape travels in the decode-encode time, as shown in Fig. 3.36.

Editing can be complicated by the presence of a convolutional interleave. It is not possible simply to switch between two interleaved data

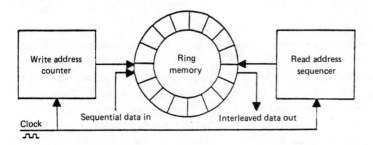

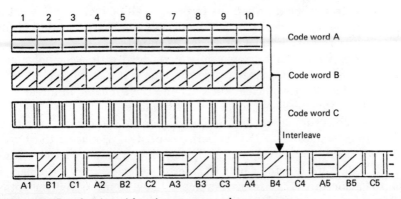

Figure 3.35 Interleaving with a ring memory and a sequencer.

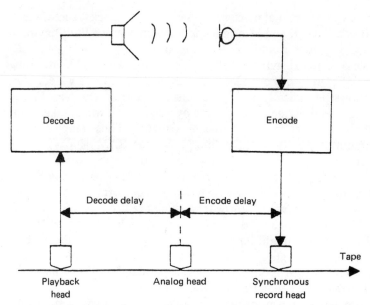

Figure 3.36 Separation between heads is required for synchronous recording owing to encode and decode delays.

streams, as this produces a confused set of samples over the constraint length. In order to edit a recording with convolutional interleave, it is necessary to use the arrangement of Fig. 3.37. The recording on the tape is reproduced and deinterleaved and then reinterleaved and rerecorded. This results in no change to the tape content. When this has been taking place for a least a constraint length, the deinterleaved reproduce samples can be faded out and samples from some other source faded in. A punch-in will commence in the same way, and will end in a reversed fade-back to deinterleaved reproduce samples. These will then be re-recorded for at least a constraint length, when the recording will be identical to what is already on the tape. Recording can then cease at a block boundary. Only in this way can convolutional interleave be edited.

The correction of burst errors can be performed with cyclic polynomials, but in general, processing of the syndrome is necessary to locate and define the burst, and sufficient time may not be available in a real-time system. It is often necessary to increase redundancy to obtain a correcting system which is simple and hence rapid. Figure 3.38 shows such a system where simple exclusive-or (XOR) parity is used to produce a third code word from two sample code words. The cyclic redundancy check (CRC) in each code word functions only as an error detector; the syndrome will be zero or nonzero. A nonzero syndrome is taken to mean that the entire code word is suspect, and it will be re-created by taking the XOR of the

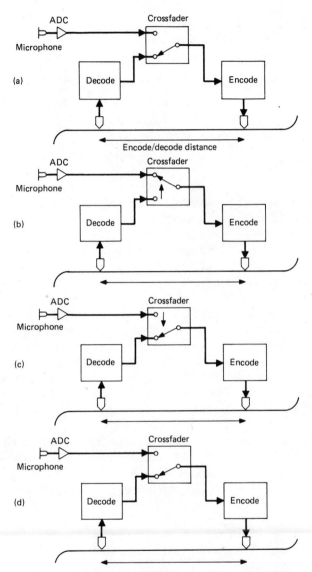

Figure 3.37 The sequence necessary to edit a convolutional interleave without corruption. *(a)* Re-record existing samples for at least one constraint length; *(b)* cross-fade to incoming samples (punch-in point); *(c)* cross fade to existing replay samples (punch-out point) *(d)* re-record existing samples for at least one constraint length.

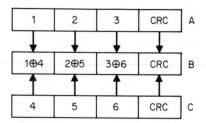

Figure 3.38 A simple exclusive-or (XOR) parity system allows rapid correction but has a large overhead.

remaining two. The redundancy is as extensive as the data, so the system is not very economical on storage, but it is used in PCM adaptors, where the channel bandwidth predetermined by the video standard is more than adequate.

The example of Fig. 3.38 may be considered to form the data into an array and form code words at right angles. This can be done only by using block interleaving. If the system is to be applied to convolutional interleaving, one of the code words must be diagonal. A smaller overhead will be necessary if the XOR term is produced from a larger number of CRC code words, as in Fig. 3.39. The use of interleaving will permit burst correction, but the presence of a random error near a burst will result in two errors in one XOR code word, which is uncorrectable. This can be overcome using cross interleaving. Code words pass through the array of data in three directions: two XOR and one CRC on the interleaved block to be recorded as shown in Fig. 3.40.

3.3.6.2 The correction mechanism. On reproduction, block CRC errors become pointers which accompany the samples through deinterleaving. Should there be two samples in error in one deinterleaved XOR code word, the data can be reinterleaved to the cross-interleave diagonal, where

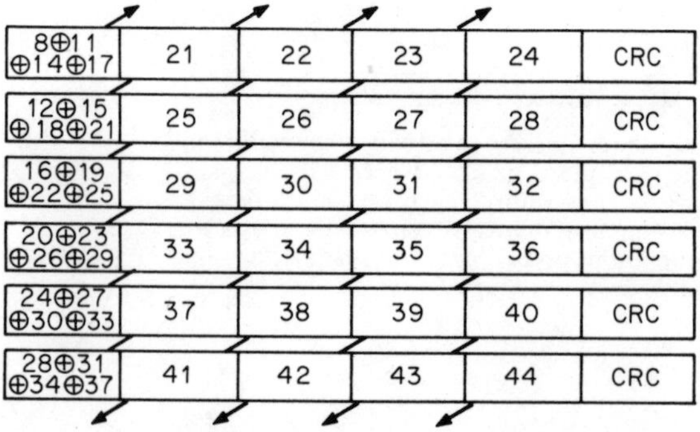

8⊕11⊕14⊕17	21	22	23	24	CRC
12⊕15⊕18⊕21	25	26	27	28	CRC
16⊕19⊕22⊕25	29	30	31	32	CRC
20⊕23⊕26⊕29	33	34	35	36	CRC
24⊕27⊕30⊕33	37	38	39	40	CRC
28⊕31⊕34⊕37	41	42	43	44	CRC

Figure 3.39 The code words in convolutional interleave.

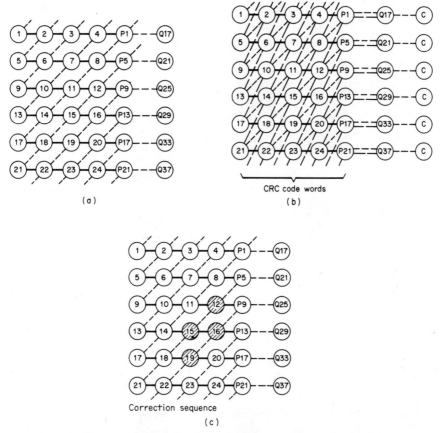

Figure 3.40 (a) The construction of code words and cross-code words; (b) the construction of cyclic redundancy check (CRC) code words used as error flags; cross interleaving permits the correction of multiple code-word errors; (c) a correction sequence using CRC errors as pointers.

one of the errors can be corrected. The samples are again deinterleaved, and the remaining error can be corrected. Cross interleaving reduces the amount of redundancy necessary to achieve a given correction performance. The technique is highly suitable for stationary-head recorders, where the data tracks are continuous. For rotary-head recorders, the tracks are not continuous, and it is necessary to interrupt the data stream at the head switch point. A convolutional interleave may operate across these interruptions, and this is done in the EIAJ standard for consumer PCM adaptors. It is possible, however, to use cross interleaving where each track is self-contained. Figure 3.41 shows that the diagonal code words may wrap around from one end of the block to the other to produce what is known as a block-completed code (Doi, 1983a).

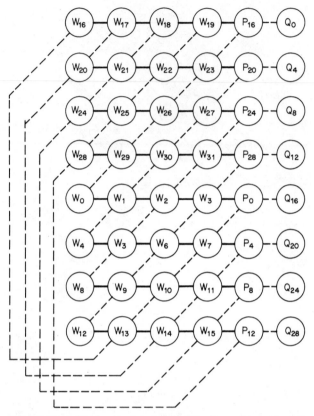

Figure 3.41 A block-completed cross interleave wraps around code words from one end of the block to the other.

3.3.6.3 Splice handling.

Tape splicing is often performed in professional digital audio recording even though its use seems anachronistic. All-electronic audio editing is possible, allowing a new tape to be assembled from source material without cutting; but the required hardware is currently complex and expensive. Also, all but disk-based editors tend to be slow to use. When an electronic editor cannot be afforded, or where editing must be done quickly, splicing is used. Splice handling causes timing problems, and special precautions have to be taken to maintain phasing between recorded samples and the timing code tracks, as these will be used when more than one machine are synchronized together. This section illustrates how these problems are overcome.

Cross interleaving is a very powerful technique for burst-error correction, but it conflicts with the requirements for splice handling. Figure 3.42 shows that the use of interleave in the area of a splice destroys code words for the entire constraint length of the interleave. The longer the constraint length, the greater the resistance to burst errors, but the greater the damage due to a splice. In order to handle splices, odd-even interleaving has to be used to interpolate when code words are overwhelmed. The odd-even distance has to be greater than the cross-interleave constraint length. In the Digital Audio Stationary Head (DASH) format, the constraint length is 1428 samples, and the odd-even delay is 2448 samples. Figure 3.43 shows the result of a perfect splice. Samples are destroyed for the duration of the constraint length, but this occurs at different times for odd and even samples.

Using interpolation, it is possible to obtain simultaneously the end of the old recording and the beginning of the new. A cross fade is made in the digital domain between the old and new recordings. The cross fade requires a multiplier and a coefficient generator. The samples from the outgoing recording are multiplied by descending coefficients, and the samples from the incoming recording are multiplied by ascending coefficients. Summing the products results in a cross fade. In the twin DASH system, all data are duplicated in a second track with reverse interleave. This permits splice handling without interpolation.

A tape splice will result in a random jump in control-track phase of up to plus or minus one-half a sector. Following a splice, the capstan will have to rephase, and in order to do so, it will have to accelerate and decelerate the tape. When tape speed is changing, the data separators have a greater chance of losing lock and causing block CRC errors. In order to reduce locking time, one solution is to control the capstan using block phase, since a block is one-quarter the size of a sector. This results in rapid locking, but there can be four relationships between block phase and sector phase. In order to restore the correct relationship between block and sector phase, the replay data delay in the time-base corrector will be changed.

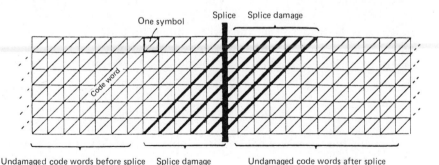

Figure 3.42 A convolutional interleave suffers corruption for a constraint length in the presence of a splice.

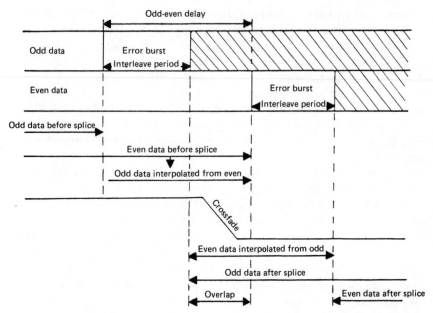

Figure 3.43 A splice-handling format has an odd-even sample interleave greater than the constraint length of the main interleave. Gross errors due to splicing cause two separate error bursts following deinterleaving, which can be concealed by interpolation.

This process is important because synchronism between two machines is achieved by sector lock; thus samples must have a fixed relationship to sector timing. Therefore tape splicing always operates in sector steps.

When a synchronous recording is made over a splice, it is necessary to duplicate exactly the change in block to sector phase (Watkinson, 1985). Thus a time-base corrector is necessary in the record channel, which will experience a timing change at the instant the splice passes the record head. The timing change on replay occurs at the instant the splice passes the playback head. The delay change parameter must be stored for the decode-encode delay so that the record timing shift occurs at the correct instant. A cross fade is neither necessary nor desirable in a synchronous recording made over a splice. Thus the decision to operate the cross-fade process must be made on a channel-by-channel basis.

A synchronous recording over a splice will suffer only a relatively small number of block CRC errors, which will be correctable after deinterleaving. A splice between two recordings will suffer a similar number of block CRC errors, but these will not be correctable, since code words have been destroyed over the constraint length. The system must use interpolation. Thus any channel which is not interpolating continuously in the area of a sector-address jump does not cross-fade. Where interpolation occurs for a constraint length, the cross fade will take place.

3.3.7 Practical digital audio machines

The options are reasonably open when the design of a digital audio machine is contemplated, provided it can accommodate the established data rate of about 1 Mb/s per audio channel. What is equally important, although less often stated, is that the working signal-to-noise ratio of the digital channel can be low. In principle, bandwidth can be obtained readily in the digital domain by operating several channels in parallel if necessary.

In the early days of digital audio recording, the need for high channel bandwidth was a major obstacle, and devices which could record such a bandwidth were restricted to videotape recorders and disk drives. Efforts centered on the use of video recorders for digital audio storage because of their much greater capacity. However, the random access of disk drives enabled them to be used as digital audio editors (Ingebretsen and Stockham, 1984; McNally et al., 1985).

3.3.7.1 Pulse code modulation adaptors. PCM adaptors convert analog audio, first into a digital data stream and then into a waveform known as *pseudovideo,* which can be recorded on a video recorder. The block diagram of such a device is shown in Fig. 3.44. PCM adaptors are available

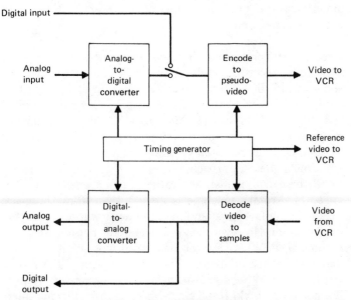

Figure 3.44 A pulse-code modulation (PCM) adapter uses a video recorder to record digital audio. The timing system locks the sampling rate to the line scan and frame rates; the ADC feeds an encoder which produces a serial data stream embedded in a pseudovideo waveform, which is recorded. On replay, the decoder returns the video waveform to sample form for the digital-to-analog converter.

for consumer video cassette recorders (Beta or VHS), and generally use the television line standard of the country of sale, so that the consumer can still use the recorder for its original purpose. Professional machines, such as those used for optical digital disk (Compact Disc) mastering, use industrial grade video recorders (U-Matic or VHS). These recorders are standardized on 525 lines and 60 Hz to permit international interchange.

The pseudovideo waveform uses some 60 percent of peak white for a binary 1 to avoid activating preemphasis circuits in standard recorders. For consumer units with auto level control, a peak-white pulse is incorporated periodically to stop the recorder from increasing the average level above 60 percent. This is unnecessary on professional machines. On all video recorders used for PCM, it is essential that any built-in dropout compensation be disabled, as this can reduce the efficiency of the error-correction system. The noncontinuous nature of video requires time compression on recording to fit samples into the active period of unblanked lines. In digital audio, the corresponding time expansion on replay can be part of the deinterleaving process, and the same time-base correction will eliminate speed variations.

The signal-to-noise ratio of even a domestic video recorder is greatly in excess of that necessary for digital recording; consequently the tape consumption will be considerably higher than strictly necessary although, by the same argument, the random-error rate will be lower, putting lower demand on the error-correction system.

3.3.7.2 Comparison of stationary and rotary-head recorders.
Progress in short-wavelength digital recording and channel codes has permitted the design of stationary-head digital audio recorders having reasonably low linear tape speed. The DASH format, for example, allows one digital audio channel per track at 760 mm/s (30 in/s). However, digital recording can also be combined with a rotary-head system, to produce a direct digital recorder, as opposed to an adapted video recorder.

Rotary- and stationary-head approaches have quite different characteristics which make them complementary rather than competitive. The head-to-tape speed of a rotary machine is greater than in a stationary head machine; thus the replay signal and medium noise will be higher for a given recording medium. However, head noise, circuit noise, and interference remain unchanged. Where a machine is head-noise limited, which is increasingly common at high densities, the greater speed of the rotary head results in a higher working signal-to-noise ratio, or conversely permits the use of a narrower track. Rotary machines determine track pitch by linear tape speed, which enables dense packing of the narrow tracks. It is also easy to employ azimuth recording, in which there are no guard bands, and the recorded tracks are actually narrower than the head poles (as described in Chap. 2). In stationary-head machines, tracks must be wider to obtain adequate signal-to-noise ratio at low speed, and a significant space must be left between tracks when recording with ferrite heads because of the difficulty of fabricating parallel head stacks. In any case, compensation will usually be necessary to reduce the crosstalk between adjacent magnetic circuits. Using thin film-head technology, crosstalk between heads can be reduced, and more magnetic circuits can be fabricated across a given width of tape. The proposed S-DAT format has 20 digital tracks on a tape only 1.9 mm wide. It remains to be seen whether such narrow tracks can be made to register with heads consistently. On the grounds of areal density, rotary-head recorders are superior by a factor of about 10. The effect of two-dimensional dropouts, which cover several tracks, is to produce several evenly spaced error bursts, and the effect of edge damage is distributed, because all tracks share the width of the tape. Both factors ease the requirements of the error-correction system. The drawbacks of a rotary-head system are that tape splice editing is impossible, and a dc-free channel code is needed since the signal has to pass a rotary transformer. The higher head speed will cause more rapid wear, and in high humidity environments, condensation can cause the tape to adhere to the rotor.

For professional use, it is often necessary to have a multitrack recorder in which tracks can be recorded independently. This causes no difficulty in a stationary-head machine; whereas in a rotary-head recorder, all of the audio channels are time multiplexed into a small number of heads. Independent channel operation requires the use of a block structure along the track with tolerance gaps between so that the heads can update a single block in the middle of the track if necessary. The added requirements of synchronous recording, variable-speed operation, and punch-in can be met in a rotary machine by fitting extra heads (necessary because of the interleaving structure) in the rotor (see Sec. 3.3.6.1). This results in a complex and expensive rotating assembly. Where tape splice editing is necessary, the stationary-head machine is mandatory. For these reasons, the stationary-head recorder is currently favored for professional multitrack applications. Such machines can still be more economical in tape use than their analog counterparts. A typical 24-track analog recorder uses 2-in-wide tape at 760 mm/s (30 in/s). The DASH format uses ferrite heads, in the single density version, to achieve 24 tracks on half-inch tape at the same speed. The double-density version will use thin film heads to offer 48 tracks on the same tape. Another digital format, by comparison, offers 32 tracks on 1-in tape. Professional machines are necessarily open-reel designs in which the tape is subject to handling damage. At the same time, recording takes place at short wavelengths to give reasonable playing time. Data errors due to fingerprints will be severe, and the error-correction system will be complex, especially since splice handling is necessary. With stationary heads, however, the dc-code content is less of a problem, and this is sometimes sacrificed to optimize density ratio, for example. With stationary heads it is possible to multiplex several tracks from one audio channel with a corresponding reduction in tape speed. The disadvantage of having record and reproduce circuits for each track must be weighed against the complication of the drum and servos of a rotary-head machine.

For consumer applications, the overriding concern is to minimize tape consumption; tape splicing is not an issue. In this case, it is sensible to use a cassette owing to the greater protection afforded from contamination, and because the high-density rotary-head can be employed. In the rotary head digital audio tape recorder (R-DAT) format, a recording density of 129 Mb/in^2 has been achieved by a combination of azimuth recording and a track-following servo (Nakajima and Odaka, 1983; Shimpuku et al., 1986). This permits 2 to 3 hours of stereo recording in a cassette measuring $73 \times 54 \times 10.5$ mm. The high recording density means that the length of tape used for a given recording time is short; thus the machine appears to rewind very quickly. In order to simplify the threading procedure, the tape is only partially wrapped around the drum, which is only 30 mm in diameter. For about half of the time, neither head is in contact with the tape, and the recording is time-compressed into the short periods of head contact.

3.3.8 Conclusions

This section has introduced the concept of sound recording in which the basic criteria of quality are set at the design stage. The potential quality of a recording thus defined can never be improved upon without fundamental change. This places a greater responsibility upon those who propose standards to cater to future demands as well as present requirements. International standards now exist for digital audio recording for the Compact Disc, consumer PCM adaptors, R-DAT, and the digital audio interface, although standardization of stationary-head multitrack machines is proving elusive.

The performance of recorders themselves will improve slightly as filters and converters approach the limit of the chosen word length. There has been discussion of the use of 20-bit word length for recorders, but there is no converter technology to support it.

The use of oversampling is likely to grow. The standardization of formats permits the quantity production necessary for use of LSI technology to reduce costs and power consumption, making portable recorders a reality. For stereo digital recorders, the economics of LSIs are such that professional machines will almost certainly be forced to use the same chip sets and formats as their consumer relatives. There may cease to be a distinction between consumer and professional formats, and the machines will differ only in the facilities offered.

References

Adams, R. W., "Companded Predictive Delta Modulation: A Low-Cost Technique for Digital Recording," *J. Audio Eng. Soc.*, **32**, 659 (1984).

Adams, R. W., "Design and Implementation of an Audio 18 Bit A/D Converter Using Oversampling Techniques," *Audio Eng. Soc. Conf.*, Hamburg, Preprint No. 2182, 1985.

Bertram, H. N., "Long Wavelength AC Bias Recording Theory," *IEEE Trans. Magn.*, **MAG-10**, 1039 (1974).

Bertram, H. N., "Wavelength Response in AC Biased Recording," *IEEE Trans. Magn.*, **MAG-11**, 1176 (1975).

Bertram, H. N., M. Stafford, and D. Mills, "The Print-Through Phenomenon," *J. Audio Eng. Soc.*, **28**, 690 (1980).

Bertram, H. N., and R. Niedermeyer, "The Effect of Spacing on Demagnetization in Magnetic Recording," *IEEE Trans. Magn.*, **MAG-18**, 1206 (1982).

Betts, J. A., *Signal Processing Modulation and Noise*, Hodder and Stoughton Ltd., Seven-oaks, England, 1970, pp. 119–136.

Blesser, B. A., "Advanced A/D Conversion and Filtering: Data Conversion," in B. A. Blesser, B. Locanthi, and T. G. Stockham (eds.), *Digital Audio*, Audio Eng. Soc., New York, 1983.

Daniel, E. D., "A Preliminary Analysis of Surface-Induced Tape Noise," *IEEE Trans. Commun. Elect.*, **83**, 250 (1964).

Daniel, E. D., "Tape Noise in Audio Recording," *J. Audio Eng. Soc.*, **20**, 92 (1972).

Daniel, E. D., and P. E. Axon, "Accidental Printing in Magnetic Recording," *BBC Quarterly*, **5**, 7 (1950).

Doi, T. T., "Error Correction of Digital Audio Recordings," in B. A. Blesser, B. Locanthi, and T. G. Stockham (eds.), *Digital Audio*, Audio Eng. Soc., New York, 1983*a*.

Doi, T. T., "Channel Codings for Digital Audio Recordings," *J. Audio Eng. Soc.*, **31**, 224 (1983*b*).

Dolby, R. M., "An Audio Noise Reduction System," *J. Audio Eng. Soc.*, **15**, 383 (1967).

Dolby, R. M., "A Noise Reduction System for Consumer Applications," *Conv. Audio Eng. Soc.*, Paper No. M-6, New York, 1970.

Dunlop, D., and J. Stirling, "Hard Viscous Remanent Magnetization in Fine Grain Hematite," *Geophys. Research Letters*, **4**, 163 (1977).

Eldridge, D. F., "DC and Moduation Noise in Magnetic Tape," *IEEE Trans. Commun. Electron.*, **83**, 585 (1964).

Eldridge, D. F., and E. D. Daniel, "New Approaches to AC-Biased Recording," *IRE Trans. Audio*, **AU-10**, 72 (1962).

Fujiwara, T., "Nonlinear Distortion in Long Wavelength AC Bias Recording," *IEEE Trans. Magn.*, **MAG-15**, 894 (1979).

Gaskell, P. S., "A Hybrid Approach to the Variable Speed Replay of Digital Audio," *Audio Eng. Soc. Conf.*, Hamburg, Preprint No. 2202, 1985.

Ingebretsen, R. B., and T. G. Stockham, "Random Access Editing of Digital Audio," *J. Audio Eng. Soc.*, **32**, 114 (1984).

Kishimoto, M., T. Sueyoshi, S. Kotaoka, K. Wakai, and M. Amemiya, "Chronological Changes in Coercivity and Printing Effect of Cobalt-Substituted Iron Oxides," *J. Japan. Soc. Powder and Powder Metallurgy*, **26**, 49 (1979).

Kogure, T., T. T. Doi, and R. Lagadec, "The DASH Format: An Overview," *Audio Eng. Soc. Conf.*, New York, Preprint No. 2038, 1983.

Köster, E., "A Contribution to Anhysteretic Remanence and AC Bias Recording," *IEEE Trans. Magn.*, **MAG-11**, 1185 (1975).

Köster, E., et al., "Switching Field Distribution and AC Bias Recording Parameters," *IEEE Trans. Magn.*, **MAG-17**, 2550 (1981).

Lagadec, R., "Digital Sampling Frequency Conversion," in B. A. Blesser, B. Locanthi, and T. G. Stockham (eds.), *Digital Audio*, Audio Eng. Soc., New York, 1983.

Mallinson, J. C., "One-Sided Fluxes: A Magnetic Curiosity?" *IEEE Trans. Magn.*, **MAG-9**, 678 (1973).

McKnight, J. G., "The Distribution of Peak Energy in Recorded Music," *J. Audio Eng. Soc.*, **7**, 65 (1959).

McKnight, J. G., "Erasure of Magnetic Tape," *J. Audio Eng. Soc.*, **11**, 223 (1963).

McKnight, J. G., "Flux and Flux-Frequency Measurements in Magnetic Recording," *J. SMPTE*, **78**, 457 (1969).

McNally, G. W., P. S. Gaskell, and A. J. Stirling, "Digital Audio Editing," *Audio Eng. Soc. Conf.*, Hamburg, Preprint No. 2214, 1985.

Meyer, J., "Time Correction of Anti-Aliasing Filters Used in Digital Audio Systems," *J. Audio Eng. Soc.*, **32**, 132 (1984).

Mori, T., M. Kasuga, M. Kikuchi, Y. Tsuchikane, and T. Matsushige, "System Design of Professional Digital Audio Mastering System," *Audio Eng. Soc. Conf.*, Eindhoven, Netherlands, Preprint No. 1959, 1983.

Nakajima, H., and K. Odaka, "A Rotary Head High Density Digital Audio Tape Recorder," *IEEE Trans. Consumer Electronics*, **CE-29**, 430 (1983).

Naumann, K. E., and E. D. Daniel, "Audio Cassette Chromium Dioxide Tape," *J. Audio Eng. Soc.*, **19**, 822 (1971).

Ohtsuki, T., S. Kazami, M. Watari, M. Tanaka, and T. T. Doi, "Digital Audio Editor," *Audio Eng. Soc. Conf.*, Hamburg, Preprint No. 1743, 1981.

Olsen, H. F., *Acoustical Engineering*, Van Nostrand, New York, 1957.

Preis, D., "Phase Distortion and Phase Equalization in Audio Signal Processing," *J. Audio Eng. Soc.*, **30**, 774 (1983).

Ragle, H. U., and P. Smaller, "An Investigation of High-Frequency Bias-Induced Noise," *IEEE Trans. Magn.*, **MAG-1**, 105 (1965).

Sawada, T., and K. Yamada, "AC Erase Head for Cassette Recorder," *IEEE Trans. Magn.*, **MAG-21**, 2104 (1985).

Shannon, C. E., "A Mathematical Theory of Communication," *Bell Syst. Tech. J.*, **27**, 379 (1948).

Shimpuku, Y., S. Fukuda, H. Suguki, and K. Odaka, "A New Digital Audio Recording Technique," *Video, Audio and Data Recording Conference*, IERE Pub. No. 67, 1986.

Sivian, L. J., H. K. Dunn, and S. D. White, "Absolute Amplitudes and Spectra of Certain Musical Instruments and Orchestras," *J. Acoust. Soc. Amer.* **2**, 330 (1931).

Tjaden, D. L. A., and J. Leyten, "A 5000:1 Scale Model of the Magnetic Recording Process," *Philips Tech. Rev.*, **25**, 319 (1963).

Tjaden, D. L. A., and A. M. A. Rijckaert, "Theory of Anhysteretic Contact Duplication," *IEEE Trans. Magn*, **MAG-7**, 532 (1971).

Van de Plassche, R. J., and D. Goedhart, "A Monolithic 14 Bit D/A Converter," *IEEE J. Solid State Circ.*, **SC-14** (1979).

Van de Plassche, R. J., and E. C. Dijkmans, "A Monolithic 16 Bit D/A Conversion System for Digital Audio," in B. A. Blesser, B. Locanthi, and T. G. Stockham (eds.), *Digital Audio*, Audio Eng. Soc., New York, 1983.

Vanderkooy, J., and S. P. Lipshitz, "Resolution below LSB in Digital Systems with Dither," *J. Audio Eng. Soc.*, **32**, 106 (1984).

Wallace, R. L., "The Reproduction of Magnetically Recorded Signals," *Bell Syst. Tech. J.*, **30**, 1145 (1951).

Watkinson, J. R., "Error Correction Techniques in Digital Audio," *Video and Data Rec. Conf.*, Southampton, England, 157 (1984).

Watkinson, J. R., "Splice Handling Mechanisms in DASH Format," *Audio Eng. Soc. Conf.*, Hamburg, Preprint No. 2214, 1985.

Westmijze, W. K., "Studies in Magnetic Recording," *Philips Res. Rep.*, **8**, 344 (1953).

Instrumentation Recording

James U. Lemke

Center for Magnetic Recording Research,
University of California, San Diego,
California

Instrumentation recording covers the broad field of data recording in which analog or encoded information is stored in other than standard audio, video, or computer formats. In the United States, the practice traces its origins to the early 1950s when flight-test programs required capacities and rates beyond those possible with oscillographic recorders, and the national security effort required ever-increasing surveillance bandwidths. The earliest instrumentation recorders were adapted from professional audio machines to accommodate frequency modulation (FM) and other higher-than-audio-frequency electronics. The bandwidth of the audio reproducing heads was increased by replacing the center conductor of the coaxial head cables with fine piano wire, thereby reducing the parallel capacity and raising the resonant frequency. In time, the audio-equipment manufacturers developed recorders specifically for the requirements of the instrumentation users, and standards evolved such as those of the Inter-Range Instrumentation Group (IRIG).

Although some instrumentation recording is done on disks, the dominant formats utilize tapes. Fixed-head recorders are usually multitrack and use wide tapes, typically $\frac{1}{2}$ or 1 in, although some are as wide as 2 in. Tapes are wound on precision reels having aluminum hubs and flanges and can have lengths exceeding 3000 m.

Rotary-head machines allow the highest areal packaging density and are now primarily of the helical-scan configuration similar to those used for professional video recording. Early rotary-head machines for wide bandwidth analog signals were adapted from transverse-scan broadcast video recorders, with signal processing to provide time alignment of the signal overlap where it was interrupted at the end of one head scan and the beginning of another. Rotary-head machines today use high-density digital recording and the interruption is easily accommodated by digital buffering.

Various modes of recording are used for instrumentation, depending upon the type of data to be stored. Pulse-code modulation (PCM) is becoming the dominant technique for telemetry, satellite downlink, and laboratory measurements, in many cases replacing FM recording. Baseband linear recording with ac bias is done in two modes: constant-flux recording for maximum information capacity per track, and constant noise-power-ratio (NPR) recording for surveillance signals where the minimum intermodulation plus additive noise is needed across a wide band. The NPR is described in more detail below.

Instrumentation recorders differ from other recorders in a number of ways but markedly so in the method of optimizing the record current and in equalizing the reproduced signals. In analog machines, the record head and reproduce head are separate structures, and each is optimized for its function. The value of the ac bias is optimized at band-edge—the highest signal frequency of interest—to obtain the maximum bandwidth. Audio recorders also use separate record and reproduce heads but have their bias current optimized for maximum low-frequency dynamic range. Bulk erasure is commonly used in instrumentation recording since the necessity for overwriting is seldom encountered, although some high-density digital recording machines exhibit excellent overwrite characteristics. Linear densities up to 3000 fr/mm (a wavelength of 0.67 μm) are achieved with partial-penetration recording of the relatively thick magnetic tape coating. Small record gaps of fractions of a micron create little phase shift during the recording process and, although not common in multitrack recorders, are increasingly being used as their benefits become better understood. Phase equalizers are used to remove much of the linear peak shift during reproduction. The tapes utilize high-coercivity, small-particle oxides and are calendered to a high finish to reduce modulation noise and spacing loss. Tape speeds can be very high, with multitrack recorders having the capability of storing data at gigabits per second. Burst-efficient error-detection and -correction codes, such as the Reed-Solomon code, are used to correct errors down to 10^{-12}.

4.1   Heads and Tapes

4.1.1   Interface

The shortest wavelengths employed by instrumentation recorders are extremely small; therefore, special care must be taken to ensure intimate contact between the gap region of the head and the surface of the tape coating. At 3000 fr/mm, for example, a spacing between the reproducing head and the tape of only 0.67 μm will produce an attenuation of the signal of almost 60 dB. This comes from the reproducing spacing-loss law (Wallace, 1951).

$$\frac{E_d}{E_0} = 20 \log e^{-2\pi d/\lambda} \tag{4.1}$$

where E_d/E_0 is the output voltage ratio at spacings of d and zero at wavelength λ. Equation (4.1) is frequently given in the form $54.6d/\lambda$ dB. An empirical record loss factor has been found to be about $45d/\lambda$ dB (Bertram and Niedermeyer, 1982).

The effective depth of recording can be found by integrating the contribution from strata distances y from the head and finding that depth δ_{eff} that contributes 90 percent of the reproduce signal. For a recording uniform in depth, the solution

$$\int_0^{\delta_{eff}} e^{2\pi y/\lambda} dy = 0.9$$

gives

$$\delta_{eff} \approx \frac{\lambda}{e} \tag{4.2}$$

Equation (4.2) points out the necessity for maintaining a magnetically dense, smooth surface on the tape. At a density of 3000 fr/mm, 90 percent of the signal comes from a layer that is only 0.25 μm thick—about half the wavelength of visible light.

Tape speeds are relatively high on many instrumentation recorders; therefore, it is frequently a challenge to maintain head-tape contact in view of the likelihood of an entrapped-air film. As a head wears, its radius of contact increases, which exacerbates this problem. Such "tape flying" is evident from observing the envelope of the reproduced-signal amplitude for instability. Although some forms of frictional polymers can form on the head surface and considerably increase head life, without being thick enough to affect output, head buildups are generally detrimental and must be removed for acceptable performance.

The impossibility of consistently achieving the near-perfection in the head-tape interface that is required for high-density recording has led the instrumentation field to use very powerful interleaved error-detection and correction codes. The strategy in instrumentation recording is to accept long bursts of errors as the penalty for using a very high-density channel and then to correct the errors to the degree desired. Philosophically, this is in contrast to the field of computer recording, where relatively low linear densities requiring little error correction are used.

4.1.2 Heads

Single-track heads for rotary recorders are fabricated in very much the same way as video heads (see Chap. 2). The optical gap lengths are in the range of 0.3 to 1.0 μm. Typically, the cores are made from single-crystal or hot-pressed manganese–zinc ferrite with sputtered ceramic or glass gaps and the poles assembled by glass bonding. Track widths are controlled by notching the cores in the region of the gap prior to glass bonding. Usually the same head is used for recording and reproducing. Such heads are used on channels with relatively narrowband modulation schemes, and as a consequence, their pole faces need not be very long. The small cores result in fairly high efficiency with low-permeability ferrite even at frequencies of up to 100 MHz. The ferrite materials must be lapped with great care and the glass bonding done under tight process control to ensure good gap definition. Aside from these considerations, the heads are relatively easy to make and, being single-track, are free from crosstalk, gap alignment, and many other problems encountered with multitrack heads.

Multitrack heads have record gap lengths of from 0.5 to 2.0 μm, reproduce gap lengths of from 0.3 to 0.5 μm, and are fabricated in halves from machined half-brackets into which the individual prewound core halves are bonded. The brackets may be made from a nonmagnetic metal such as aluminum or from a nonmagnetic ferrite in the case of the all-ferrite head. Frequently, a tip plate containing thin metal-alloy poles is bonded to the ferrite cores in such a way as to cap the cores and thereby serve as the contact surface of the head. This creates a hybrid structure with the bulk of the magnetic circuit made up of low-loss ferrite but with the gap region, where the flux density is highest, composed of thin poles of hard, high-saturation-magnetization alloy. The alloy tips may be laminated to decrease their eddy-current losses and are usually made from a soft magnetic material such as aluminum-silicon-iron.

Shields between the tracks are inserted into slots provided for them, after the two half-brackets have been gapped with a sputtered or evaporated gap material, or they are bonded into the half-brackets before gapping in the case of all-ferrite heads. Both types of heads require precise fabrication of diverse materials to submicron tolerances over dimensions that may exceed 50 mm in some heads.

The tolerances associated with track width, track position, gap length, gap scatter, gap depth, and electrical parametrics are all difficult to control; one of the most critical tolerances relates to individual head gap azimuth and the relative azimuth track-to-track. The equation relating azimuth misalignment loss L (in dB) to track width w, wavelength λ, and misalignment angle β is

$$L = 20 \log \frac{\sin[(\pi w \tan \beta)/\lambda]}{(\pi w \tan \beta)/\lambda} \tag{4.3}$$

For a high-density, 28-track instrumentation head operating at a wavelength of 0.67 μm and a track width of 457 μm, an azimuth misalignment scatter track-to-track of 2 minutes of arc will result in a 14-dB difference in signal level between tracks if the overall azimuth alignment is optimized for one track. Since azimuth scatter exists in both record and reproduce heads, and tapes must be interchangeable between recorders, tolerances of 1 minute of arc must be maintained to keep the azimuth loss to less than 2.5 dB in the worst case. Metal heads that are bonded together by epoxy resins are more susceptible to this problem than are the all-ferrite heads.

Head designs are, of necessity, compromises between core size, inductive crosstalk, efficiency, gap depth (head life), turns, and resonance frequency. The problem is lessened somewhat when the record and reproduce heads are separate and their designs may be optimized separately. Asymmetries are often designed into the head pole-tip shapes to create compensating long-wavelength response bumps caused by the outer dimensions of the head. With care, useful response over 12 octaves can be achieved.

4.1.3 Tapes

Instrumentation tapes are in many respects similar to video tapes; they differ primarily in their width, length, packaging (reel), and in the smaller amount (or absence) of abrasive material formulated into the binder system for controlled head cleaning. Abrasivity is a property of major concern in selecting an instrumentation tape, since multitrack heads are very expensive and may have an allowable wear depth of only 10 or 12 μm. Usually, head manufacturers will warrant head life for a limited number of hours with only a few specific types of tape.

Low-density tapes use conventional gamma ferric oxide of about 28 kA/m (350 Oe) coercivity, but most high-density tapes use very small particles of cobalt-modified gamma ferric oxide of about 52 kA/m (650 Oe) coercivity. A matte finish back coating containing a conductive material is usually used to prevent electrostatic buildup and to facilitate tracking and uniform winding onto the reels. The very long lengths of some tapes—sometimes exceeding 3000 m—necessitate controlled winding tensions to obtain a stable pack that can be accelerated to high angular rates on a tape transport without cinching or blocking and that can also be shipped without damage. Tight tolerances are maintained on slitting dimensions, surface finish, and, in the case of high-density digital recording tapes, dropouts.

Acicular tape particles are oriented along the longitudinal axis of the tape during manufacture, but the process is imperfect and most tapes will support components of magnetization perpendicular to the tape plane that are 30 percent, or more, of the saturation magnetization. This affects the equalization requirements, as will be seen later.

4.2 Recording Modes

4.2.1 The channel

In the usual communication-theory sense, a transfer function does not exist for direct recording without bias. The hysteretic properties of the tape cause the input-output function to be double-valued; in fact, even the shapes of the response and group-delay curves are dependent upon the input-signal amplitude. Furthermore, the noise of the channel is sig-

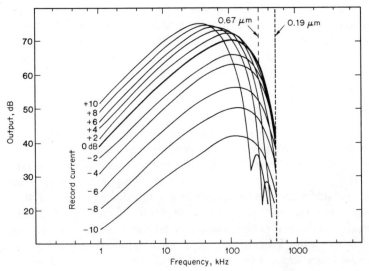

Figure 4.1 Constant-current response curves. A level of 0 dB represents the maximum output at $\lambda = 0.67$ μm. (Tape speed: 19 cm/s; gaps = 0.25 μm.)

nal-dependent. Figure 4.1 shows a family of constant-current response curves with the 0 dB curve obtained by maximizing the output at 0.67-μm wavelength. An increase in record current of 10 dB reduces the bandwidth of the channel by almost two-thirds and causes a phase shift of π radians beyond the null in the response curve. For binary signals, this channel is usable since it has been found that a class of preemphasis exists that allows a detectable invariant during reproduction—the zero crossing (Mallinson and Steele, 1969). The proper preemphasis is that which allows the record-head flux at the gap to reach the tape-saturating level within each bit cell. This mode of recording is used for pulse-code-modulation (PCM) signals.

4.2.2   AC-biased recording

The channel can be linearized by employing ac bias in which a high-frequency current (typically four, or more, times the band-edge frequency) is added to the signal current during recording. The bias current is much larger than the signal current and is adjusted to yield the maximum signal-to-noise ratio during reproduction. For most instrumentation tapes, the bias level is about 2 dB over the level giving maximum output at band-edge. The resulting transfer function is linear over about 30 percent of the dynamic range of the channel for a 1 percent distortion level. The recording sensitivity of the channel is increased by the use of bias when the signal currents are only about 10 percent of the bias current. The transfer function is actually a member of a family of transfer functions determined by the bias level. High bias currents restrict the bandwidth of the channel but permit high signal levels for a given linearity. In high-density recorders, high signal-to-noise ratio at long wavelengths is sacrificed for bandwidth by the use of a relatively low bias current.

The channel is set up by recording at constant flux (head losses equalized) with the bias optimized at band-edge frequency. The maximum record level for linear operation is found by adjusting the record current to yield 1 percent third-harmonic distortion at one-tenth the band-edge frequency. The reproduce channel is then equalized for flat response and the wideband signal-to-noise ratio measured. The bias is increased slightly, after which the record level for 1 percent distortion is reset, the channel reequalized, and the signal-to-noise ratio again measured. By this process, the proper bias for maximum signal-to-noise ratio is found. The constant-current channel is commonly used for telemetry recording, but it exhibits a nonflat signal-to-noise ratio with a maximum at midband. Another approach to setting up the linear channel results in a flat signal-to-noise ratio and a flat response. This technique is called flat noise-power-ratio recording and will be discussed in Sec. 4.3.

4.2.3 Frequency-modulation recording

Frequency-modulation recording was originally developed for telemetry channels where dc response, linearity, and low intermodulation were primary considerations. A relatively large modulation index is used for enhanced signal-to-noise ratio since this ratio increases with the square of the modulation index. In some cases, pilot tones are demodulated from spare channels to allow first-order correction of flutter-induced noise. Frequency modulation is used primarily for the recording of laboratory data when great accuracy is not required but dc response is necessary. Most other applications have been replaced with pulse-code-modulation (PCM) recording.

4.2.4 Pulse-code-modulation recording

The bulk of instrumentation recording, including that done on multitrack recorders, uses the binary mode without bias. Rotary-head recorders employ this technique exclusively and have achieved areal densities greater than 10^5 bits per square millimeter. Many different modulation codes are used in PCM, including run-length-limited (RLL) codes such as the (2, 7) code widely used in computer storage, and duobinary codes, also known as class IV partial response.

Instrumentation recording has pioneered the area of very high linear densities where the channel is dominated by burst errors. In multitrack records it was convenient to interleave the data across the tape width to reduce burst error lengths, since the tracks were statistically independent with regard to dropouts. It is now very common to use Reed-Solomon parallel-interleaved codes for multitrack recorders and such codes in serial interleave for rotary-head instrumentation applications. Raw bit-error rates of 10^{-5} or 10^{-6} are corrected to 10^{-12} with low overhead.

4.2.5 Vector field recording

Considerable evidence exists for the proposition that recording on conventional particulate media involves both longitudinal and perpendicular components of magnetization. The vector nature of the magnetization is detectable in instrumentation recording where small record gaps and relatively low write currents are used, but it also can be seen in the erasure process when an erase head is used. Prominent resonances in erasure as a function of erase-head field have been known for some time (McKnight, 1963) but were not susceptible to analysis with a longitudinal-recording model alone. The inclusion of the perpendicular component is necessary to create interference, the source of the resonances.

The field from a write head is vector in nature and may be broken down into longitudinal and perpendicular components as shown in Fig. 4.2. The

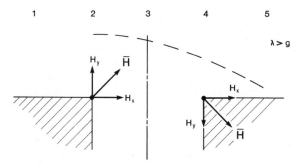

Figure 4.2 Record-head field vectors at gap edges.

peak field at the gap edges has values H_y and H_x as shown. Although the resultant fields are at $\pm 45°$ at the corners, they are always orthogonal and rotate to $0°$ and $90°$ a small distance from the gap edges at the head surface. The magnetic coating of the tape consists of acicular particles, each with a preferred axis of magnetization. The particles are imperfectly oriented so that a significant population has axes canted toward the perpendicular direction. Magnetization measurements on typical tapes reveal about a 30 percent perpendicular magnetization component when the proper demagnetization compensation is taken into account.

It is instructive to consider the field history of an initially unmagnetized particle ensemble near the surface of the tape as it traverses the gap region, as shown in Fig. 4.3. Five regions are indicated: (1) far from the left gap edge, (2) the left gap edge, (3) the gap centerline, (4) the right gap edge, and (5) far from the right gap edge. Considering a wavelength that is long relative to the gap length, the M-H curves are shown for the longitudinal field component and the perpendicular field component. The longitudinal field component is unidirectional, decreasing only slightly at the gap centerline near the head surface. Consequently, the longitudinal magnetization is established by the peak longitudinal field at the left gap edge. The perpendicular magnetization, established by the peak perpendicular field, is created at the right gap edge. Thus the perpendicular and longitudinal magnetizations are separated by a distance approximately equal to the record-head gap length. This leads naturally to interference effects, as will be shown.

The vector fields emanating from a magnetized region of the recording medium are different for longitudinal magnetizations and perpendicular magnetizations, as shown in Fig. 4.4. Both types of magnetization generate longitudinal and perpendicular external fields that are related to each other as Hilbert transform pairs (defined below). Ideally, ring heads respond only to the longitudinal external fields from either longitudinal magnetizations or perpendicular magnetizations (similarly, pole heads respond only to perpendicular external fields).

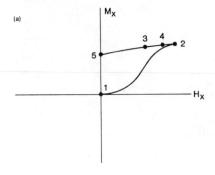

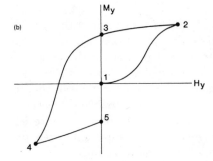

Figure 4.3 Remanent magnetizations near the medium surface, M_x and M_y, induced by (a) the left gap edge and (b) the right gap edge. (See Fig. 4.2 for points 1 through 5.)

The longitudinal-only sensitivity of the ring head derives from the fact that, in a high-efficiency head, the core reluctance is small relative to the gap reluctance, and consequently, most of the magnetomotive force from the external field appears across the gap. The head is equivalent, therefore, to a thin rectangular coil outlining the gap face and lying with its plane in the plane of the gap. Such a coil is insensitive to perpendicular external fields since they do not link the coil, whereas the longitudinal external fields do. A ring head thus senses an even function of external-field intensity ($\cos kx$) from the longitudinal magnetization and an odd function ($-\sin kx$) from the perpendicular magnetization (see Fig. 4.4).

The Hilbert transform of a function is obtained by the convolution of that function with $-1/\pi t$

$$
\begin{aligned}
\hat{e}_L(t) &= -\frac{1}{\pi t} * e_p(t) \\
&= \frac{1}{\pi} \int_{-\infty}^{+\infty} \frac{e_p(t)}{t' - t} \, dt'
\end{aligned}
\tag{4.4}
$$

The Hilbert transform of a function is also equivalent to its Fourier transform with fixed coefficients but with each component rotated by $-\pi/2$.

If separate longitudinal and perpendicular sinusoidal magnetizations were recorded by the same gap edge and, therefore, at the same site in the recording medium, the reproduce head flux would be of the form ($A \cos kx - B \sin kx$), where A and B are the relative magnitudes of the longitudinal and perpendicular fields, respectively. If, however, the components are separated in the direction of tape travel by a distance equal to the gap length, as described above, constructive or destructive interference will result as a function of wavelength. In fact, the optimum record gap length is $\lambda/4$ for maximum signal-to-noise ratio, where λ is the wavelength at band-edge (Lemke, 1982). That is the spacing that puts the cosine function recorded at the leading edge of the record-head gap in phase with the $-\sin$e function recorded at the trailing gap edge. The effect is very pronounced at low recording intensities, as encountered in partial-penetration recording, but less so at higher recording intensities where some of the leading-edge magnetization is altered by the intense trailing-edge field.

As noted above, erasure with an erase head exhibits characteristics that are explainable only though the employment of both magnetization vectors. Figure 4.5 shows the obvious presence of re-recording of a signal with marked resonances during the process of erasure (McKnight, 1963). For the particular recording conditions reported, certain frequencies (2100 Hz, 5000 Hz, and 8000 Hz) exhibit a high degree of erasure that peaks at a critical erase current and diminishes at higher currents. This is due to re-recording the original signal out of phase at the peaks, with the erase field serving as a bias field. The wavelengths at which resonance occurs

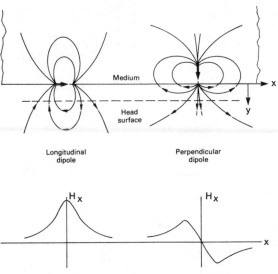

Figure 4.4 Idealized longitudinal field from dipoles rotated $\pi/2$.

are unaffected by erase current and are related to the dimensions of the head gap and the recorded signal.

The source of the re-recorded signal field is the unerased layer that lies deeper in the coating than the depth of penetration of the erase field. The erase field erases a stratum on the surface of the tape and then serves as a bias field for the underlying layers that then function as a signal source. As can be seen schematically in Fig. 4.6, the surface layer is magnetized

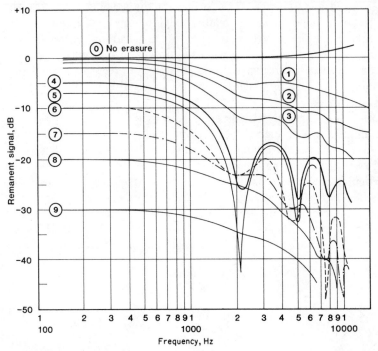

Figure 4.5 AC erasure. Erasure ac current increasing from 0 to 9 *(McKnight, 1963)*.

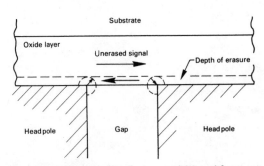

Figure 4.6 Re-recording by overwriting, without ac erasure.

in a sense opposite to that of its source. It is tempting to argue that the out-of-phase flux from the surface layer that is in close proximity to the read head is weak and cancels the contribution of flux from the deeper unerased layer. That is, in fact, the mechanism of overwrite in unbiased recording with thick media using partial-penetration recording, where the peak in overwrite occurs at a wavelength dependent upon the overwriting current. However, such a model fails to account for the fixed critical wavelengths that are independent of the erasure field magnitude and for the multiple peaks shown in Fig. 4.5.

In Fig. 4.2, the vector fields from the gap edges at $\pm 45°$ angles are orthogonal and generate Hilbert-transform-pair magnetizations in the same way that was considered earlier. At low fields near the surface of the head, the field lines are concentrated at the gap edges, as shown in Fig. 4.7, and two distinct erasure zones exist having orthogonal peak field directions. The layer of recording that is not erased by the two zones has an external field that links the erase head. The erase head acts as a very large-gap reproduce head in that flux circulates in its core from the

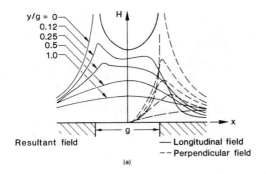

(a)

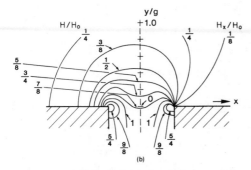

(b)

Figure 4.7 Fields in the vicinity of the recording gap. (a) Field distribution; (b) contours of equal field strength (Duinker, 1957).

recorded tape, but it also acts as a record head in the biased mode with two zones of recording. The magnetization thus recorded can have a large magnitude since the flux is concentrated at the gap corners and the anhysteretic process has a susceptibility of typically about 3. At wavelengths that are long relative to the erase-head gap length, the flux in the core causes a re-recording that is in phase with the unerased signal, and at the wavelength equal to the gap length, no flux links the core. At wavelengths less than the erase-head gap length, the flux in the core can have a reversed polarity and cause gap-edge signals to be recorded that oppose the originating signal, and interference occurs. At wavelengths such that

$$g = \frac{3\lambda}{4}, \frac{7\lambda}{4}, \frac{11\lambda}{4}, \ldots$$

the two Hilbert component signals are in destructive interference, as seen in Fig. 4.8, and a null in the response results. This is, in effect, the condition of zero re-recording and, therefore, the condition of maximum erasure. The erasure peaks will occur at wavelengths such that

$$\lambda_n = \frac{4g}{4n - 1} \tag{4.5}$$

where n is an integer. In the McKnight reference (1963), the tape speed was 38.1 cm/s, the gap of the erase head was 133 μm (108 μm of mica and two 12.7-μm glue lines), and the critical frequencies were 2100 Hz, 5000 Hz, and 8000 Hz. The first three values of n give frequencies of 2142 Hz, 5000 Hz, and 7857 Hz, in very good agreement with the experimental results. Thus, the erasure process gives evidence of the importance of both

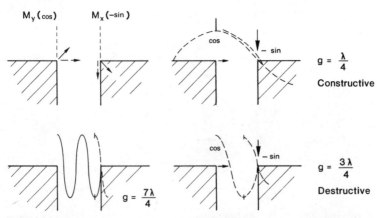

Figure 4.8 Constructive and destructive interferences from Hilbert pairs of magnetization.

longitudinal and perpendicular components in analyzing the operation of a recording channel.

The simultaneous presence of phase-separated Hilbert transform pairs in instrumentation recording has a significant impact on a number of performance characteristics of the recording channel. It affects the shape of the response curve in constant-current recording, and it is a free parameter that can be used to increase the band-edge signal-to-noise ratio. It poses a problem, however, in equalizing linear channels, as will be discussed in Sec. 4.3, and in analyzing the spectral response of a channel using a single-transition magnetization change and Fourier analysis. The perpendicular components are masked by demagnetization of the semi-infinite magnetization pattern of the single transition, with the result that the channel does not follow the model at short wavelengths where the perpendicular demagnetization is much reduced. The analysis problem can be obviated by employing a triplet or missing-pulse response rather than a single transition.

4.3 Equalization and Signal-to-Noise Ratio

4.3.1 Head losses

Record-head loss refers to the reduction in available flux at the gap as a function of frequency for constant-current head excitation. According to the reciprocity principle, reproduce heads suffer the same loss in sensitivity as they would as record heads. These losses are easily measured and are compensated for by preequalization and postequalization.

Reproduce losses are measured by recording a tape at low speed with two frequencies, band-edge and one-half band-edge, and then reproducing it at speeds that are doubled over many octaves. With each speed-doubling, the one-half-band-edge signal reproduces at the frequency of the former band-edge signal and should exhibit no frequency losses. Losses that do appear are due to misalignment at the higher speed or, more usually, to spacing loss induced by the tape flying. The head loss is the residual loss after correction for misalignment and spacing loss at each frequency. Record-head losses are measured similarly—by reproducing at a fixed low speed and doubling the record speed over the same range.

Typical instrumentation multitrack heads with ferrite cores and metal tips exhibit about 3 dB of loss at 2 MHz and about 10 dB at 10 MHz. Although bias currents are about 10 times the signal current, the bias-induced increase in recording sensitivity is only about a factor of 3 when head losses are accounted for; this is the value of typical anhysteretic susceptibility for oxide media.

4.3.2 Perpendicular and longitudinal field components

A constant-flux recording results in a response that typically rises somewhat less than 6 dB per octave to peak, where it then rolls off in an ever-increasing slope to a band-edge that is usually chosen such that the peak-to-band-edge loss is less than 20 dB. The frequency equalization of this response to one that is flat is not difficult, but it does introduce phase shift that must be removed. The classical equation for reproduction does not contain a phase-shift term since it assumes a constant magnetization as a function of depth into the tape (Wallace, 1951). Real recordings, however, generate a leading phase with frequency that must also be equalized. The leading phase arises from the wavelength-dependent effective depth of recording and the curvature of the recording field produced by a ring-type head. The shortest-wavelength signals reside on the surface of the tape, whereas the centroid of the longer-wavelength signals is deeper and, therefore, retarded.

A channel equalized for phase and amplitude yields an unexpected square-wave response, as seen in Fig. 4.9. The rounding of the trailing edge is not derivable from the band-limited Fourier components of the square-wave but, rather, is due to the Hilbert transform components, discussed earlier, that are generated by the perpendicular magnetization. The effect of the Hilbert transform is to rotate each Fourier component through $\pi/2$ while keeping the coefficients unchanged. A Fourier sine expansion of a square wave (Fig. 4.10a) will lead to a Hilbert expansion in cosine terms (Fig. 4.10b) if each component is rotated by $\pi/2$. If the Hilbert terms are weighted by 30 percent of the Fourier terms and the components are band-limited, a waveform results that is similar to Fig. 4.9.

In some instrumentation recorders, the effect of the perpendicular component is compensated somewhat by differentiating the signal, inverting it, and adding it to the original signal to compensate for the unwanted trailing-edge droop. This technique works well at only one frequency but

Figure 4.9 Square waves showing in-phase Fourier and Hilbert components with properly phase-equalized channel and conventional tape ($M_y/M_x \approx 0.4$).

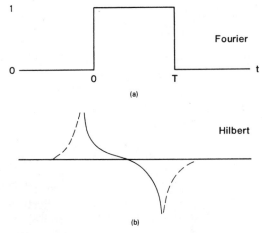

Figure 4.10 Hilbert transform of a square wave.

gives acceptable results over a modest range of frequencies. A channel equalized in this way can be misleading when its phase response is judged from its square-wave waveform. The phase response of the channel is actually highly distorted to accommodate at one square-wave frequency the Fourier and Hilbert components that cannot, in general, be equalized simultaneously to give a square wave at all frequencies. A properly equalized channel yields a waveform as in Fig. 4.9.

The existence of both longitudinal and perpendicular magnetization components also results in a modification to the response curve that affects equalization (Lemke, 1982). Interferences between the leading and trailing edges of the record-head gap change the shape of the response curve and are responsible for the fact that the unequalized response initially rises at 5.5 dB per octave, rather than the 6 dB per octave expected from simple theory.

4.3.3 Multiplicative noises

The dominant sources of noise in most instrumentation channels are multiplicative rather than additive. Multiplicative noise is caused by coating inhomogeneities, variations in spacing and azimuth loss, and variations in time-base error (flutter). Multiplicative errors are more important to instrumentation recording than to computer recording owing to the much shorter wavelengths typically used in the instrumentation field. Also, error-concealment techniques cannot be used, as they are in audio or video recording.

Modulation noise derives its name from the fact that it is evident only in the presence of a signal and appears as a modulation of that signal. It exists in few recording systems, photographic film being the primary other medium that exhibits the effect. Its existence complicates the analysis of a channel since information theory assumes that signal and noise are independent functions, an assumption which is grossly incorrect for a magnetic recording channel.

Modulation noise has two sources associated with the magnetic coating: bulk inhomogeneities and surface finish (including asperities). The bulk inhomogeneities can be controlled somewhat through the use of dispersants and careful milling of the magnetic dispersion prior to coating. The spectrum of this kind of modulation noise is predominantly long-wavelength. The surface-induced noise is much more difficult to control and has a severe effect on signals at short wavelengths where the highest information density resides. As noted in Eq. (4.2), 90 percent of the signal is contributed by a thin layer of depth λ/e. For a 0.67 μm wavelength, 90 percent of the signal resides in a layer of thickness corresponding to the wavelength of ultraviolet light. Even the best optical surfaces, if achievable on the tape surface, would have measurable modulation noise that would have to be considered in designing an error-detection and error-correction method.

Magnetic recording at wavelengths usual in instrumentation recorders represents an unusual channel from a communication-theory point of view. As system design is currently practiced, the channels enjoy an embarrassment of riches in average signal-to-noise ratio but suffer from the worst-case modulation-noise limit. Much more must be done to push the instrumentation-channel density further to an additive-noise limit, and a portion of the increased capacity should then be spent to implement error-detection and error-correction codes tailored to the peculiarities of instrumentation recording.

4.3.4 Noise-power ratio

The noise-power ratio (NPR) is a critical measurement in telephone systems since it measures not only the thermal noise in a channel but also all of the intermodulation components present from other channels. It was adapted to instrumentation recording for use in surveillance recorders which are required to detect weak signals in the presence of strong signals over a wide bandwidth. Equal probability of error per root-Hertz is achieved by making the NPR constant over the channel bandwidth. The resultant channel satisfies the Shannon equation in its simplest form

$$C = 2B \log_2 (1 + \text{SNR}) \tag{4.6}$$

where C = information capacity in bits per second
  B = bandwidth of the channel
  SNR = the signal-to-noise ratio.

Flat noise-power ratio in magnetic recording is achieved solely through preemphasis and is an iterative process. It is measured by recording white noise in a biased channel that has a frequency-selectable, narrow-notch filter. During reproduction, a narrower-band selective voltmeter is used to measure the thermal noise and the intermodulation products that have been created in the band from all of the noise sources outside the band. The ratio of the noise signal in the vicinity of the notch to the noise in the notch is the noise-power ratio at that frequency (Fig. 4.11).

Preemphasis at band-edge to levels above the record current that causes 1 percent third-harmonic distortion, will result in a greater signal-to-noise ratio at band-edge, but the harmonics of that signal, although not reproducible, will intermodulate with in-band frequencies and lower the signal-to-noise ratio at midband. Some loss at midband can be tolerated, since that is the region of maximum signal-to-noise ratio, but too much band-edge preemphasis must be avoided. At low frequencies, considerable preemphasis is possible since the intermodulation products are also of low frequency and do not affect the higher frequencies. By moving the selective filters across the band and iterating the adjustments in bias and preemphasis, it is possible to equalize a channel optimally. It is also possible, in principle, by this technique to adapt the signal-to-noise ratio (including intermodulation products) of a channel to match closely the optimum channel-response curve for a particular code. The transfer characteristics of a flat-noise-power-ratio channel, relating noise-power ratio to input signal level, is given (Tant, 1974) by

$$\text{NPR} = 10 \log \frac{P}{P_0} - 10 \log \left[\left(\frac{P}{P_0} \right)^{\gamma} + (\gamma - 1) \right] + 10 \log \gamma \qquad (4.7)$$

where P_0 = power at the maximum value
P = relative power
γ = order of distortion

The value of γ is 3 for magnetic recording which is dominated by third-order harmonic distortion. As shown in Fig. 4.12, the noise-power ratio rises 1 dB for each decibel of input below the maximum and falls 2 dB

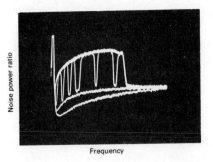

Noise power ratio

Frequency

Figure 4.11 Flat noise-power ratio showing multiple notches. The NPR is constant for all frequencies independent of any subsequent postequalization for flat response.

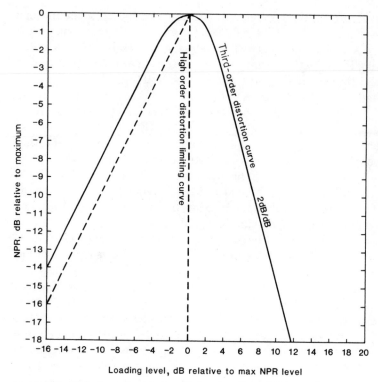

Figure 4.12 Noise-power ratio for third-order distortion in an instrumentation tape channel.

per decibel of input above the maximum, as can be seen in the equations below.

Linear region: $(P/P_0)^\gamma \ll (\gamma - 1)$

$$\text{NPR} = 10 \log \frac{P}{P_0} + 10 \log \frac{\gamma}{\gamma - 1} \tag{4.8}$$

Nonlinear region: $(P/P_0)^\gamma \gg (\gamma - 1)$

$$\text{NPR} = 10 \log \gamma - (\gamma - 1)\, 10 \log \frac{P}{P_0} \tag{4.9}$$

The signal-to-noise ratio relative to that at the maximum for the linear and nonlinear regions, respectively, is

$$\text{SNR} = -10 \log \frac{\gamma}{\gamma - 1} \tag{4.10}$$

and

$$\text{SNR} = \gamma \left[10 \log \frac{P}{P_0} \right] - 10 \log \gamma \tag{4.11}$$

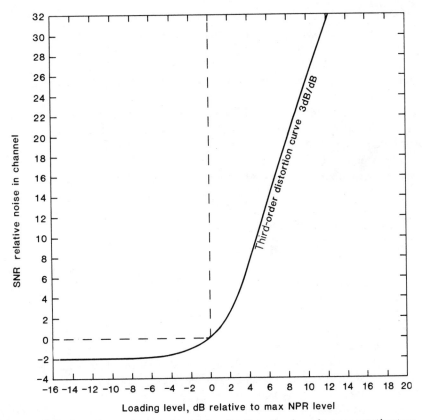

Figure 4.13 Signal-to-noise ratio for third-order distortion in an instrumentation tape channel.

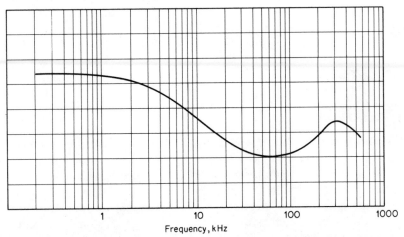

Figure 4.14 Flat noise-power-ratio preemphasis curve for a record gap length of 0.25 μm (300 kHz; 19 cm/s).

In the linear region, the signal-to-noise ratio is unaffected by adjacent channels and is limited only by thermal noise. Above the maximum noise-power ratio, however, the signal-to-noise ratio deteriorates 3 dB for each decibel of input. At the maximum, the thermal-noise power is twice the intermodulation power. Figure 4.13 shows input-output curves for signal-to-noise ratio.

A typical preemphasis curve for flat noise-power ratio is shown in Fig. 4.14. It is interesting to note that much of the total equalization is in the record channel, in contrast to constant-flux recording where it is all in the reproduce channel. For flat noise-power-ratio recording, the reproduce channel is equalized only to facilitate detection of the signal.

4.3.5 High-density digital recording

High-density digital recording (HDDR) is usually equalized by integrating the reproduced signal and then phase-compensating the channel for the best eye pattern and maximum detection-window margin. Final window-margin adjustments are made with a time-interval analyzer which plots histograms of the zero-crossing times. The channel is then tested by recording a pseudorandom sequence which is reproduced through a bit synchronizer and analyzed for bit- and burst-error rates. Raw bit-error rates are typically on the order of 10^{-6} for narrow-track high-rate channels. Error detection and correction is then used to correct to the desired final error rate, which may be as low as 10^{-12}. Overheads of from 5 percent to 20 percent are used.

Some recorders use playback equalization to achieve class IV partial response characteristics (Kretzmer, 1966). The transfer function of such a channel matches somewhat the general response of magnetic recording with a null at zero frequency and at the bit rate. It results in a three-level response at the detector.

4.4 Tape Transport

4.4.1 Fixed heads

The earliest instrumentation tape-transport designs adapted from professional audio recorders have been replaced with new transport concepts to meet the more stringent requirements of data recording. Although many formats have been tried in the quest for error-free tape transport, all of them are flawed to various degrees. Unfortunately, the uniform transport of an elastic medium such as tape is a very difficult task, and no format is as free from disturbances as would be desired. Some of the formats that have been built are shown schematically in Fig. 4.15.

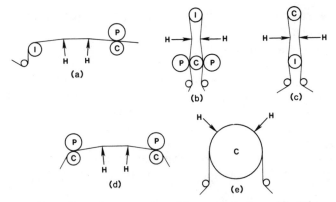

Figure 4.15 Several common tape-drive configurations. All of them transmit input-reel flutter disturbances to the head region. (C = capstan; H = head; I = idler; P = pinch roller.)

The professional audio configuration of Fig. 4.15a isolates the head area from the take-up-reel disturbances, but the tension variations from the supply reel are magnified since the tension at the input to the capstan (and, therefore, at the last head) is increased by the wrap around the heads. The friction at the head-tape interfaces, μ, increases the tape tension over a combined head-wrap angle of θ by a factor of $\exp(\mu\theta)$. For typical wrap angles of 10° per head and a friction coefficient of 0.2, the input-reel tension disturbances are increased by about 15 percent across the head stack.

Several other configurations employed in instrumentation recorders are shown in Fig. 4.15b, c, d, and e. Figure 4.15b shows the closed-loop approach that was developed as an attempt to isolate the heads from reel disturbances by clamping the tape to the capstan surface by two pinch rollers. Variations of this approach used a stepped capstan and pinch rollers to establish a fixed-tension differential. Figure 4.15c has no pinch rollers, since the capstan has a large wrap angle. The free idler is coated with a high-friction layer so that there is no slippage between it and the tape after the tape is at speed.

Figure 4.15d shows a two-capstan drive with differential speed between the two capstans to maintain tension at the heads. Usually, pinch rollers are used to prevent tape slippage. Figure 4.15e shows a typical capstan arrangement designed to permit high-speed servoing of the tape speed for time-base error correction. Frequently the tape is lifted off the vacuum capstan locally to permit improved head-to-tape contact relative to what is possible with the head urged against tape in contact with the capstan surface. Various devices for achieving compliance have been interposed between the capstan-head region and the reels, including vacuum columns and moving arms. Frequently the reel speed is servoed from the compliance structure to maintain constant tape tension. The introduction of compliance is only partially effective in preventing reel disturbances from entering the head area.

The operation of a tape transport may be analyzed simply by utilizing the conservation of mass in considering the flow of tape into and out of imaginary closed regions of the transport. For example, consider the dashed-line boundary in Fig. 4.16. The mass flow into region 1 must equal the mass flow out of region 3. Other regions can be placed in the tape path, and a series of simultaneous equations can be arrived at for complete analysis of the tape motion.

Defining a modulus of strain per unit length and tension as $\Omega = \Delta L / LT$, the density per unit length of tape relative to the initial density ρ_0 is

$$\rho = \frac{\rho_0}{1 + \Omega T} \tag{4.12}$$

The tape velocity entering the capstan at region 1 has the capstan velocity of V_1 and the rate of mass flow into the region is

$$\frac{dm}{dt_1} = \frac{V_1 \rho_0}{1 + \Omega T_1} \tag{4.13}$$

Similarly, the mass flow out of region 3 is

$$\frac{dm}{dt_3} = \frac{V_3 \rho_0}{1 + \Omega T_3} \tag{4.14}$$

Since the two tape paths are driven by the same capstan surface, $V_1 = V_3$ and $T_3 = T_1$. Thus, the tension inside the closed loop is governed by the input tension outside the loop and is, in fact, exactly equal to T_1,

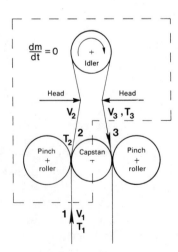

Figure 4.16 Closed-loop tape-drive configuration. Mass flow into the dashed region equals the mass flow out of the region.

including any disturbances on that tension. The differential tape length that has been created by $T_1 + \Delta T_1$ in the immediate vicinity of the input side of the capstan is clamped to the capstan by static friction generated by the pinch roller and is transported to region 2, where it is released into a new tension environment. A transition from static friction to dynamic friction occurs between the tape and the capstan surface at region 2 as the pinch-roller force is reduced. This can excite longitudinal vibrations in the tape that appear in the form of high-frequency flutter.

All tape-drive configurations experience flutter at the head sites from outside stimuli independently of the method of capstan drive, unless the tape tension is forced to go to zero just prior to the input capstan that precedes the head site. Some configurations are worse than others. Consider the design shown in Fig. 4.17. The output capstan is designed to have a surface velocity slightly higher (by a factor δ) than that of the input capstan; this establishes tape tension in the head region. The two regions are where the symbols V_1, T_1 and V_2, T_2 appear in Fig. 4.17, and the velocities in these regions are related by $V_2 = V_1(1 + \delta)$. The mass flow through the dashed region requires that

$$\frac{V_1 \rho_0}{1 + \Omega T_1} = \frac{V_1(1 + \delta)\rho_0}{1 + \Omega T_2} \tag{4.15}$$

With an input tension at region 1 of $T_1(1 + \Delta)$, where Δ is the fractional amount of tension disturbance, the tension at region 2 is

$$T_2 = \frac{\delta}{\Omega} + T_1(1 + \delta)(1 + \Delta) \tag{4.16}$$

$$\Delta T_2 = \Delta T_1(1 + \delta) \tag{4.17}$$

This indicates that variations in the input tension T_1 are transmitted into region 2 and magnified by $1 + \delta$.

Thus, any disturbance at the input capstan propagates through the head region undiminished. If the tape tension does not go to zero in the

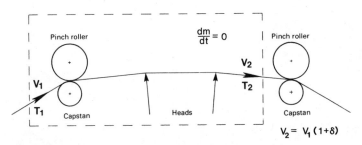

Figure 4.17 Dual capstan drive configuration.

tape path between the supply reel and the input capstan, the relatively large sources of disturbances associated with the reel and tape pack will appear at the heads, although diminished somewhat by any compliance inserted in that path.

The action of the compliance can be seen from examining a simple pulley and tape arrangement, as shown in Fig. 4.18. The massless pulley is attached to a spring with force constant k that exerts a tension in the tape at a position x of $kx/2$. The tape is assumed to be delivered to the pulley from the reel at an average velocity V_r and removed from the pulley by the capstan at a fixed velocity V_c equal to V_r. A small sinusoidal velocity component in V_r would cause a peak change in the length of tape of

$$\Delta L = \int_0^{T/4} \Delta V_4 \cos \omega t \, dt \tag{4.18}$$

The mechanical advantage is 2, therefore $\Delta L = kx/2 = \Delta V_r/2\omega$ and

$$\Delta T = \frac{\Delta V_r k}{\omega} \tag{4.19}$$

The disturbances are, therefore, reduced by a small spring constant k and are largest at low frequencies, decreasing at 6 dB per octave. In practice, the inertia of the compliant arm permits disturbances at frequencies above several tens of hertz to pass unattenuated by the compliance. Also, the compliant device is usually a rotating arm with mechanical advantage much less than shown in Fig. 4.18.

Frequently, the position of compliant arms or tape positions in vacuum columns are used to servo the supply and take-up reel speeds to achieve constant tape tension, particularly on the supply side. Except for the lowest-frequency components that can be handled by the servo bandwidth,

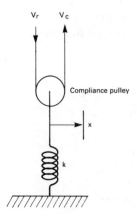

Figure 4.18 Compliance model.

some of the reel disturbances still get through, since the vacuum column has an effective spring constant associated with its loop-position servo.

Vacuum columns serve a very useful function in edge-guiding tape; this function is unrelated to their role in tape tensioning. During manufacture, all tapes are slit from wide webs by rotary knives. The tapes exhibit a small weave related to the circumference (typically 20 cm) of the knives. If the column is tapered to encourage the tape to ride on one edge, and if the length of tape in the column exceeds twice the knife circumference, the tape in the column, which is folded and relatively stiff, will ride continuously on the two peaks of the slitting weave. Reproducible tracking to several microns, as evidenced by output envelope, can be maintained between transports of the same type using this technique.

Low-frequency disturbances that enter the head region can be removed through servoing the speed of a low-inertia capstan with a reference signal on the tape. On sophisticated transports, such servos are used to reduce the time-base error (TBE) to the order of 10 ns. The time-base error is the time integral of the velocity flutter and is often of more interest than is the instantaneous velocity of the tape. Most high-frequency disturbances contribute little to the time-base error but can cause signal errors by modulating signals with unwanted sidebands. So-called scrape flutter is induced by abrupt large tension transitions, as noted above, and by friction at fixed surfaces such as the heads.

4.4.2 Rotary heads

The primary advantage of rotary-head transports over fixed-head transports is their greater areal information packing density. For tape-noise-limited systems, the signal-to-noise ratio is reduced 3 dB for each track-width halving, but the bandwidth is doubled since two tracks can be accommodated in the same area. The net gain is 3 dB in information-handling capacity each time the trackwidth is halved. Rotary head transports currently handle well over 6×10^5 bits per square millimeter, and are theoretically capable of densities well beyond optical limits. They are restricted by head constraints in the bandwidth they can handle in one track. Bandwidths beyond about 300 Mb/s per track are difficult to achieve and will probably require multiple rotary heads.

Typically, two servos (and sometimes a third—a tension servo) are employed to synchronize the linear tape advance and the head position in its rotary mount to ensure that the tracks are positioned on the tape in a reproducible, nonoverlapping pattern that can be duplicated during reproduce. The basic head-tape geometry is very similar to the video formats, and, in fact, many of the rotary-head instrumentation transports have been adapted from professional video recorders.

References

Bertram, H. N., and R. Niedermeyer, "The Effect of Spacing in Magnetic Recording," *IEEE Trans. Magn.* **MAG-18,** 1206 (1982).

EMI engineering staff, *Modern Instrumentation Recording,* EMI Technology, London, England, 1978.

Howard, J. A., and L. N. Ferguson, "Magnetic Recording Handbook," Hewlett Packard Application Note 89, Palo Alto, Calif., 1970.

Kretzmer, E. R., "Generalization of a Technique for Binary Communication," *IEEE Trans. Commun.,* **14,** 67 (1966).

Lemke, J. U., "An Isotropic Particulate Magnetic Medium with Additive Hilbert and Fourier Field Components," *J. Appl. Phsy.,* **53,** 2561 (1982).

Lemke, J. U., "Ferrite Transducers," *Ann. N. Y. Acad. Sci.,* **189,** 171 (1972).

Mallinson, J. C., and C. W. Steele, "Theory of Superposition in Tape Recording," *IEEE Trans. Magn.* **MAG-5,** 886 (1969).

McKelvey, J. P., "General Kinematics of Winding Processes," *Am. J. Phys.,* **53,** 1156 (1985).

McKnight, J. G., "Erasure of Magnetic Tape," *J. Audio Eng. Soc.,* **11,** 223 (1963).

NASA, *High Density Digital Recording,* NASA Reference Publication, 1111, Washington, D.C., 1985.

Star, J., "Specifying Phase Response and Envelope Delay in an Instrumentation Tape Recorder," *Telemetry J.,* **5,** 17 (1970).

Tant, M. J., *The White Noise Book,* White Crescent Press, Luton, England, 1974.

Wallace, R. L., "The Reproduction of Magnetically Recorded Signals," *Bell Syst. Tech. J.,* **30,** 1145 (1951).

5

Signal and Error-Control Coding

Arvind M. Patel

IBM Corporation, San Jose, California

5.1 Introduction

A magnetic recording channel is designed to accept data for storage and deliver the same on demand at a later time, with reasonable access time and without errors. These data may be in the form of fixed- or variable-length records corresponding to user-identified blocks which can be updated or redefined at any time. In a typical magnetic recording channel, the incoming data are first encoded with an error-correction code with a well-defined data format. The data format is designed to fit the requirements imposed by the device architecture, and the error-correction code is designed to protect the data against commonly encountered error modes. A second encoder then converts data into an analog signal in accordance with a waveform encoding technique or a modulation code. This code is designed to fit the requirements of the magnetic recording channel for high packing density and error-free detection of the recorded data.

Error-control coding and modulation techniques have played a significant role in the design and development of digital magnetic recording products. The trend towards higher densities and data rates has presented continuing demands for innovations and improvements in signal coding and error-control techniques. The ability to operate at a lower signal-to-noise ratio and to tolerate an increased number of correctable errors has become part of the design strategy. As a result, the cost of storage is reduced significantly through defect tolerance and higher yield in manufacturing, while very high data integrity, as seen by the user, has been provided.

In this chapter, we present summary results on signal encoding techniques and the theory of error-correction coding in relation to the requirements of digital magnetic recording channels. We also present examples of the modulation codes and error-correction schemes used in various magnetic tape and disk products. The chapter is not meant to be a survey of all the coding results used in magnetic recording. However, it will present a variety of results and help the reader to gain a sufficient hold on many useful concepts without requiring a background in the mathematical development of the theory.

5.2   Signal-Coding Methods

Binary information is recorded on magnetic media as magnetic transitions in the form of a coded binary waveform. For example, a 1 and a 0 of the binary data can be identified by an up and a down level in the waveform, respectively. This mapping of the waveform is known as the *non-return-to-zero* (NRZ) data-encoding method. Alternatively, a 1 and a 0 of the binary data may be represented by a presence and an absence of a transition in the binary waveform, respectively. This mapping of the waveform is known as the *modified non-return-to-zero* (NRZI) data-encoding method. In digital magnetic recording, each magnetic transition on the medium generates a corresponding electric pulse in the readback signal. An error in detecting this pulse will result in a 1-bit error in NRZI-coded data. In contrast, the same error will propagate indefinitely in NRZ-coded data since a transition in the NRZ waveform represents a switch from a string of 1s to a string of 0s, or vice versa. Thus, NRZI is the commonly accepted encoding method in digital magnetic recording.

The principal drawback of NRZI is that long strings of 0s are recorded as long periods with no magnetic transitions, and hence no pulses in the readback signal; the read clock can lose synchronization during these periods. Furthermore, circuits with dc response are required for signal processing and data detection. The NRZI method is also subject to a more subtle difficulty. At high densities, the readback pulses produced by a string of consecutive 1s tend to interfere with one another. When such a string of 1s is preceded or followed by a string of consecutive 0s, this interference is asymmetrical, causing the first or last few pulses to have larger amplitudes; thus the baseline appears to drift. A similar shift can arise from a string of consecutive 0s if the dc response of the electronic circuitry is less than adequate. Interference also causes the pulses to seem to slide into the signal-free zone occupied by the 0s, creating a displacement in time of the pulse peaks. This peak shift can be a significant fraction of the nominal time between bits. These problems have led to a variety of coding modifications being made to the basic NRZI method, as shown in Fig. 5.1.

Phase encoding (PE) was devised to alleviate some of these problems. It is used on $\frac{1}{2}$-in tapes recorded at 1600 bits per inch. In PE, a 1 corresponds to an up-going transition and a 0 to a down-going transition at the center of the bit cell. Where two or more 1s or 0s occur in succession, extra transitions are inserted at the bit-cell boundaries. The resulting waveform is self-clocking and has no dc component, since the waveform in each bit cell has up (positive) and down (negative) signal levels of equal duration. However, PE requires twice the transition density of the NRZI method for a random data pattern. Thus recording efficiency is poor.

Frequency modulation (FM) has transitions at every bit-cell boundary. FM is similar to PE in all waveform properties, although the 1 and 0 correspond to a presence or absence of a transition rather than to an up or down transition at the center of the corresponding bit cell. In certain applications, FM has been supplanted by delay modulation, or MFM, which provides enough clocking transitions without doubling the transition density. In MFM, as in FM, a 1 and a 0 correspond to the presence and absence, respectively, of a transition in the center of the corresponding bit cell. Additional transitions at the cell boundaries occur only between bit cells that contain consecutive 0s. This method retains an adequate minimum rate of transitions for clock synchronization without exceeding the maximum transition density of NRZI. A disadvantage is the

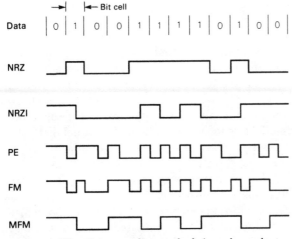

Figure 5.1 Waveform encoding methods in early products.

dc component, with indefinitely large accumulated dc charge for some data patterns, such as 0110110110. . . .

Modified forms of NRZI include synchronized NRZI (NRZI-S) and group-coded recording (GCR). In these methods, the data stream is precoded by adding bits to break up long strings of 0s. In NRZI-S, a 1 extends each 8-bit group to 9 bits and establishes a synchronization transition. In GCR, each 4-bit group is mapped into 5 bits using a fixed assignment that guarantees that the coded data stream never contains more than two consecutive 0s. The detection process for both methods is the same as that for NRZI, but maximum transition density is necessarily higher. These methods marked the beginning of precoding the binary data with run-length-limiting codes for desired properties in the resultant waveform. These run-length-limiting codes are also called *modulation codes*.

5.3   Modulation Codes

A modulation code for magnetic recording is a one-to-one mapping of binary data into a constrained binary sequence which is then recorded on the magnetic recording medium in the form of an NRZI waveform. In this waveform, the maximum and minimum spaces between consecutive transitions correspond to the maximum and minimum run lengths of 0s betweend two consecutive 1s in the corresponding binary sequence. Thus, the modulation codes for magnetic recording fall into the class of run-length-limited (RLL) codes. These codes are characterized by the parameters (d, k) where d represents the minimum and k represents the maximum number of 0s between two consecutive 1s in the coded sequence (Franaszek, 1970). The parameter d controls the highest transition density and the resulting intersymbol interference. The parameter k controls the lowest transition density and ensures adequate frequency of transitions for synchronization of a read clock.

The run-length constraints, however, are not the only consideration if the channel uses ac-coupling network elements in processing the read-write signals. For example, in a rotary-head system, the head is coupled through a transformer. In that case, a dc component in the waveform results in a nonzero average value of the amplitude and causes charge accumulation at the ac-coupling element. A constraint on the maximum accumulated charge is an effective means of reducing the signal distortion caused by the ac-coupling network element. In the case of a binary coded sequence, the accumulated charge increases or decreases by one unit at each digit depending on an up or down level, respectively, in the corresponding waveform. A run-length-limiting code with charge constraint is characterized by the parameters $(d, k; c)$ where d and k are the previously described run-length constraints, and the accumulated charge at any digit position in the coded sequence is bounded by $\pm c$ units (Patel, 1975).

If, on average, x data bits are mapped into y binary digits in the coded sequence ($y \geq x$), the ratio x/y is called the *rate of the code*. A run-length-limiting code is completely described by its rate and the code parameters, written as "x/y (d, k)" code, or "x/y ($d, k; c$)" code. The following are some of the important characteristics of these codes.

1. *Density ratio.* A higher value of d provides greater separation between consecutive transitions in the recorded waveform, although with some tradeoff in the rate of the code. The ratio of the data density versus the highest density of recorded transitions is a measure of recording efficiency and is called the *density ratio* (DR) given by:

$$\text{DR} = \frac{\text{data density}}{\text{transitions density}} = \frac{x}{y}(d + 1)$$

2. *Detection window.* The read waveform is processed through the channel where the presence or absence of a transition within each signaling element is detected using a variable-frequency clock. The available time for detection, called the *detection window*, is determined by the width of the signaling element in terms of the data bit cell. This is completely determined by the rate of the code.

$$\text{Detection window} = \frac{x}{y} \text{ (width of the data bit cell)}$$

3. *Maximum/minimum pulse width ratio.* The ratio of maximum pulse width to minimum pulse width is related to the resolution or uniformity in the resulting read waveform and the corresponding peak-shift problem when a series of long-wavelength pulses follow a short-wavelength pulse or vice versa. This ratio is also related to the worst-case crosstalk between two adjacent tracks, which occurs when long-wavelength pulses in one track produce interference into short-wavelength pulses in the next track, or vice versa. This ratio is given by:

$$\frac{\text{Maximum pulse width}}{\text{Minimum pulse width}} = \frac{k + 1}{d + 1}$$

The rate x/y of the code is bounded above by the information capacity of the constrained sequences (Shannon, 1948). The information capacity of the unconstrained binary sequences is 1 bit per digit, and that of constrained sequences is necessarily less than 1 bit per digit. This capacity is a strong function of the constraints and can be determined for any given (d, k) or ($d, k; c$) parameters. Table 5.1 lists the capacities and possible rates corresponding to some (d, k) and ($d, k; c$) parameters (Franaszek, 1970; Patel, 1975). Figure 5.2 provides an insight into the interrelationship of various code parameters; the figure shows that the density ratio progressively increases with larger values of the parameter d, although the capacity decreases. It also shows how the density ratio is affected as one allows a wider range of pulse widths. The ratio $(k + 1)/(d + 1)$ determines this range and is strongly related to the amount of peak shift of adjacent transitions, crosstalk between adjacent tracks at high linear and track densities, and overwrite noise. The maximum rate and density ratio are not always realizable with a practical algorithm in all (d, k) codes. A good algorithm realizes a rate that is close to the information capacity of the constrained sequences, uses a simple implementation, and avoids the propagation of errors in the decoding process.

Codes with a fixed word length can be developed through an exhaustive search of code words which can be catenated without violating the desired constraints. A look-up table may be used; however, in some cases a comprehensive word assignment can be obtained to create logic equations for encoding and decoding, as is the case with an $\frac{8}{9}$ (0, 3) block code (Patel, 1985a). Sometimes a design with variable-length code words provides shorter word lengths with reduced error propagation, as occurs with the $\frac{1}{2}$ (2, 7) code (Franaszek, 1972). Low-rate codes can be designed by creating isomorphic state diagrams (Patel, 1975) to represent the constrained and unconstrained sequences. In general, this provides an iterative algorithm with some look-ahead or look-back requirements, or both, as occurs with a $\frac{1}{2}$ (1, 3; 3) code, otherwise known as *zero modulation code*. Zero modulation also provides a case study of a code with infinite look-ahead; one in which the rate approaches the information capacity of the sequences, and yet the error propagation remains zero. When the look-ahead requirement

TABLE 5.1 Information Capacity of Coded Sequences

Parameters			
d	k	c	Capacity $\geq$ rate
0	3	$\cdots$	$0.947 > \frac{8}{9}$
1	3	$\cdots$	$0.552 > \frac{1}{2}$
1	3	3	$0.500 = \frac{1}{2}$
1	6	$\cdots$	$0.669 > \frac{2}{3}$
1	7	$\cdots$	$0.679 > \frac{2}{3}$
1	7	10	$0.668 > \frac{2}{3}$
2	7	$\cdots$	$0.517 > \frac{1}{2}$
2	7	8	$0.501 > \frac{1}{2}$
2	8	7	$0.503 > \frac{1}{2}$
3	7	$\cdots$	$0.406 > \frac{2}{5}$
4	9	$\cdots$	$0.362 > \frac{1}{3}$
5	17	$\cdots$	$0.337 > \frac{1}{3}$

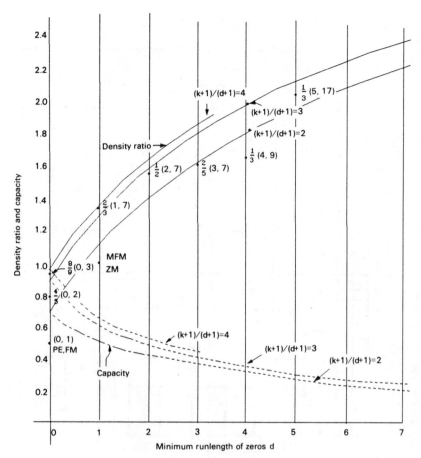

Figure 5.2 Density ratio and capacity of coded sequences.

is finite and reasonable, the isomorphism of the state diagrams can be obtained easily through a step-by-step procedure (Adler et al., 1983).

Many different modulation codes are in use in various magnetic tape and disk products. These codes are the result of a steady evolution of waveform design along with that of the high-performance magnetic recording channel including the clocking and signal-detection process. In modern machines, the choice of code parameters involves trade-offs (MacKintosh, 1980) between the various properties of a code and the characteristics of a particular device. Table 5.2 lists these codes with their parameters. The earlier data-encoding methods NRZI-S, PE, FM, and MFM are also identified in this table with (d, k) parameters in order to put them in perspective. GCR is a $\frac{4}{5}$ (0, 2) block code used in tape machines recording at 250 bits per millimeter (6250 bits per inch) in 9-track standard data format. This code is replaced by a new, more efficient

TABLE 5.2 Modulation Codes Used in Magnetic Recording Products

Code	RLL d,k	Rate	Range of pulse widths in waveform [†]	PW ratio (max/min)	Density ratio	Detection window	Max DC charge	Usage
NRZI	0,∞	1		∞	1	1	∞	800 bpi tape
NRZI-S	0,8	8/9		9	0.89	0.89	∞	13xx disk
PE	0,1	1/2		2	0.5	0.5	T	24xx tape 23xx disk
MFM (Miller)	1,3	1/2		2	1	0.5	∞	33xx disk
GCR	0,2	4/5		3	0.8	0.8	∞	3420 tape
8/9	0,3	8/9		4	0.89	0.89	∞	3480 tape
ZM	1,3	1/2		2	1	0.5	3T/2	3850 MSS
Miller[2]	1,5	1/2		3	1	0.5	3T/2	
3PM	2,11	1/2		4	1.5	0.5	∞	ISS-8470
2,7	2,7	1/2		2.67	1.5	0.5	∞	3370-80 disk
1,7	1,7	2/3		4	1.33	0.67	∞	

[†]The left-hand vertical bar denotes the pulse width of the clock.

254

$\frac{8}{9}$ (0, 3) block code in the newest tape subsystem. Zero modulation (ZM) was specifically designed for a mass storage system which uses a rotary head and requires dc-free waveform. ZM is a $\frac{1}{2}$ (1, 3; 3) code with an iterative algorithm and a strong error-detection capability. High-performance disk files have shown preference for codes with higher density ratios, such as the $\frac{1}{2}$ (2, 7) code with variable-length code words, and the $\frac{2}{3}$ (1, 7) code.

5.3.1   The $\frac{8}{9}$ (0, 3) code

Recent tape-storage subsystems record data at a density of 24,000 bits per inch in an 18-track data format on $\frac{1}{2}$-in tape using the $\frac{8}{9}$ (0, 3) group code. It is a one-to-one mapping of the set of all 8-digit binary sequences (data bytes) onto a set of 9-digit binary code words as given in Table 5.3 (Patel, 1985a). The coded binary sequence corresponding to any sequence of data bytes possesses the following two properties: (1) The run lengths of consecutive 0s between any two 1s are limited to zero, one, two, and three only; thus it is called $\frac{8}{9}$ (0, 3) code. (2) The pattern S = 100010001 is not a code word and also does not occur anywhere in the coded sequence with original or shifted code-word boundaries; thus S can be used as a synchronization pattern at selected positions in the data stream to identify format boundaries.

The code-word assignment of Table 5.3 provides simple and inexpensive encoder and decoder logic. The encoder logic for the mapping is derived using four functions, M, N, R, and S. These are the controlling functions which identify the subsets in the mapping depending on the types of partitions in the code-word assignment. Any 9-bit code word W is partitioned into three parts: W_1 of four digits, a connecting digit C, and W_2 of four digits.

$$W = [W_1], C, [W_2]$$

Any 8-bit data group is partitioned into two parts; V_1 of four digits and V_2 of four digits.

$$V = [V_1], [V_2]$$

The function M represents the subset in the mapping where the partitions W_1 and W_2 are the same as V_1 and V_2, respectively. The connecting digit C is 1, and there are 154 code words in this subset. Function N represents the subset in the mapping where the partition W_1 is a special identifying pattern for V_1 and the partition W_2 is the same as V_2. The connecting digit C is 0, and there are 20 code words in this subset. Function R represents the subset in the mapping where the partition W_1 is a special identifying pattern for V_2 and the partition W_2 is the same as V_1. The connecting digit C is 0, and there are 50 code words in this subset. Function

TABLE 5.3   The ⅞ (0, 3) Modulation Code†

X	P	X	P	X	P	X	P
00000000	011001011	01000000	010001011	10000000	111001011	11000000	110001011
00000001	011001001	01000001	010001001	10000001	111001001	11000001	110001001
00000010	001001101	01000010	010010010	10000010	100010010	11000010	110010010
00000011	101100011	01000011	010010011	10000011	100010011	11000011	110010011
00000100	011001010	01000100	010001010	10000100	111001010	11000100	110001010
00000101	101100101	01000101	010010101	10000101	100010101	11000101	110010101
00000110	101100110	01000110	010010110	10000110	100010110	11000110	110010110
00000111	101100111	01000111	010010111	10000111	100010111	11000111	110010111
00001000	011001111	01001000	010001111	10001000	111001111	11001000	110001111
00001001	101101001	01001001	010011001	10001001	100011001	11001001	110011001
00001010	101101010	01001010	010011010	10001010	100011010	11001010	110011010
00001011	101101011	01001011	010011011	10001011	100011011	11001011	110011011
00001100	011001110	01001100	010001110	10001100	111001110	11001100	110001110
00001101	101101101	01001101	010011101	10001101	100011101	11001101	110011101
00001110	101101110	01001110	010011110	10001110	100011110	11001110	110011110
00001111	101101111	01001111	010011111	10001111	100011111	11001111	110011111
00010000	001001011	01010000	011100101	10010000	011101001	11010000	011101101
00010001	001001001	01010001	001100101	10010001	001101001	11010001	001101101
00010010	011001101	01010010	010110010	10010010	100110010	11010010	110110010
00010011	100100011	01010011	010110011	10010011	100110011	11010011	110110011
00010100	100100010	01010100	010100010	10010100	010100010	11010100	010100010
00010101	100100101	01010101	010110101	10010101	100110101	11010101	110110101
00010110	100100110	01010110	010110110	10010110	100110110	11010110	110110110
00010111	100100111	01010111	010110111	10010111	100110111	11010111	110110111
00011000	001001111	01011000	111100110	10011000	111101001	11011000	111101101
00011001	101101001	01011001	010111001	10011001	100111001	11011001	110111001
00011010	100101010	01011010	010111010	10011010	100111010	11011010	110111010
00011011	101101011	01011011	010111011	10011011	100111011	11011011	110111011
00011100	001001110	01011100	110100110	10011100	110100011	11011100	110101101
00011101	100101101	01011101	010111101	10011101	100111101	11011101	110111101
00011110	100101110	01011110	010111110	10011110	100111110	11011110	110111110
00011111	100101111	01011111	010111111	10011111	100111111	11011111	110111111
00100000	101001111	01100000	011100110	10100000	011101010	11100000	011101110
00100001	101001101	01100001	001100110	10100001	001101010	11100001	001101110
00100010	001010010	01100010	011010010	10100010	101010010	11100010	111010010
00100011	001010011	01100011	011010011	10100011	101010011	11100011	111010011
00100100	101001110	01100100	010100110	10100100	010101010	11100100	010101110
00100101	001010101	01100101	011010101	10100101	101010101	11100101	111010101
00100110	001010110	01100110	011010110	10100110	101010110	11100110	111010110
00100111	001010111	01100111	011010111	10100111	101010111	11100111	111010111
00101000	101001011	01101000	111100110	10101000	111101010	11101000	111101110
00101001	001011001	01101001	011011001	10101001	101011001	11101001	111011001
00101010	001011010	01101010	011011010	10101010	101011010	11101010	111011010
00101011	001011011	01101011	011011011	10101011	101011011	11101011	111011011
00101100	101001010	01101100	110100110	10101100	110101010	11101100	110101110
00101101	001011101	01101101	011011101	10101101	101011101	11101101	111011101
00101110	001011110	01101110	011011110	10101110	101011110	11101110	111011110
00101111	001011111	01101111	011011111	10101111	101011111	11101111	111011111
00110000	011100011	01110000	011100111	10110000	011101011	11110000	011101111
00110001	001100011	01110001	001100111	10110001	001101011	11110001	001101111
00110010	001110010	01110010	011110010	10110010	101110010	11110010	111110010
00110011	001110011	01110011	011110011	10110011	101110011	11110011	111110011
00110100	010100011	01110100	010100111	10110100	010101011	11110100	010101111
00110101	001110101	01110101	011110101	10110101	101110101	11110101	111110101
00110110	001110110	01110110	011110110	10110110	101110110	11110110	111110110
00110111	001110111	01110111	011110111	10110111	101110111	11110111	111110111
00111000	111100011	01111000	111100111	10111000	111101011	11111000	111101111
00111001	001111001	01111001	011111001	10111001	101111001	11111001	111111001
00111010	001111010	01111010	011111010	10111010	101111010	11111010	111111010
00111011	001111011	01111011	011111011	10111011	101111011	11111011	111111011
00111100	110100011	01111100	110100111	10111100	110101011	11111100	110101111
00111101	001111101	01111101	011111101	10111101	101111101	11111101	111111101
00111110	001111110	01111110	011111110	10111110	101111110	11111110	111111110
00111111	001111111	01111111	011111111	10111111	101111111	11111111	111111111

†Synchronization pattern: 100010001; X = data byte; P = code word.

S represents the subset in the mapping where partitions W_1 and W_2 both are special identifying patterns for V_1 and V_2, respectively. The connecting digit C is 0, and there are 32 code words in this subset.

The functions M, N, R, and S are generated by logic expressions describing the corresponding subsets in the mapping. Each digit of the 9-bit code word corresponding to an 8-bit data byte is then obtained using logic expressions involving M, N, R, and S. The decoder logic is derived in the same manner, using the partitioning functions M, N, R, and S in reverse.

Let $[X_1, X_2, X_3, X_4, X_5, X_6, X_7, X_8]$ and $[P_1, P_2, P_3, P_4, P_5, P_6, P_7, P_8, P_9]$ denote the 8-bit data byte and the corresponding 9-bit coded pattern, respectively. The encoder and decoder functions are given by the binary logic equations shown in Tables 5.4 and 5.5, respectively. The decoder logic also includes an error-checking function E which checks for an invalid pattern in place of a code word. Each code word also possesses opposite parity relation with the corresponding data byte; this can be used as a diagnostic check for the encoder and decoder hardware.

5.3.2   Zero modulation: $\frac{1}{2}$ (1, 3; 3) code

Zero modulation (ZM) (Patel, 1975) features waveforms with zero dc component. It was created especially for a rotary-head storage system (IBM

TABLE 5.4   The Encoder Function: $\frac{8}{9}$ (0, 3) Code

$$[X_1, X_2, X_3, X_4, X_5, X_6, X_7, X_8] \rightarrow [P_1, P_2, P_3, P_4, P_5, P_6, P_7, P_8, P_9]$$

8-digit data byte $\rightarrow$ 9-digit code word

$M = AH$ where $A = X_1 + X_2 + X_3$

$N = \overline{A}G$ $H = (X_5 + X_6 + X_7)(X_7 + X_8)$

$R = B\overline{H}$ $B = (X_1 + X_2)(X_3 + X_4) + X_3 X_4$

$S = \overline{A}\,\overline{G} + \overline{B}\overline{H}$ $G = (X_5 + X_6)(X_7 + X_8) + X_7 X_8$

$$P_1 = X_1 M + N + X_5 R + (X_1 + X_3) S$$
$$P_2 = X_2 M + \overline{X}_7 \overline{X}_8 R + \overline{X}_3 (\overline{X}_4 \oplus X_7) S$$
$$P_3 = X_3 M + \overline{X}_4 N + \overline{X}_6 \overline{X}_7 R + \overline{X}_2 S$$
$$P_4 = X_4 M + N + R$$
$$P_5 = M$$
$$P_6 = X_5 M + X_5 N + X_1 R + S$$
$$P_7 = X_6 M + X_6 N + X_2 R + [(X_3 \oplus X_5) + X_7] S$$
$$P_8 = X_7 M + X_7 N + X_3 R + \overline{X}_7 \overline{X}_8 S$$
$$P_9 = X_8 M + X_8 N + X_4 R + \overline{X}_6 S$$

TABLE 5.5   The Decoder Function: $\frac{2}{3}$ (0, 3) Code

$$[P_1, P_2, P_3, P_4, P_5, P_6, P_7, P_8, P_9] \rightarrow [X_1, X_2, X_3, X_4, X_5, X_6, X_7, X_8]$$

9-digit code word → 8-digit data byte

$$M = P_5$$

$$N = P_1 \, \overline{P}_2 \, P_4 \, \overline{P}_5$$

$$R = (\overline{P}_1 + P_2) \, P_4 \, \overline{P}_5$$

$$S = \overline{P}_4 \, \overline{P}_5$$

$$X_1 = P_1 \, P_5 + P_6 \, R + P_1 \, P_2 \, S$$

$$X_2 = P_2 \, P_5 + P_7 \, R + \overline{P}_3 \, S$$

$$X_3 = P_3 \, P_5 + P_8 \, R + P_1 \, \overline{P}_2 \, S$$

$$X_4 = P_4 \, P_5 + \overline{P}_3 \, N + P_9 \, R + \overline{P}_1 \, (\overline{P}_2 \oplus P_7 \, \overline{P}_8) \, S$$

$$X_5 = P_6 \, P_5 + P_6 \, N + P_1 \, R + (P_1 \, \overline{P}_2 \oplus P_7) \, P_8 \, S$$

$$X_6 = P_7 \, P_5 + P_7 \, N + \overline{P}_3 \, R + \overline{P}_9 \, S$$

$$X_7 = P_8 \, P_5 + P_8 \, N + \overline{P}_1 \, P_7 \, \overline{P}_8 \, S$$

$$X_8 = P_9 \, P_5 + P_9 \, N + \overline{P}_2 \, R + (P_1 + \overline{P}_7) \, \overline{P}_8 \, S$$

Error check

$$E = \overline{P}_1 \, \overline{P}_2 \, \overline{P}_3 + \overline{P}_2 \, \overline{P}_3 \, \overline{P}_4 \, \overline{P}_5 + \overline{P}_4 \, \overline{P}_5 \, \overline{P}_6 + \overline{P}_5 \, \overline{P}_6 \, \overline{P}_7 \, \overline{P}_9 + \overline{P}_6 \, \overline{P}_7 \, \overline{P}_8$$

$$+ \, \overline{P}_8 \, \overline{P}_9 + \overline{P}_3 \, \overline{P}_4 \, \overline{P}_5 \, P_7 \, \overline{P}_8 + P_1 \, \overline{P}_4 \, \overline{P}_5 \, \overline{P}_8 \, (P_2 \oplus \overline{P}_7)$$

Synchronization pattern

$$P_1 \, \overline{P}_2 \, \overline{P}_3 \, \overline{P}_4 \, P_5 \, \overline{P}_6 \, \overline{P}_7 \, \overline{P}_8 \, P_9$$

3850). The read-write signal in this tape system is transmitted to a rotary transducer by coupling through a transformer whose primary and secondary coils are in continuous relative motion.

In the coded ZM binary sequence, any two consecutive 1s are separated by at least one and at the most three 0s. This sequence is converted into a waveform using a transition for a 1 and no transition for a 0 in the binary sequence. Consequently, the narrowest pulse in the ZM waveform spans two digits in the coded sequence, thus keeping the ratio of the data density to the highest-recorded transition density close to 1. Similarly, the widest pulse spans four digits, thus limiting the range of different pulse widths. In the ZM waveform, the accumulated dc charge value always remains within ±3 units, and it always returns to a zero value at specific boundaries.

The stringent coding constraints of ZM also provide a powerful check on errors in decoding the read data. In particular, the commonly encountered bit-shift errors in magnetic recording are always detected by the bound of ±3 units on the accumulated change.

The ZM algorithm maps every data bit sequentially into two binary digits. The mapping is described in terms of a data bit to be encoded, one preceding data bit, and the two coded digits corresponding to the preceding data bit and in terms of two parity functions that look ahead and back relative to the bit being encoded. Look-ahead parity $P(A)$ is the count, modulo 2, of 1s in the data stream, beginning with the data bit being encoded and counting forward to the next 0 bit; look-back parity $P(B)$ is the count, modulo 2, of all 0s in the data stream from its beginning up to the present bit. For example, in the data sequence 01011110, with bit positions considered from left to right, $P(A) = 1$ at the second, fifth, and seventh bits and $P(B) = 1$ at the first, second, and eighth bits.

The encoding and decoding rules are expressed in the form of binary logic functions. The encoding function is

$$a_0 = \overline{d}_0\overline{d}_{-1} + d_0\overline{d}_{-1}\,\overline{P(A)}P(B) + d_{-1}\overline{a}_{-1}\overline{b}_{-1} \tag{5.1}$$

$$b_0 = d_0[\mathrm{P}(A)\overline{d}_{-1} + \overline{P(B)} + b_{-1}] \tag{5.2}$$

and the decoding function is

$$d_0 = b_0 + a_0\overline{a}_1\overline{b}_1 + a_0a_{-1}\overline{b}_{-1} \tag{5.3}$$

where the symbol d represents a data bit; a and b represent coded digits; and subscripts -1, 0, and 1 signify preceding, current, and succeeding bits, respectively. For convenience, the nonexistent bit preceding the first data bit is assumed to be 1 and its look-back parity is 0; the nonexistent bit following the last bit is 0. Figure 5.3 shows a typical data sequence encoded into ZM waveform.

Look-back *parity* $P(B)$ can be obtained by updating a 1-bit storage cell for every 0 in the data as data bits are encoded. Look-ahead parity $P(A)$ depends on the length of a string of 1s in the succeeding data sequence. When the algorithm imposes no limit on the length of this string, the com-

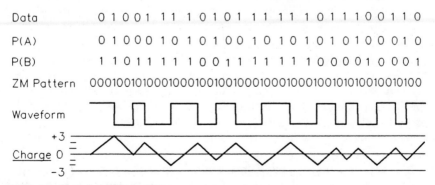

Figure 5.3 Encoded ZM waveform.

putation of $P(A)$ requires an encoder with infinite memory. In order to limit the amount of memory, a parity is inserted at the end of every section of f data bits, which makes $P(B)$ equal to 0 at position $f + 1$ at the end of each section. When $P(B)$ is 0, the encoding functions no longer depend on $P(A)$. Thus, $P(A)$ has no effect on ZM mapping at the boundary of every section of $f + 1$ bits in the data sequence with parity bits, and the computation of $P(A)$ at any data bit need not extend farther than f bits. Then $P(A)$ is given by the binary logic function of the data stored in f bits of memory

$$P(A) = d_0\overline{d}_1 + d_0 d_2 \overline{d}_3 + d_0 d_2 d_4 \overline{d}_5 + \cdots$$
$$+ d_0 d_2 d_4 \cdots d_{t-4}\overline{d}_{t-3} + d_0 d_2 d_4 \cdots d_{t-4}d_{t-2} \tag{5.4}$$

where $t = f$ if f is even and $t = f - 1$ if f is odd.

The encoding process is delayed by f bit periods in a continuous stream of data for computing $P(A)$, but the decoding errors in ZM do not propagate, and the decoding process always terminates at the section boundary.

In the particular rotary-head system under discussion, the value of f is 128. A known unique pattern is inserted at the end of each section and is used as a synchronization pattern. This pattern is

$$S = 01000100101000101000100101001.$$

The waveform corresponding to S satisfies the ZM pulse-width constraints. It has zero dc component, but does not satisfy the ± 3 charge constraint. The pattern S contains the sequence 00101000101000, which is the shortest among those that never occur in the valid ZM pattern in its original or shifted position. Thus, when the synchronization is lost, the sequence S can still be identified from the shifted data, which then reestablishes the synchronization.

The reverse of the waveform corresponding to S also makes a good synchronization mark. This distinction may be used to mark the beginning of the data in a segment by means of the reverse sync waveform in contrast with the regular sync waveform at the end of each section.

One interesting property of the ZM-coded waveform is read-backward symmetry. A properly terminated ZM waveform, when read backward, is a ZM waveform corresponding to the same data in reverse. In particular, when a parity bit is appended to the data to make $P(B)$ equal to 0 at the end, the encoding process terminates the dc charge value at zero at the end of the corresponding waveform. Such a waveform, which has a zero dc component, can be decoded forward or backward by means of the ZM decoding algorithm.

The zero-modulation algorithm is unique in that it achieves a rate that comes arbitrarily close to the channel capacity with corresponding increase in look-ahead, as was promised by the Shannon theorem. Other codes with relaxed parameters, such as $\frac{1}{2}$ (1, 4; 3), called *modified zero modulation* (Ouchi, 1976), and $\frac{1}{2}$ (1, 5; 3), called *Miller squared code* (Miller, 1977), are also available for waveforms with zero dc component and with a very small amount of look-ahead.

5.3.3   The $\frac{1}{2}$ (2, 7) code

The $\frac{1}{2}$ (2, 7) code (Franaszek, 1972) is used in high-performance disk files (such as IBM 3370, 3375, and 3380). It has variable-length codewords. The main advantage this code offers is that the minimum spacing between transitions is three clock periods, as compared to only two for MFM with the same clock and detection window. There are two encoded bits corresponding to each data bit. Thus the detection window is one-half the bit cell, and recorded-transitions density is lower by a factor of 1.5. The mapping of variable-length code words is given in Table 5.6.

The encoding of the data can be done by partitioning the data sequence into two-, three-, and four-bit partitions to match the entries in the code table and then mapping them into corresponding code words. The decoding process is similar but in reverse. An equivalent but more convenient iterative algorithm for this code is also available (Eggenberger et al., 1978). For this purpose, let d_i denote the ith data bit and a_i and b_i denote the corresponding two encoded digits. The preceding and following data bits are indicated by the subscripts $i - 1$ and $i + 1$, respectively, and $i = 0$ indicates the digits in process. The encoding equations use an intermediate variable p_i which provides the necessary look-back and look-ahead to determine the word boundaries. The equations are given by

$$a_0 = \overline{d}_0 d_1 \overline{d}_2 \overline{p}_{-2} + d_0 d_1 \overline{p}_{-2} \tag{5.5}$$

$$b_0 = \overline{d}_1 p_{-1} \tag{5.6}$$

where

$$p_0 = \overline{d}_0 d_1 + d_0 d_1 \overline{p}_{-1} + \overline{d}_0 \overline{d}_1 \overline{d}_2 \overline{p}_{-1} \overline{p}_{-2} \tag{5.7}$$

TABLE 5.6   Code Table for $\frac{1}{2}$ (2, 7) Code

Data	Code word
10	0100
11	1000
000	000100
010	100100
011	001000
0010	00100100
0011	00001000

The value $p_0 = 1$ indicates that d_2 is the last digit of the word. The decoding equation is independent of the word boundaries and is given by

$$d_0 = a_{-1} + a_0\overline{b}_1 + b_0(a_{-2} + b_{-2}) \tag{5.8}$$

The sequential encoding and decoding process can be initialized and terminated by means of a known preamble and postamble of 0 values. As can be seen from the decoding equation, any error in a coded bit may cause a decoding error in up to two following data bits, the current data bit, and up to one preceding data bit. Thus, the error propagation is limited to four consecutive data bits. Another code used, $\frac{1}{2}$ (2, 11), is called the *three-position modulation* (3PM) code (Jacoby, 1977); it has similar parameters (see Table 5.2) and fixed word length. This design also requires look-ahead and look-back in encoding.

5.3.4   The $\frac{2}{3}$ (1, 7) code

It is known that $(d, k) = (1, 6)$ sequences have the information capacity of 0.669 bits per symbol; however, a practical encoding and decoding algorithm for a rate $\frac{2}{3}$ code is not available. The known algorithm (Horiguchi and Morita, 1976) is unattractive in terms of complexity and error propagation. A code with the next best set of constraints is $\frac{2}{3}$ (1, 7), which is available in various forms. Here we present the code that is used in a particular magnetic disk product (Jacoby and Kost, 1984). In this code, the minimum spacing between transitions is two clock periods as in MFM; however, the clock period and the detection window are two-thirds of a bit cell, as compared to one-half of a bit cell in MFM. The recorded-transitions density is lower by a factor of $(2 \times \frac{2}{3}) = 1.33$, the density ratio.

The mapping rules are described in Tables 5.7 and 5.8 by means of a basic code table and a substitution table. The basic encoding table shows a mapping for the four 2-bit data words. When the data are encoded according to the basic encoding table, some catenations of 3-bit code words violate the $d = 1$ constraint. Whenever that happens, an appropriate substitution is made, using double-word entries from the substitution table, that removes the violations. In decoding the 3-bit partitions,

TABLE 5.7   Basic Encoding Table for $\frac{2}{3}$ (1, 7) Code

Data	Code
00	101
01	100
10	001
11	010

**TABLE 5.8 Substitution Encoding
Table for $\frac{2}{3}$ (1, 7) Code**

Data	Code
00.00	101.000
00.01	100.000
10.00	001.000
10.01	010.000

the cases in which the substitution was made during encoding will be identified using look-ahead or look-back. Then all words can be decoded without ambiguity.

Here we create a new more convenient iterative algorithm for this code, in which the encoding and decoding rules are expressed in the form of binary logic equations. In these equations the symbols a, b represent a 2-bit data word; the symbols x, y, z represent a corresponding 3-digit code word; and subscripts -1, 0, and 1 signify preceding, current, and succeeding bits, respectively. The encoding function is given by

$$x_0 = \overline{a}_0\overline{b}_0\overline{z}_{-1} + \overline{a}_0 b_0(b_{-1} + \overline{x}_{-1}\overline{y}_{-1}) \tag{5.9}$$

$$y_0 = a_0(b_0 + \overline{a}_1 b_1) \tag{5.10}$$

$$z_0 = \overline{b}_0(a_1 + \overline{b}_1)\,(a_0 + \overline{z}_{-1}) \tag{5.11}$$

and the decoding function is given by

$$a_0 = \overline{x}_0(y_0 + z_0) \tag{5.12}$$

$$b_0 = \overline{x}_0\overline{y}_0\overline{z}_0\overline{z}_{-1} + (x_0 + y_0)(x_1 + y_1 + z_1)\overline{z}_0 \tag{5.13}$$

The sequential encoding or decoding operation processes groups of two data bits and three coded bits at each cycle. For convenience, the nonexistent preceding bits at the first cycle and following bits at the last cycle are assumed to be zero. As can be seen from the decoding equations, any error in a coded bit may cause a decoding error in the preceding b bit, the current a and b bits, and the following b bit. Thus, the error propagation is limited to five consecutive data bits.

This concludes the discussion of data encoding methods and modulation codes. When these codes are used in recording and reading the data at high data rates, the detection window is a very important parameter. For example, at the data rate of 6 MB/s, the detection window for a half-rate code is 10.4 ns. Thus lower rate codes with larger value of d provide an increasingly difficult trade-off between the detection window and the density ratio in applications with very high density and data rates.

5.4 Error-Correction Coding

The standard $\frac{1}{2}$-in 9-track tape drives have progressively increased their dependence on error detection and correction coding in order to achieve higher densities and data rates. The development of error-correction coding (ECC) in tape-storage drives is summarized in Table 5.9. In earlier tape machines, a vertical byte-parity check (VRC) and a longitudinal track-parity check (LRC) were used. Later, as the density was increased to 31.5 bits per millimeter (800 bits per inch), a cyclic redundancy check (CRC) byte was added (Brown and Sellers, 1970) at the end of the record to provide one-track error correction on rereading of the record. Subsequently, the recording tape with 250 bits per millimeter (6250 bits per inch) used a powerful scheme, called optimal rectangular code (ORC), with a byte-processing algorithm which provided on-the-fly correction of up to two erased tracks (Patel and Hong, 1974). All these tape products used parallel-track format and provided correction of track erasures. The mass storage system with a single-element rotary head used a single-track data format and provided correction of long sections of data (Patel, 1980a). The latest 18-track machines use a new code called *adaptive-cross-parity* (AXP) code and provide correction of up to four erased tracks (Patel, 1985b).

Contemporary magnetic disk products use error-correction coding schemes that correct one burst of error in a variable-length block of up to one full track. Earlier products used the Fire code (Fire, 1959), in which processing is slow, one bit at a time. The later products used byte-oriented

TABLE 5.9   Error-Correction Coding in Tape Drives

Year	IBM product	Density, b/mm	No. of tracks	Data code	ECC capability
1953–8	72X	4–22	9	NRZI	Parity track and LRC
1962	729	31.5	9	NRZI	Parity track and CRC, one-track correction on reread
1966	24XX	63	9	PE	Parity track and CRC, one-track correction on-the-fly
1973	3420	250	9	GCR (0, 2)	Parity track and ORC 1 check byte/7 bytes, two-track correction on-the-fly
1975	3850 MSS	275†	†	ZM (1, 3; 3)	Interleaved subfield code, word = 15 sections, section = 16 bytes, redundancy 2 sections/15 sections, on-the-fly correction of two sections
1985	3480	880	18	(0, 3)	Multitrack correction, adaptive cross-parity code

†Rotary head.

TABLE 5.10   Error-Correction Coding in Disk Drives

Year ship	IBM product	Density		Data code	ECC capability	
		b/mm	t/mm			
1957	350	4	0.8	NRZI	Parity check	
196X	13XX	20.4	2	NRZI	CRC check	
		40	2	NRZI	Multiple error detection	
	23XX	866	3.9	FM	Multiple error detection	
					Modified Fire code single-burst correction	
					Correct/ detect	Check bytes
1971	3330	159	7.6	MFM	11 bits/22 bits	7/record
1973	3340	222	11.8	MFM	3 bits/11 bits	6/record
1976	3350	253	18.7	MFM	4 bits/10 bits	6/record
					Modified Reed-Solomon code single-burst correction	
					Correct/ detect	Check bytes
1979	3370	478	25	(2, 7)	9 bits/17 bits	9/block
1980	3375	478	31.5	(2, 7)	9 bits/17 bits	12/record
1981	3380	598	31.5	(2, 7)	17 bits/33 bits	12/record
1985	3380E	638	54.6	(2, 7)	17 bits/33 bits	12/record

structures using the algebra of Reed-Solomon codes (Hodges et al., 1980). Table 5.10 presents a summary of the development of ECC in disk-storage products.

Encoding for error detection and correction consists of creating a well-defined structure on random data by introducing data-dependent redundancy. Presence of errors is detected when this structure is disturbed. Errors are corrected by making a minimum number of alterations to reestablish the structure. A good error-correction coding scheme provides this function with a reasonable amount of cost in terms of redundancy, processing time, and hardware. For this purpose, the operations in finite fields are found to be most convenient. The simplest finite field is the binary field of two elements. The sum and product operations in this field are modulo-2 sum and modulo-2 product as given by the following tables:

+	0	1		×	0	1
0	0	1		0	0	0
1	1	0		1	0	1

For the sake of coding, binary sequences are interpreted as polynomials with binary coefficients. Modulo-2 sum and product operations on these polynomials are used in creating coding structures which are easy to implement with binary shift registers and logic hardware.

5.5   Polynomial Code Structure

A polynomial with binary coefficients can be used to represent a sequence of binary bits. For example, message B with k information bits is represented by a polynomial $B(x)$ as follows

$$B = [b_0, b_1, b_2, \ldots, b_{k-1}] \tag{5.14}$$

$$B(x) = b_0 + b_1 x + b_2 x^2 + \cdots + b_{k-1} x^{k-1} \tag{5.15}$$

When we attach r check bits at the low end of this message, the resultant code word W is a sequence of $n = k + r$ bits given by

$$W = [c_0, c_1, \ldots, c_{r-1}, b_0, b_1, \ldots, b_{k-1}] \tag{5.16}$$

Let a polynomial $C(x)$ represent the r check bits

$$C(x) = c_0 + c_1 x + \cdots + c_{r-1} x^{r-1} \tag{5.17}$$

Then, attaching the check bits at the low end can be simply written in polynomial form as

$$W(x) = C(x) + x^r B(x) \tag{5.18}$$

The r check bits are determined by a coding process which requires that every code word $W(x)$ is divisible by a preselected polynomial $G(x)$ with zero remainder. $G(x)$ is called the *generator polynomial* of the code. Thus we have

$$W(x) = 0 \text{ modulo } G(x) \tag{5.19}$$

Then, from Eqs. (5.18) and (5.19) we have

$$C(x) = -x^r B(x) \text{ modulo } G(x) \tag{5.20}$$

In the binary field, we can ignore the negative sign. Thus, $C(x)$ represents the r-bit remainder when one divides $x^r B(x)$ by $G(x)$. Equation (5.20) provides the means for computing r check bits for a k-bit information sequence $B(x)$ such that the resultant n-bit code word $W(x)$ is divisible by $G(x)$.

The code thus generated posseses a cyclic property: For a specific code word length n, any cyclic shift of a code word is also a code word when the high-order bit corresponding to x^n after the shift is wrapped around to the low-order bit corresponding to x^0. This code-word length n is the smallest positive integer e that satisfies the relation

$$x^e = 1 \text{ modulo } G(x) \tag{5.21}$$

Alternatively, e is the minimum number of cyclic shifts in a modulo-$G(x)$ register after which it resets to its original content. The integer e is called the *cycle length* or *exponent* of $G(x)$. The codes are called *cyclic codes*.

The code-word length may be shortened to fit the requirements of a particular application. In that case, a number of information digits at the high-order end of the code word are physically absent and are assumed to be zero in processing. We will see that the cyclic structure is very useful in mechanizing the encoding and decoding operations by means of shift registers and modulo-2 summing (exclusive-or) circuits.

On readback, the received word, denoted by $\hat{W}(x)$, may be different from $W(x)$ in some positions. Thus on computing the remainder in accordance with (5.19), we may not get a result of zero. Let $S(x)$ denote this remainder computed from $\hat{W}(x)$ as

$$S(x) = \hat{W}(x) \text{ modulo } G(x) \tag{5.22}$$

Let $E(x)$ denote the difference between the received word $\hat{W}(x)$ and the original code word $W(x)$; it represents the errors in $\hat{W}(x)$ in polynomial form

$$E(x) = \hat{W}(x) - W(x) \tag{5.23}$$

$E(x)$ is called the *error pattern* or *error polynomial*. Note that presence of an error corresponds to a one value for the coefficient of $E(x)$ in the corresponding position. From Eq. (5.19), (5.22), and (5.23) we have

$$S(x) = E(x) \text{ modulo } G(x) \tag{5.24}$$

Thus, $S(x)$ as computed from the received word in accordance with Eq. (5.22) depends only on the errors. $S(x)$ is called the *syndrome of errors*. Absence of errors will result in a zero value for $S(x)$. A nonzero syndrome will need further interpretation in order to determine the exact nature of the errors. This process of interpreting the syndrome is called the *decoding process*.

If the generator polynomial $G(x)$ is a product of two or more polynomials, then the code word $W(x)$ must be divisible by each of the factors in order to be divisible by $G(x)$. In some applications, this fact is used to simplify the decoding process by obtaining and interpreting the separate syndromes corresponding to each of the factors of $G(x)$. The decoding process and the range of error detection and correction capability depend on the choice of the generator polynomial and other practical considerations. We will describe these in detail in the discussion of various practical applications.

As seen so far, the encoding and decoding operations in cyclic codes require computation of the remainder when a given polynomial is divided by $G(x)$. In particular, the encoding operation is given by Eq. (5.20), the syndrome generation is described by Eq. (5.22), and the decoding of the errors is an interpretation of Eq. (5.24). All these computations can be carried out easily and conveniently by means of a linear-feedback shift register.

Consider a shift register as shown in Fig. 5.4 with feedback connection according to the coefficients of a binary generator polynomial $G(x)$:

$$G(x) = g_0 x^0 + g_1 x^1 + g_2 x^2 + \cdots g_{r-1} x^{r-1} + x^r \qquad (5.25)$$

where $g_i = 1$ means a connection is present and $g_i = 0$ means a connection is absent.

The binary content of the shift register is viewed as a degree-$(r - 1)$ polynomial, and the feedback connection effectively performs

$$x^r = g_0 x^0 + g_1 x^1 + g_2 x^2 + \cdots g_{r-1} x^{r-1} \text{ modulo } G(x) \qquad (5.26)$$

Thus an upward-shifting operation in this shift register corresponds to multiplying the contents by x and reducing the results modulo $G(x)$.

We can use this shift register for encoding the data $B(x)$ to compute check bits $C(x)$ in accordance with Eq. (5.20). The data bits are entered at the feedback input (see Fig. 5.4), which is equivalent to premultiplying the entering digit by x^r. Each shifting operation, onwards, multiplies the content of the shift register by x. The high-order coefficient b_{k-1} of $B(x)$ is entered first, followed by $b_{k-2}, b_{k-3}, \ldots, b_1, b_0$, in that order, with each

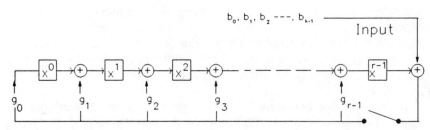

Figure 5.4 Encoding shift register.

successive shifting operation. The contents of the shift register at each shift successively represent

$$b_{k-1}x^r \text{ modulo } G(x)$$

$$b_{k-2}x^r + b_{k-1}x^{r+1} \text{ modulo } G(x)$$

$$\vdots$$

$$b_0x^r + b_1x^{r+1} + \cdots + b_{k-1}x^{r+k-1} \text{ modulo } G(x)$$

Thus the final content, at the end of k shifts, is the check polynomial $C(x)$ as defined in Eq. (5.20). The encoding progresses in synchronism with bit-by-bit transmission of the k data bits. After k shifting operations, the check bits are ready in the shift register and may be shifted out during the following r shifts. The feedback connections of the shift register are disabled during these last r shift cycles.

On readback, we can use a similar shift register to process the received word $\hat{W}(x)$ for generation of the syndrome $S(x)$ in accordance with Eq. (5.22). In this case the entire received word (including check bits) is entered at the x^0 position of the shift register, as shown in Fig. 5.5. The high-order digit is entered first, followed by other digits in succession. When the last digit is entered, the content of the shift register is the syndrome $S(x)$ as defined in Eq. (5.22).

A nonzero value of $S(x)$ indicates the presence of errors in accordance with Eq. (5.24). Since $S(x)$ is the remainder, there are many error polynomials $E(x)$ which will satisfy Eq. (5.24). Here, an assumption will be made regarding the most likely error patterns, such as single error, single-burst error of smallest length, etc. In fact, code design will specify the range of most likely error patterns that will provide a unique solution to Eq. (5.24). These are called *correctable error patterns*. Various techniques are used to process the syndrome $S(x)$ to arrive at a correctable error position and determine its error pattern. We will examine some of these techniques in the discussion of various practical applications.

5.5.1 Cyclic codes for error detection

One of the most widely used binary cyclic codes is called the *cyclic redundancy check* (CRC); it is used for detection of errors in transmission and

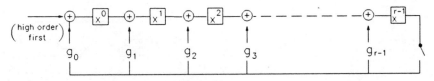

Figure 5.5 Shift register for syndrome generation.

storage of serial data. Any polynomial with long cycle length can be used as a generator for CRC. The error-detection capability is largely determined by the degree r of $G(x)$ or the number of CRC bits (Brown and Peterson, 1961). In particular, any single-burst error of length r or less is detected. Also, the fraction of undetected single-burst errors of length $r + 1$ is $1/2^{r-1}$ and that of length greater than $r + 1$ is $1/2^r$. For example, the commonly used 16-bit CRC has a cycle length of $2^{15} - 1$ and has the generator

$$G(x) = 1 + x^2 + x^{15} + x^{16} = (1 + x)(1 + x^2 + x^{15})$$

It will provide detection of all single-burst errors of length 16 or less and detect better than 99.998 percent of all other burst errors in a binary record of any length up to 32,767 bits. A linear-feedback shift register for this generator is shown in Fig. 5.6. This is essentially the only hardware needed for encoding and decoding variable-length binary records for error detection. The CRC character is generated by serially shifting the binary information in the feedback shift register as it is transmitted. The CRC character is then transmitted at the end of the information sequence. On readback, the received sequence is processed in the same manner. The generated CRC character is compared with the received check character for detection of any errors in the received message.

5.5.2 Interleaving cyclic codes

A burst error can be effectively handled by spreading its effect over many code words. This is easily done by physically interleaving m code words to create an m-times-longer message. The interleaving operation can be viewed as arranging m code words of the original code into m rows of a rectangular block and then sending the digits column by column. The resulting code is an m-way interleaved code. A burst error of length m or less will affect at the most one digit in each of the m code words in the interleaved code. Thus if the original code corrects single-bit errors, then the m-way interleaved code corrects single-burst errors of length m or less.

The interleaving can be accomplished simply by replacing each binary storage element of the encoding shift register with m binary storage elements connected in tandem. The feedback connections are not changed.

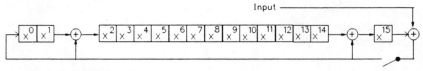

Figure 5.6 Serial CRC register.

In fact, if the original code is a cyclic code then the interleaved code is also a cyclic code. If the generator of the original code is a degree-r polynomial $G(x)$, then the generator of the interleaved code is a degree-mr polynomial $G(x^m)$. The decoding process for the interleaved code follows from the decoding of the original code. Interleaving is illustrated later in various applications of error-correction coding in magnetic recording products.

5.5.3 Fire code

The Fire code (Fire, 1959) is commonly used in magnetic disk products for detection and correction of single-burst errors in variable-length records. This is a cyclic code with generator polynomial of the form

$$G(x) = (x^c + 1)G_1(x) \tag{5.27}$$

where the factor $G_1(x)$ determines the position of the error, and the factor $(x^c + 1)$ relates to the pattern of the error burst. A Fire code is capable of correcting a burst error of length b when $c \geq 2b - 1$. Simultaneous detection of a longer burst of length up to d bits can be included in the capability when $c \geq b + d - 1$. The polynomial $G_1(x)$ must be of degree $m \geq b$. The full length of the code words is determined by the cycle length of $G(x)$ which is the least common multiple of the number c and the cycle length of $G_1(x)$.

The encoding of a Fire code is carried out with a shift register connected for modulo $G(x)$ operation in accordance with Eq. (5.20) as described previously. The syndrome of error in the Fire code is computed as two separate remainders $S_0(x)$ and $S_1(x)$ corresponding to the two factors $(x^c + 1)$ and $G_1(x)$ of the generator $G(x)$. Thus in place of syndrome Eq. (5.22) we have two equations

$$S_0(x) = \hat{W}(x) \text{ modulo } (x^c + 1) \tag{5.28}$$

$$S_1(x) = \hat{W}(x) \text{ modulo } G_1(x) \tag{5.29}$$

Similarly, the error pattern of Eq. (5.24) will be written in the form of two equations

$$S_0(x) = E(x) \text{ modulo } (x^c + 1) \tag{5.30}$$

$$S_1(x) = E(x) \text{ modulo } G_1(x) \tag{5.31}$$

The most likely and correctable error pattern is a burst of length b or less. Hence, we can express the error polynomial $E(x)$ more specifically as an error burst at location p as follows

$$E(x) = x^p E_b(x) \tag{5.32}$$

where $E_b(x)$ is a degree-$(b - 1)$ polynomial representing a b-bit burst, and p is the location of the low-order bit of this burst in the error polynomial $E(x)$. All other unspecified coefficients in $E(x)$ are zero. From Eqs. (5.30), (5.31), and (5.32) we get

$$S_0(x) = x^p E_b(x) \text{ modulo } (x^c + 1) \tag{5.33}$$

$$S_1(x) = x^p E_b(x) \text{ modulo } G_1(x) \tag{5.34}$$

Since n is the cycle length of $G(x)$, we can use Eq. (5.21) and rewrite Eq. (5.33) and (5.34) as

$$x^{n-p} S_0(x) = E_b(x) \text{ modulo } (x^c + 1) \tag{5.35}$$

$$x^{n-p} S_1(x) = E_b(x) \text{ modulo } G_1(x) \tag{5.36}$$

Now we can describe the decoder hardware and algorithm. The hardware consists of two shift registers SR_0 and SR_1 similar to the one shown in Fig. 5.5. A shifting operation in SR_0 and SR_1 multiplies the contents in each by x modulo $(x^c + 1)$ and x modulo $G_1(x)$, respectively. These registers are used to compute syndromes $S_0(x)$ and $S_1(x)$ from the received code word $\hat{W}(x)$ in accordance with Eqs. (5.28) and (5.29), respectively. The computed syndromes are then decoded by means of the same two shift registers by a method called the *error-trapping technique* (Peterson, 1961).

Notice from Eqs. (5.35) and (5.36) that, after $(n - p)$ shifts of registers SR_0 and SR_1, the original contents $S_0(x)$ and $S_1(x)$ will both reduce to the burst error $E_b(x)$. Uniqueness of the burst pattern is assured, since SR_0 merely circulates the pattern and $c \geq 2b - 1$ (see Fig. 5.7). Since p is unknown, we will have to examine the contents after each shifting operation. The exact error pattern is recognized from the fact that the contents of SR_0 and SR_1 must be equal, and the contents of SR_0 may be a b-bit burst in the low-order (b) bit positions; the high-order ($c - b$) bit positions must be zero. When these two requirements are satisfied we have captured the error pattern. Since n is known, the number of shifts $(n - p)$ determines the position of the burst, starting with its low-order bit.

In some applications, when the cycle length n is large and the practical length of any code word is relatively short, one may use backward-shifting

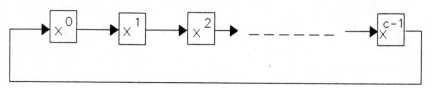

Figure 5.7 Shift register SR_0; multiplication by x modulo $(x^c + 1)$.

registers in which a shifting operation amounts to multiplication of the contents by x^{-1} modulo $x^c + 1$ and x^{-1} modulo $G_1(x)$ respectively. In that case p instead of $n - p$ shifting operations will be required, since

$$x^{n-p} = x^{-p} \text{ modulo } G(x) \tag{5.37}$$

Later we will see an application of Fire codes in magnetic disk products, which include a design modification that further reduces the number of shifting operations. Fire codes are limited in their capability to the correction of only one burst, and the bit-by-bit processing is relatively slow. Thus Fire codes will remain popular in low-data-rate applications where error correction can be deferred until the end of a variable-length record. In contrast, the codes in extended binary fields, in particular, the Reed-Solomon codes (Reed and Solomon, 1960) provide more flexibility in terms of capability options and fast byte-parallel processing.

5.6 Operations in Extended Binary Fields

The binary base field can be extended to a finite field of 2^m elements, also known as the Galois field $GF(2^m)$ in which the field elements are represented by m-bit binary bytes. As an example, we will consider the Galois field $GF(2^8)$ in which the 256 distinct elements are represented by the 256 eight-bit bytes. Each element can be written as an eight-bit column vector A or a polynomial $A(x)$ with eight binary coefficients as follows

$$A = \begin{bmatrix} a_0 \\ a_1 \\ a_2 \\ a_3 \\ a_4 \\ a_5 \\ a_6 \\ a_7 \end{bmatrix} \tag{5.38}$$

$$A(x) = a_0 + a_1 x + a_2 x^2 + a_3 x^3 + a_4 x^4 + a_5 x^5 + a_6 x^6 + a_7 x^7 \tag{5.39}$$

The sum and product of the elements in $GF(2^8)$ are computed as the modulo-2 sum and product of the corresponding polynomials with the results reduced modulo $P(x)$, where $P(x)$ is a prime polynomial of degree 8. (A polynomial is prime if it cannot be factored into two or more polynomials of lower degrees. It is also called an *irreducible polynomial*.) Thus the sum and product of two field elements $A(x)$ and $B(x)$ are defined as

$$\text{Sum} = A(x) + B(x) \text{ modulo } P(x) \tag{5.40}$$

$$\text{Product} = A(x) \times B(x) \text{ modulo } P(x) \tag{5.41}$$

Since $P(x)$ is a degree-8 polynomial, the result of the reduction modulo $P(x)$ is always a polynomial of degree less than 8, thus an element of the field. Furthermore, $P(x)$ is irreducible, which ensures that the result of the sum or product of field elements is always a unique field element. The 0 element of this field is the byte with $b_i = 0$ for all i; the 1 element is the byte with $b_i = 1$ for $i = 0$ and $b_i = 0$ for all other values of i.

The field operations can be expressed in a more convenient form as the matrix-sum and matrix-product operations using the companion matrix T of the polynomial $P(x)$. The companion matrix T is defined in terms of the coefficients of $P(x)$ as

$$T = \begin{bmatrix} 0 & 0 & 0 & 0 & 0 & 0 & 0 & p_0 \\ 1 & 0 & 0 & 0 & 0 & 0 & 0 & p_1 \\ 0 & 1 & 0 & 0 & 0 & 0 & 0 & p_2 \\ 0 & 0 & 1 & 0 & 0 & 0 & 0 & p_3 \\ 0 & 0 & 0 & 1 & 0 & 0 & 0 & p_4 \\ 0 & 0 & 0 & 0 & 1 & 0 & 0 & p_5 \\ 0 & 0 & 0 & 0 & 0 & 1 & 0 & p_6 \\ 0 & 0 & 0 & 0 & 0 & 0 & 1 & p_7 \end{bmatrix} \tag{5.42}$$

where $P(x)$ is given by

$$P(x) = p_0 + p_1 x + p_2 x^2 + p_3 x^3 + p_4 x^4$$
$$+ p_5 x^5 + p_6 x^6 + p_7 x^7 + x^8 \tag{5.43}$$

For any field element $A(x)$ we can compute a corresponding 8×8 matrix T_A by substituting $x = T$ in Eq. (5.39). Thus T_A is given by

$$[T_A] = a_0[I] + a_1[T] + a_2[T^2] + \cdots + a_7[T^7] \tag{5.44}$$

where T^i denotes T multiplied by itself i times, I is the 8×8 identity matrix, and all results are reduced modulo 2. Now the sum and product of field elements $A(x)$ and $B(x)$ can be computed through modulo-2 matrix expressions

$$[\text{sum}] = [A] + [B] \tag{5.45}$$

and $$[\text{product}] = [T_A][B] \tag{5.46}$$

where A and B are vector forms of $A(x)$ and $B(x)$ respectively and T_A is the matrix given by Eq. (5.44). The operations defined by the polynomial equations, Eq. (5.40) and (5.41), and those defined by the matrix equations, Eq. (5.45) and (5.46), are equivalent. This follows from the fact that the vector multiplication by the matrix T is equivalent to the polynomial multiplication by x modulo $P(x)$.

If the 8-bit field element expressed as x modulo $P(x)$ is denoted by α, then the field element expressed by x^i modulo $P(x)$ is α^i, the element α

multiplied by itself i times. The corresponding matrix T^i is the matrix T multiplied by itself i times. A shift-upward operation (without new input) in a modulo-$P(x)$ shift register is equivalent to multiplying the contents, an 8-bit vector, by the matrix T. This shift register is similar to the one shown in Fig. 5.5 for modulo-$G(x)$ computations. If e is the minimum number of shift-upward operations after which the shift register resets to its original contents, then e is called the cycle length of the shift register. It is the smallest positive integer such that

$$x^e \text{ modulo } P(x) = x^0 \tag{5.47}$$

Thus, the 8-bit field elements α^i expressed as x^i modulo $P(x)$ for $i = 0, 1, 2, \ldots, e - 1$ are distinct nonzero elements of $GF(2^8)$. The corresponding matrices T^i for $i = 0, 1, 2, \ldots, n - 1$ are also distinct, and $T^e = T^0 = I$ where I is an 8×8 identity matrix.

Now the 8-bit column vectors in the companion matrix T of Eq. (5.42) can be identified as the elements of $GF(2^8)$ as follows

$$T = [\alpha, \alpha^2, \alpha^3, \alpha^4, \alpha^5, \alpha^6, \alpha^7, \alpha^8] \tag{5.48}$$

Since
$$[T][\alpha^i] = [\alpha^{i+1}] \tag{5.49}$$

we can obtain T^2, T^3, $\ldots$ iteratively and it follows that the matrix T^i is given by

$$T^i = [\alpha^i, \alpha^{i+1}, \alpha^{i+2}, \alpha^{i+3}, \alpha^{i+4}, \alpha^{i+5}, \alpha^{i+6}, \alpha^{i+7}] \tag{5.50}$$

TABLE 5.11 Irreducible Polynomials of Degree 8

No.	p_0 / p_8	p_1 / p_7	p_2 / p_6	p_3 / p_5	p_4 / p_4	p_5 / p_3	p_6 / p_2	p_7 / p_1	p_8 / p_0	Cycle length
					Coefficients					
1	1	0	0	0	1	1	1	0	1	255
2	1	0	1	1	1	0	1	1	1	85
3	1	1	1	1	1	0	0	1	1	51
4	1	0	1	1	0	1	0	0	1	255
5	1	1	0	1	1	1	1	0	1	85
6	1	1	1	1	0	0	1	1	1	255
7	1	0	0	1	0	1	0	1	1	255
8	1	1	1	0	1	0	1	1	1	17
9	1	0	1	1	0	0	1	0	1	255
10	1	1	0	0	0	1	0	1	1	85
11	1	0	1	1	0	0	0	1	1	255
12	1	0	0	0	1	1	0	1	1	51
13	1	0	0	1	1	1	1	1	1	85
14	1	0	1	0	1	1	1	1	1	255
15	1	1	1	0	0	0	0	1	1	255
16	1	0	0	1	1	1	0	0	1	17

Table 5.11 lists all the irreducible polynomials of degree 8 and their cycle lengths. The reciprocal of each of the listed polynomials is also an irreducible polynomial with the same cycle length.

The maximum cycle length for a degree-8 polynomial is $(2^8 - 1)$. A polynomial with maximum cycle length is called a *primitive polynomial.* A primitive polynomial is necessarily an irreducible polynomial (but not vice versa). When $P(x)$ is a degree-8 primitive polynomial, we have $e = 255$, and hence all 255 nonzero 8-bit field elements can be expressed as remainders x^i modulo $P(x)$ for $i = 0, 1, 2, \ldots, 254$. In that case, the element α, expressed as x modulo $P(x)$, is called a *primitive element.* Any nonzero 8-bit field element $A(x)$ and corresponding matrix T_A can be expressed as α^i and T^i, respectively, for some specific value i, where $0 \leq i \leq 254$. This unique integer i can be regarded as the logarithm to the base α of the field element $A(x)$. (For the case of field operations that use a nonprimitive polynomial $P(x)$, the element α expressed as x modulo $P(x)$ is not primitive; thus the base for the logarithm will have to be some element other than α. Any finite field has at least one primitive element. Any two finite fields with the same number of elements are isomorphic. The one-to-one correspondence of this isomorphism is established through the one-to-one correspondence between the primitive elements of the two fields.)

Now we are ready to describe some practical coding schemes based on the operations in $GF(2^m)$. These codes are often referred to as *byte-correcting* or *symbol-correcting codes.* Since modern data-processing machines almost always use bytes in their data flow it is altogether fitting that operations in $GF(2^m)$, and concepts in coding theory which use the byte as a basic element for processing, provide a more convenient and efficient method of processing errors.

5.7   Reed-Solomon Codes

Now, we extend the concepts of cyclic codes to the codes in an extended binary field $GF(2^m)$, where a field element is represented by an m-bit binary byte. A polynomial with coefficients in $GF(2^m)$ can be used to represent a sequence of bytes. In fact, all processing of these polynomials is done byte-parallel using the sum and product operations defined in the extension field $GF(2^m)$. This may look overwhelmingly complex now, but it turns out to be simple, convenient, and efficient. In this system, the generator $G(x)$, the information sequence $B(x)$, the remainder $C(x)$, and the code word $W(x)$ are all polynomials whose coefficients are elements in $GF(2^m)$. In other words, the information symbols, the check symbols, and the errors in these codes are in the form of m-bit bytes. Among these, the Reed-Solomon codes (Reed and Solomon, 1960) are of particular interest. They are convenient in implementation and most efficient in code capability.

The Reed-Solomon codes are cyclic codes with symbols in $GF(q)$. In particular, we use 8-bit binary bytes as symbols in $GF(2^8)$. In these codes, r check bytes provide correction of t error bytes and u erasure bytes, when $r \geq 2t + u$. (An erasure byte is an erroneous byte with known location.)

The degree-r generator polynomial of this code is given by

$$G(x) = (x - \alpha^0)(x - \alpha^1)(x - \alpha^2) \cdots (x - \alpha^{r-1}) \tag{5.51}$$

where α is any element of $GF(2^8)$. The main characteristic† of this $G(x)$ is that the roots $\alpha^0, \alpha^1, \alpha^2, \ldots, \alpha^{r-1}$ are the r consecutive powers of an element of $GF(2^8)$. A full-length code word will have n symbols, where n, the cycle length of $G(x)$, is the least integer for which $\alpha^n = \alpha^0$. When α is the primitive element of $GF(2^8)$ then n has the largest value $(2^8 - 1)$. The actual length may be shorter or variable from $r + 1$ up to n.

The generator polynomial $G(x)$ is used in computing the check bytes in a manner analogous to that used in binary cyclic codes. In particular, on multiplying out the factors in Eq. (5.51), we can rewrite $G(x)$ in the form of Eq. (5.25) with coefficients g_i in $GF(2^8)$

$$G(x) = g_0x^0 + g_1x^1 + g_2x^2 + \cdots + g_{r-1}x^{r-1} + x^r$$

Then the encoding shift register is similar in form to that shown in Fig. 5.4, although here we process bytes instead of bits. All lines in the data flow are 8 bits in parallel, each x^i storage device holds 8 bits, and the feedback connections require multiplication by the field element g_i in $GF(2^8)$. This process is illustrated in an application in Sec. 5.8.

Next, we examine the coding relations implied by the roots of the generator polynomial. By definition of cyclic codes, a code word $W(x)$ is divisible by $G(x)$. Thus with an appropriate quotient polynomial $Q(x)$ we can write

$$W(x) = Q(x)G(x) \tag{5.52}$$

From Eq. (5.51) we see that

$$G(x) = 0 \qquad \text{at } x = \alpha^0, \alpha^1, \alpha^2, \ldots, \alpha^{r-1} \tag{5.53}$$

Substituting this in Eq. (5.52) we have

$$W(x) = 0 \qquad \text{at } x = \alpha^0, \alpha^1, \alpha^2, \ldots, \alpha^{r-1} \tag{5.54}$$

---

†$G(x) = (x - \alpha^a)(x - \alpha^{a+1}) \cdots (x - \alpha^{a+r-1})$ provides the most general definition, where a is any integer. However, a nonzero value of a can be viewed as a constant scale factor. Specific choice of a, in some cases, provides symmetry or other useful structure.

The code word polynomial $W(x)$ can be written as

$$W(x) = B_0 + B_1x + B_2x^2 + \cdots + B_{n-1}x^{n-1} \qquad (5.55)$$

where $B_0, B_1, \ldots, B_{r-1}$ represent the r check bytes, and $B_r, B_{r+1}, \ldots, B_{n-1}$ represent the k data bytes. From Eq. (5.54) and (5.55) we obtain the coding rule corresponding to each of the r roots of the generator polynomial by substituting $x = \alpha^j$

$$B_0 + B_1\alpha^j + B_1\alpha^{2j} + \cdots + B^{n-1}\alpha^{(n-1)j}$$
$$= 0 \qquad \text{for } j = 0, 1, 2, \ldots, r - 1. \quad (5.56)$$

The product of field elements B and α^i can be expressed as the modulo-2 matrix product T^iB, where T is the companion matrix corresponding to α and B is an 8-digit column vector. Thus the coding relations of Eq. (5.56) can be expressed as the modulo-2 matrix equations

$$B_0 \oplus B_1 \oplus B_2 \oplus \cdots \oplus B_{n-1} = 0$$
$$B_0 \oplus TB_1 \oplus T^2B_2 \oplus \cdots \oplus T^{n-1}B_{n-1} = 0 \qquad (5.57)$$
$$B_0 \oplus T^2B_1 \oplus T^4B_2 \oplus \cdots \oplus T^{2(n-1)}B_{n-1} = 0$$
$$B_0 \oplus T^jB_1 \oplus T^{2j}B_2 \oplus \cdots \oplus T^{(n-1)j}B_{n-1} = 0 \qquad j < r$$

The coding relations of Eq. (5.57) represent r simultaneous equations in $\text{GF}(2^m)$ that each code word must satisfy. The r unknown check bytes can be determined by solving these equations; however, this is not necessary. Instead, we will use a shift-register circuit in $\text{GF}(2^m)$ to obtain the r check bytes analogous to r check bits for the binary cyclic codes in $\text{GF}(2)$. On the other hand, the coding relations of Eq. (5.57) are very useful at the receiver end, where they can be easily checked and used for detection and correction of errors.

Reed-Solomon codes provide correction of t errors and u erasures with r check bytes, when $r \geq 2t + u$. Among these, the 1-byte-correcting Reed-Solomon code with two-way or three-way interleaving provides an interesting alternative to a Fire code for burst-error correction in magnetic-recording disk products. In two-way interleaving, we can correct one byte in each of the two phases of interleaving; thus any two adjacent bytes in the interleaved codeword are correctable. Any burst of length $(m + 1)$ bits or less is always confined to two adjacent m-bit bytes and hence is correctable by this code. Three-way interleaving allows correction of three adjacent m-bit bytes and hence any burst of length $(2m + 1)$ bits or less will be correctable. In magnetic-recording tape products, up to 16-way interleaving has been used for correction of long bursts in rotary-head systems. The byte processing in Reed-Solomon codes is fast and convenient, in contrast to the slow bit-by-bit processing in Fire codes. Furthermore,

Reed-Solomon codes provide other options, such as correction of multiple-byte or multiple-burst errors with a reasonable amount of redundancy and cost of decoding.

The code for nine-track tape systems with 6250 bits per inch (IBM 3420) is based on the decoding equations of the single-byte-correcting Reed-Solomon code. It uses 8-bit bytes as symbols in a code word with two check symbols and seven-data symbols. The code words are in the form of an eight-by-nine rectangle with two orthogonal sides as check bits. This code, called *optimal rectangular code,* possesses orthogonal symmetry around one of the diagonals of the eight-by-eight square block formed by eight horizontal bytes along or eight vertical bytes across eight tracks; the ninth track is a parity track providing the byte-parity check on all recorded bytes. This diagonal symmetry is an artifact of algebra that allows a tapelike data format for encoding and a Reed-Solomon structure for decoding rectangular code words that exhibit a geometric relationship with the errors along the tracks. The code provides correction of up to two erased tracks or one unknown track in error. Further details on this code may be found in the reference (Patel and Hong, 1974).

The error-recovery scheme for the rotary-head mass storage system (Patel, 1980a) is also based on the single-byte-correcting Reed-Solomon code. In Sec. 5.8, we present the implementation of this scheme, illustrating the encoding process and the procedures for on-the-fly correction of a single-byte error and two byte erasures.

The correction of multiple-byte errors and erasures with a Reed-Solomon code is a very interesting subject that is beyond the scope of this chapter. A complete on-the-fly decoder for multiple byte errors has been described (Patel, 1986). Decoder equations and architecture are presented in which the location of each error is computed in a cyclic order and the value of that error is determined simultaneously without explicit information regarding the unknown locations of other errors. As a result, one can begin delivery of the decoded data symbols, one at a time, in synchrony with each cycle of the decoding process, regardless of the number of errors in the code word. This decoder structure is well suited for large-scale integration in the chip design with pipelined data flow; compared to any decoder with comparable correction capability, it will provide delivery of the data with the minimum access time.

In digital audio and video recording systems, it has been demonstrated that error-control coding can be cost-effective in consumer products. One code, used in the compact-disk digital audio player, is a two-dimensional arrangement of the Reed-Solomon code with an extensive amount of interleaving (Carasso and Nakajima, 1982; Doi, 1984). It is called the *cross interleave Reed-Solomon code* (CIRC) and is designed for correction of long burst errors. It also provides interpolation of the audio signal for the detected, but uncorrectable, longer bursts of error. This interpolation of the signal, called *error concealment,* is an effective compromise solution for very long errors in audio and video applications. Error concealment is not useful for computer data.

In Sec. 5.9, we present another coding arrangement for correction of very long errors; it is used in modern tape subsystems. This code represents an iterative or convolutional method of coding using simple cross parity checks and very effective error-correction capability.

5.8   Error-Recovery Scheme for a Rotary-Head Tape Storage System

The error-recovery scheme to be described here illustrates various important concepts in coding to obtain specific features (Patel, 1980*a*). The basic building block of the scheme is a 15-byte code word of a single-byte-correcting Reed-Solomon code. The scheme demonstrates (1) the power of interleaving code words for long-burst correction; (2) how the use of subfields in a field provides interesting relationships of bit errors, byte errors, and burst errors and also provides economies in implementation; and (3) how a modulation code and an error-correction code can work together to enhance the error-correction capability. It is used in the IBM 3850 mass storage system.

This mass storage system consists of an array of data cartridges about 4.8 cm (1.9 in) in diameter and 5.9 cm (3.5 in) long, with a capacity of 50 million characters each. Each cartridge contains magnetic tape 6.9 cm (2.7 in) wide and 19.5 m (64 ft) long on which data are organized in cylinders analogous to those of a disk file; data can be transferred to the disk file one cylinder at a time. Up to 4720 cartridges are stored in hexagonal compartments in a honeycomblike apparatus that includes a mechanism for fetching cartridges from the compartments in order to read data and to write data on them and for replacing them back into the compartments.

Instead of the multitrack fixed head used in conventional tape machines, a rotary read-write head is used in this mass storage system. The rotary head creates short slanted stripes across the tape instead of long tracks along the tape. The tape follows a helical path around a mandrel and is stepped in position from one slanted stripe to the next over a circular slit in the mandrel which houses the continuously moving read-write transducer element of the rotary head.

The data are recorded as coded binary sequences corresponding to the presence or absence of magnetic flux transitions in fixed-length stripes at a density of 26.4 stripes per centimeter (67 stripes per inch) and linear density ranging from 1356 to 2712 flux transitions per centimeter (3444 to 6888 flux transitions per inch). A read-write operation always involves the processing of whole stripes, with each data stripe containing exactly 4096 net data bytes after decoding.

5.8.1 Data format

The parallel multitrack data format of the conventional tape machines is not available. Instead, the data are organized in resynchronizable sections in order to facilitate recovery from mixed-mode errors involving 1-bit errors caused by random noise, multiple 2-bit errors caused by bit shift, and clusters of errors caused by defects and dust particles—this includes the capability of resynchronizing the clock. The data format of the stripe is illustrated in Fig. 5.5. The stripe is divided into 20 segments. The segments are appended to each other, forming a continuous waveform; however, each segment is a separate entity and can be decoded without reference to the data in other segments. Each segment consists of 13 data sections followed by 2 check sections. In a write operation, the bit values for the 2 check sections are computed in accordance with an error-correction code by processing the 13 data sections as they are being recorded. The computed check sections are then appended to the 13 data sections, thus completing a segment.

Each section is 129 bits long and consists of 16 bytes of binary information with an overall odd-parity bit. This sequence of 129 bits is encoded into a 258-digit, zero-modulation waveform followed by a known unique synchronization signal (see Sec. 5.3 for details on ZM). The odd-parity bit serves the dual purpose of checking data errors and of limiting

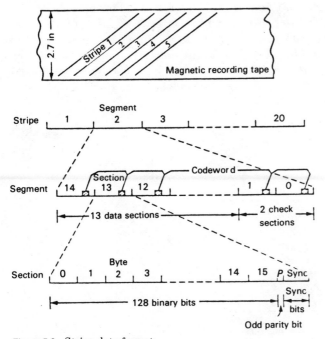

Figure 5.8 Stripe data format.

the memory requirement in ZM. The sections are appended to one another to form a continuous ZM waveform. Thus, each section is protected by the synchronization signal at both of its ends. This allows resynchronization of the decoding clock at the beginning and at the end of each section, in the event of a long error causing loss of synchronization.

The data format for the error-correction code is designed around the resynchronizable sections. The 16 bytes in each section belong to 16 different code words, as shown in Fig. 5.9. Each code word consists of 15 bytes—one from each of the 15 sections in a segment. Let k be the index for the code words and j be the index for the sections in a segment, where $0 \leq k \leq 15$ and $0 \leq j \leq 14$. Then B_{jk} denotes the byte in position k in section j. The group of bytes $B_{0k}, B_{1k}, B_{2k}, \ldots, B_{14k}$ is recorded in position k of the sections $0, 1, 2, \ldots, 14$, respectively, and form the kth code word W_k. Thus, the 16 code words are interleaved in the data format of 15 sections in a segment. This interleaving of the code words facilitates correction of mixed-mode errors. When a defect or dust particle affects up to 2 full sections, the resultant error is recoverable by correcting the corresponding 2 bytes in each of the 16 code words. On the other hand, many combinations of multiple 1-bit and 2-bit errors in a segment are also recoverable, since each code word can detect and correct any one of its bytes. Any stripe with a defect length of more than 128 bits is demarked by the write operation. Every write operation is followed by a readback check. Every demarked stripe is bypassed by the read operation.

Excluding the 2 error-checking sections in each segment, the 20 segments in a stripe provide a net recording space for a data stream of 4160 bytes (Fig. 5.10). The first 2 bytes in this data stream are reserved for stripe identification. This is followed by a block of 4096 bytes of data, 60 bytes of filler 0s, and 2 bytes of cyclic redundancy check (CRC) code. The

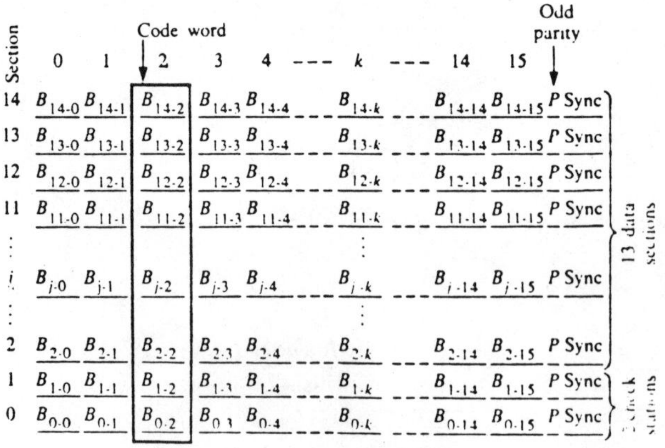

Figure 5.9 A segment: 15 sections formed with 16 interleaved code words.

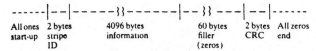

| All ones start-up | 2 bytes stripe ID | 4096 bytes information | 60 bytes filler (zeros) | 2 bytes CRC | All zeros end |

Figure 5.10 Data stream of 4160 bytes in a stripe.

2-byte CRC code provides an overall check for the data integrity of the stripe, including errors caused by malfunctions in the data-flow hardware and error-correction process. The read operation is retried both with and without a change in the read-amplifier gain setting when an uncorrectable error is encountered in any code word or when a miscorrected error is detected by the CRC at the end of the stripe.

5.8.2 Error-correction code

As just described, the error-recovery scheme is designed around the concept of resynchronizable sections. The basic building block of this scheme is a 15-byte code word of a single-byte-correcting Reed-Solomon code. The generator polynomial of this code is given by

$$G(x) = (x - \beta^0)(x - \beta^1) \tag{5.58}$$

where β is an element of $GF(2^8)$ and is also a primitive element of a 16-element subfield of $GF(2^8)$. In particular

$$\beta = \alpha^\lambda \tag{5.59}$$

where α is a primitive element of $GF(2^8)$ and $\lambda = 68$ (λ must be a multiple of 17 and a prime to 15). The elements β, β^2, β^3, ..., β^{15} are distinct non-zero elements of a 16-element subfield of $GF(2^8)$, and $\beta^{15} = \beta^0$. The sum and product of these elements are closed in the sense that the result is always one of the elements of the subfield. It will be seen later that these properties of the subfield elements facilitate implementation of the decoder and save hardware and decoding time.

As seen in the set of Eq. (5.57), the two coding relations corresponding to the roots β^0 and β^1 of the generator polynomial can be written as

$$B_0 \oplus B_1 \oplus B_2 \oplus \cdots \oplus B_{14} = 0 \tag{5.60}$$

$$B_0 \oplus T^\lambda B_1 \oplus T^{2\lambda} B_2 \oplus \cdots \oplus T^{14\lambda} B_{14} = 0 \tag{5.61}$$

where B_0 and B_1 are 2 check bytes and B_3 through B_{14} are 13 data bytes. This code possesses dual error-correction capability; namely, it detects all double-bit errors and corrects all single-bit errors (a Hamming code), or it detects and corrects all single-byte errors (a Reed-Solomon code). This dual capability provides an effective method of reducing the probability of miscorrections in a tapelike, mixed-mode error environment. In particular, the following two assertions can be made (Patel, 1980a):

1. If the code is used for correction of single-byte errors, then it will not miscorrect any combination of two 1-bit errors.
2. If the code is used for correction of single-bit errors, then it will not miscorrect any combination of one byte error with a 1-bit error in another byte.

In the particular application under discussion, the code is used for correction of single-byte errors in the absence of error pointers. In this mode, the code exhibits a high level of protection against miscorrection of noise-induced bit errors in more than one byte. In the presence of error pointers, the code corrects two erroneous bytes. Another feature of this error recovery scheme is the zero-modulation encoding and its dual-function role. Zero modulation ensures the absence of a dc component in the recording signal and also provides very reliable error-detection pointers.

In a read operation, each section is read through the ZM decoding algorithm, which also checks for errors through stringent run-length and dc-charge constraints. Error-free ZM patterns possess run lengths of at least one and at most three 0s between two 1s, and the dc-charge value is always constrained within ± 3 units. Thus, two consecutive 1s or four consecutive 0s indicate an error. The dc charge can be monitored with an up-down counter which increments for every digit position recorded with a positive level and decrements in a similar manner when the level is negative; it signals an error if the total exceeds ± 3 at any time. The charge value must also be zero at the end of the section, before the synchronization pattern. These checks and the odd parity at the end of each section detect most errors, including the 2-bit errors caused by bit shift and the dropout and synchronization errors caused by defects and dust particles. The errors in ZM decoding do not propagate, and the decoding process always terminates at a section boundary. Thus, the presence of an error is usually detected by the ZM error-detection circuits in the vicinity of the error within the same section. The resynchronization signal at the beginning and at the end of each section provides or confirms the proper phase of the ZM double-frequency clock, thereby rendering each section independent in error modes.

All detected errors are reported to the decoder of the error-correction code for error recovery. Errors in up to two full sections in a segment can be recovered by means of this error-correction code; this includes the longest error in a worst-case situation in which the defect coincides with a section boundary and affects two adjacent sections. A wide variety of shorter multiple errors are also detected and corrected by the same error-correction code.

Encoding process.   The generator polynomial of Eq. (5.58) can be rewritten as

$$G(x) = x^2 + (1 + \alpha^\lambda)x + \alpha^\lambda \tag{5.62}$$

The encoding can be performed by a shift register network built for modulo $G(x)$ operations. Figure 5.11 shows a block diagram of this shift register, which can be constructed from the conventional binary network elements. The sum of any field elements β_1 and β_2, represented by eight-digit binary vectors B_1 and B_2, can be accomplished by the modulo-2 matrix sum of B_1 and B_2. The multiplication of any field element β by the elements α^λ and $1 \oplus \alpha^\lambda$ can be accomplished by the modulo-2 matrix multiplications $T^\lambda B$ and $[I \oplus T^\lambda]B$, respectively, where the eight-digit binary vector B represents β, and matrix T is given by

$$T = \begin{bmatrix} 0 & 0 & 0 & 0 & 0 & 0 & 0 & 1 \\ 1 & 0 & 0 & 0 & 0 & 0 & 0 & 1 \\ 0 & 1 & 0 & 0 & 0 & 0 & 0 & 0 \\ 0 & 0 & 1 & 0 & 0 & 0 & 0 & 1 \\ 0 & 0 & 0 & 1 & 0 & 0 & 0 & 0 \\ 0 & 0 & 0 & 0 & 1 & 0 & 0 & 1 \\ 0 & 0 & 0 & 0 & 0 & 1 & 0 & 0 \\ 0 & 0 & 0 & 0 & 0 & 0 & 1 & 0 \end{bmatrix} \tag{5.63}$$

The resulting encoding network is, in fact, an eight-channel binary shift register, as shown in Fig. 5.12, in which each storage element of Fig. 5.11 is replaced by eight binary storage elements, and the modulo-2 matrix multiplication or addition is realized by means of a set of binary modulo-2 gates (XOR circuits). Figure 5.13 shows separately a network of modulo-2 gates for multiplication of any 8-bit vector with the matrix T^λ.

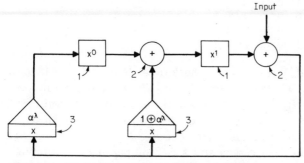

Figure 5.11   Block diagram of encoding network.

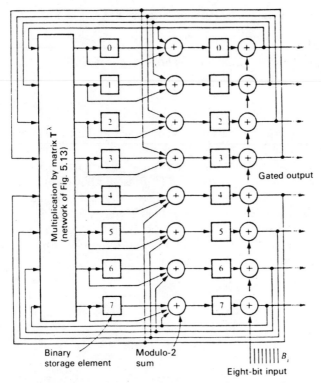

Figure 5.12 Encoding shift register $G(x) = T^\lambda \oplus (1 \oplus T^\lambda) x \oplus X^2$. All feedback connections are gated open for output.

The matrix T^λ for $\lambda = 68$ can be computed from T and is given by

$$
T^{68} = \begin{bmatrix}
0 & 0 & 0 & 0 & 1 & 0 & 0 & 0 \\
1 & 0 & 0 & 0 & 1 & 1 & 0 & 0 \\
0 & 1 & 0 & 0 & 0 & 1 & 1 & 0 \\
0 & 0 & 1 & 0 & 1 & 0 & 1 & 1 \\
1 & 0 & 0 & 1 & 0 & 1 & 0 & 1 \\
0 & 1 & 0 & 0 & 0 & 0 & 1 & 0 \\
0 & 0 & 1 & 0 & 0 & 0 & 0 & 1 \\
0 & 0 & 0 & 1 & 0 & 0 & 0 & 0
\end{bmatrix} \tag{5.64}
$$

The 15-byte code word W consists of 2 check bytes B_0 and B_1 and 13 data bytes $B_2, B_3, \ldots, B_{14}$. The check bytes B_0 and B_1 are computed by processing the data bytes $B_2, B_3, B_4, \ldots, B_{14}$ in the encoding shift register of Fig. 5.12. Initially, the storage elements of the shift register are all set to zero. The ordered sequence of data bytes $B_{14}, B_{13}, B_{12}, \ldots, B_2$ is entered into the shift register in 13 successive shifts, as 8-bit parallel vector inputs. At the end of this operation, the shift register contains the check bytes B_1

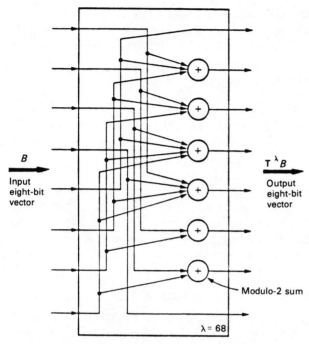

Figure 5.13 Network for multiplication by T^{λ}.

and B_0 in its high- and low-order positions, respectively. Then B_1 and B_0 are gated out without feedback and appended to the data bytes to form a 15-byte code word.

Table 5.12 illustrates this encoding process with an example showing the contents of the shift register after each shift.

5.5.3 Syndromes of error

The read data are checked for errors by means of the coding equations, Eqs. (5.60) and (5.61). All 16 code words of a segment are stored in a temporary storage pending any correction of errors. The decoding process is carried out by applying the decoding algorithm to each of the 16 code words independently. The algorithm will correct any 1 byte in an unknown position or any 2 bytes in indicated positions in each of the 16 code words. Let $\hat{B}_0, \hat{B}_1, \hat{B}_2, \ldots, \hat{B}_{14}$ denote the read bytes corresponding to the written bytes $B_0, B_1, B_2, \ldots, B_{14}$, respectively. Let S_0 and S_1 denote the results of computations when the read byte values are substituted in place of the written byte values in the left-hand side of Eqs. (5.60) and (5.61). If the read code word is error-free, then S_0 and S_1 both will be zero, as seen from these equations; however, a nonzero value in S_0 or S_1 indicates that one or more read bytes are in error. The 8-bit vectors S_0 and S_1 are called *syndromes of error* and are given by

$$S_0 = \hat{B}_0 \oplus \hat{B}_1 \oplus \hat{B}_2 \oplus \cdots \oplus \hat{B}_{14} \tag{5.65}$$

and
$$S_1 = \hat{B}_0 \oplus T\hat{B}_1 \oplus T^{2\lambda}\hat{B}_2 \oplus \cdots \oplus T^{14\lambda}\hat{B}_{14} \tag{5.66}$$

The syndrome vectors S_0 and S_1 can be computed by means of two separate 8-bit shift registers SR_0 and SR_1, respectively. These shift registers are shown in Fig. 5.14. The shifting operation in SR_0 and SR_1 corresponds to multiplying the content vector by the matrix I and by the matrix T^λ, respectively. Initially, the registers are set to contain zeros. The ordered sequence of read bytes $\hat{B}_{14}, \hat{B}_{13}, \hat{B}_{12}, \ldots, \hat{B}_1, \hat{B}_0$ is entered into both registers, SR_0 and SR_1, in 15 successive shifting operations, with $\hat{B}_{14}$ entering first. As a result, SR_0 contains the syndrome S_0, and SR_1 contains the syndrome S_1. Table 5.13 illustrates the syndrome generation process with an example showing the contents of the two shift registers after each shift.

5.8.4 Correction of two bytes

When the erroneous sections are indicated by ZM detection, this information is passed on to the decoder in the form of error pointers. Let i and j denote the position values of the two erroneous bytes in a code word, where $i < j$. The symbols E_i and E_j are used to represent the unknown error patterns in bytes B_i and B_j, respectively, so that

$$\hat{B}_i = B_i \oplus E_i \tag{5.67}$$

and
$$B_j = B_j \oplus E_j \tag{5.68}$$

TABLE 5.12 Computation of Check Bytes B_0 and B_1

	Write data input	Shift count	Contents low-order byte	Contents high-order byte
B_{14}	10010111	1	00000111	10010000
B_{13}	11101000	2	11101111	10010000
B_{12}	10101010	3	11111111	00101010
B_{11}	11111000	4	01010001	01111100
B_{10}	11011001	5	00101000	11011100
B_9	10010001	6	10000110	11100011
B_8	00010101	7	00101011	01011011
B_7	01111111	8	01111010	01110101
B_6	00000000	9	01001101	01000010
B_5	00100111	10	01000100	01101100
B_4	10000001	11	11011100	01110101
B_3	01010101	12	00010010	11101110
B_2	10111110	13	00101101 = B_0	01101111 = B_1

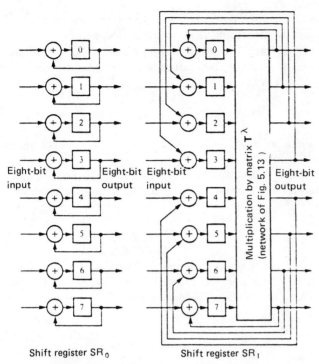

Figure 5.14 Decoding shift registers. All feedback connections are gated open and multiplication by T^λ is bypassed for output.

TABLE 5.13   Syndrome Computation: Two Known Bytes in Error

	Read data input	Shift count	Contents of SR_0	Contents of SR_1
B_{14}	10010111	1	10010111	10010111
B_{13}	11101000	2	01111111	11101111
B_{12}	10101010	3	11010101	01000010
B_{11}	11111000	4	00101101	11101000
B_{10}	01110011	5	01011110	11011101
B_9	10010001	6	11001111	01010110
B_8	00010101	7	11011010	01100100
B_7	01111111	8	10100101	00100001
B_6	00000000	9	10100101	00001000
B_5	00100111	10	10000010	11110111
B_4	00000111	11	10000101	00110110
B_3	01010101	12	11010000	00010010
B_2	10111110	13	01101110	10000011
B_1	01101111	14	00000001	00001001
B_0	00101101	15	$\boxed{00101100 = S_0}$	$\boxed{11100111 = S_1}$

When the indices i and j are known, the unknown error patterns E_i and E_j can be determined by processing the syndromes S_0 and S_1, provided all other bytes are error-free. The syndrome equations, Eqs. (5.65) and (5.66), can be reduced in terms of these unknown error patterns by combining with the coding equations, Eqs. (5.60) and (5.61), respectively. Thus we have

$$S_0 = E_i \oplus E_j \tag{5.69}$$

and
$$S_1 = T^{i\lambda}E_i \oplus T^{j\lambda}E_j \tag{5.70}$$

Since i and j are known, the two simultaneous equations, Eqs. (5.69) and (5.70), may be solved for the two unknown variables E_i and E_j to obtain

$$E_j = [I \oplus T^{(j-i)\lambda}]^{-1}[S_0 \oplus T^{-i\lambda}S_1] \tag{5.71}$$

and
$$E_i = S_0 \oplus E_j \tag{5.72}$$

Equations (5.71) and (5.72) may be implemented with simple hardware. For this, the closure property and the multiplicative inverse of the subfield elements, which were discussed before, are used. In particular, we note that

$$T^{p\lambda} = T^{-i\lambda} \tag{5.73}$$

and
$$T^{q\lambda} = [I \oplus T^{(j-i)\lambda}]^{-1} \tag{5.74}$$

where p and q depend only on the known values of i and j. The parameters p and q are precalculated for all possible values of i and j and are given in Tables 5.14 and 5.15, respectively.

Thus the decoding equations, Eqs. (5.71) and (5.72), can be rewritten into simpler form as

$$E_j = T^{q\lambda}[S_0 \oplus T^{p\lambda}S_1] \tag{5.75}$$

and
$$E_i = S_0 \oplus E_j \tag{5.76}$$

TABLE 5.14   Parameter p as a Function of i

i	0	1	2	3	4	5	6	7	8	9	10	11	12	13	14
p	15	14	13	12	11	10	9	8	7	6	5	4	3	2	1

TABLE 5.15   Parameter q as a Function of $(j - i)$

$j - i$	1	2	3	4	5	6	7	8	9	10	11	12	13	14	†
q		3	6	11	12	5	7	2	9	13	10	1	14	8	4	0

†j is undefined, and error pattern E_j is zero.

The decoder then consists of the following four simple steps:

Step 1. Multiply S_1 by the matrix $T^{p\lambda}$

Step 2. Add S_0 to the result of step 1

Step 3. Multiply the result of step 2 by $T^{q\lambda}$

Step 4. Add S_0 to the result of step 3

Table 5.16 illustrates this syndrome-decoding process with an example showing the results at each of the four steps.

The multiplication by $T^{p\lambda}$ and $T^{q\lambda}$ of steps 1 and 3 can be performed by means of the shift register SR_1 of Fig. 5.14 with p and q shifting operations, respectively. The addition of S_0 of steps 2 and 4 can be accomplished by entering the vector S_0 into SR_1 at the time of the last shifting operation of the previous step. The results of steps 3 and 4 provide the correction patterns E_i and E_j for bytes $\hat{B}_i$, and $\hat{B}_j$, respectively. When only one byte is in error, as indicated by pointer i, and the second pointer value j is undefined, the syndrome processing still determines E_i and E_j, in which E_j must result in a zero value. A nonzero value of E_j in this case indicates an uncorrectable error in one or more unknown byte positions.

5.8.5 Detection and correction of one byte

Through violations of one or more ZM constraints, almost all errors are detected. However, if any error escapes this detection, the decoder may encounter a code word with nonzero syndromes and absence of error pointers. In this case the syndromes are processed for detection and correction of a 1-byte error. Here the decoder determines the index i of the erroneous byte and the corresponding error pattern E_i. When all other bytes are error-free, the syndrome equations, Eqs. (5.65) and (5.66), in

TABLE 5.16 Syndrome Processing for Two Erroneous Bytes

Step	Operation	Contents of SR_1
1	Shift SR_1 $p = 11$ times	10111110
2	Add contents of SR_0 to SR_1	10010010
3	Shift SR_1 $q = 7$ times	$\boxed{10101010 = E_j}$
4	Add contents of SR_0 to SR_1	$\boxed{10000110 = E_i}$

$\hat{B}_i = 00000111$	$\hat{B}_j = 10101010$
$E_i = 10000110$	$E_j = 10101010$
Corrected $B_i = \overline{10000001}$	Corrected $B_j = \overline{11011001}$

TABLE 5.17   Syndrome Computation: One Byte in Error

	Read data input	Shift count	Contents of SR_0	Contents of SR_1
$\hat{B}_{14}$	10010111	1	10010111	10010111
$\hat{B}_{13}$	11101000	2	01111111	11101111
$\hat{B}_{12}$	10101010	3	11010101	01000010
$\hat{B}_{11}$	11111000	4	00101101	11101000
$\hat{B}_{10}$	01110011	5	01011110	11011101
$\hat{B}_{9}$	10010001	6	11001111	01010110
$\hat{B}_{8}$	00010101	7	11011010	01100100
$\hat{B}_{7}$	01111111	8	10100101	00100001
$\hat{B}_{6}$	00000000	9	10100101	00001000
$\hat{B}_{5}$	00100111	10	10000010	11110111
$\hat{B}_{4}$	00000111	11	10000101	00110110
$\hat{B}_{3}$	01010101	12	11010000	00010010
$\hat{B}_{2}$	10111110	13	01101110	10000011
$\hat{B}_{1}$	01101111	14	00000001	00001001
$\hat{B}_{0}$	00101101	15	$\boxed{10101010 = S_0}$	$\boxed{01001101 = S_1}$

view of the coding equations, Eqs. (5.60) and (5.61), reduce to

$$S_0 = E_i \tag{5.77}$$

and $$S_1 = T^{i\lambda}E_i \tag{5.78}$$

Thus, the error pattern E_i is determined by the syndrome S_0. Also, from Eqs. (5.77) and (5.78) we have

$$T^{-i\lambda}S_1 = S_0 \tag{5.79}$$

that is, $$T^{(15-i)\lambda}S_1 = S_0 \tag{5.80}$$

Once again, using the shift register SR_1 of Fig. 5.14, the index i can be determined. With S_1 as the initial content, SR_1 is shifted and its contents are compared with S_0 while counting down from 15 with each shift. When the contents do compare with S_0, the count indicates the index i of the erroneous byte. If the contents do not compare with S_0 even when the counter reaches zero, then this indicates the presence of two or more erroneous bytes. Tables 5.17 and 5.18 illustrate the syndrome computation and syndrome decoding processes for the case of one byte in error.

5.9   Coding in an 18-Track Tape-Storage Subsystem

A recent tape-storage subsystem (IBM 3480) introduces a tape cartridge with an 18-track data format which uses a new coding scheme called the

TABLE 5.18   Syndrome Processing for One Erroneous Byte

Shift count	Contents of SR_0	Contents of SR_1	Equal
15	10101010	01001101	No
14	10101010	10000110	No
13	10101010	00010100	No
12	10101010	01100001	No
11	10101010	00101100	No
10	10101010	10101010	Yes

Erroneous byte position = 10
$$\hat{B}_{10} = 01110011$$
$$\text{Error} = 10101010 \qquad \text{(contents of } SR_0\text{)}$$
$$\text{Corrected } B_{10} = \overline{11011001}$$

adaptive cross-parity code (AXP) (Patel, 1985*b*). In this scheme, the 18 tracks are divided into two interleaved sets of 9 tracks, and each set consists of 7 data tracks and 2 check tracks. The number of check tracks, thus, are in the same proportion as that in a prior 9-track scheme; however, by adaptive use of the checks in the two interleaved sets, the new scheme corrects up to 3 erased tracks in any one set of 9 tracks and up to 4 erased tracks in the two sets together.

The coding structure is based on the concept of interacting vertical and cross parity checks. The vertical parity checks are applied independently to each of the two sets of tracks; the cross parity checks extend over both the sets, providing adaptive usage of redundancy. The decoding procedure is iterative and uses parity equations which involve only one unknown variable at a time. The resulting implementation is simple and inexpensive.

During the iterative error-correction process, the decoder identifies an approaching new erroneous track and corrects up to two erroneous tracks in the two sets together. The third erroneous track in one set and the fourth erroneous track of the two sets together are corrected on-the-fly or on reread when they are identified as erasures by external means.

5.9.1   Encoding equations

As shown in Fig. 5.15, 18 parallel tracks are recorded along the tape. The tracks are grouped into two sets. Set A consists of any 9 parallel tracks and set B consists of the remaining 9 parallel tracks. In Fig. 5.15 the two sets are shown side by side with a symmetrically ordered arrangement of the tracks. This is done for convenience in describing the code. In actual practice, however, the tracks from two sets are interleaved and may be arranged in any other order.

Let $A_m(t)$ and $B_m(t)$ denote the mth bit in the track t of set A and set B, respectively. The track number t takes on values from 0 to 8 in each set. The bit position m takes on values from 0 to M. Tracks labeled 0 and 8 in each set are check tracks.

Each check bit in track 0 of set A provides a cross parity check along the diagonal with positive slope, involving bits from both sets, as seen in Fig. 5.15. The mth cross parity check of set A is given by the encoding equation

$$A_m(0) = \sum_{t=1}^{7} A_{m-t}(t) \oplus \sum_{t=0}^{7} B_{m-15+t}(t) \tag{5.81}$$

where the circle superimposed on the summation symbol indicates modulo-2 sum.

Each check bit in the track 0 of set B provides a cross parity check along the diagonal with negative slope, involving bits from both sets, as seen in

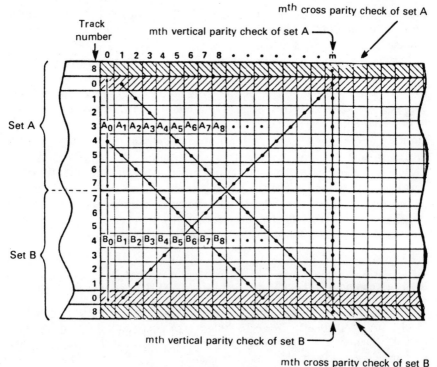

Figure 5.15 Data format; adaptive cross-parity code; 18 tracks grouped into two sets.

Fig. 5.15. The mth cross parity check of set B is given by the encoding equation

$$B_m(0) = \sum_{t=1}^{7} B_{m-t}(t) \oplus \sum_{t=0}^{7} A_{m-15+t}(t) \tag{5.82}$$

Eqs. (5.81) and (5.82) can be rewritten in a more convenient symmetrical form as

$$\sum_{t=0}^{7} [A_{m-t}(t) \oplus B_{m-15+t}(t)] = 0 \tag{5.83}$$

$$\sum_{t=0}^{7} [B_{m-t}(t) \oplus A_{m-15+t}(t)] = 0 \tag{5.84}$$

At the beginning of the record, computations of the cross-parity-check bits for positions 0 to 15 involve data-bit values from void positions (with negative position numbers). For convenience, these data-bit values will be considered to have zero binary value. At the end of the record, in order to provide diagonal checks to all bits in each track, the zero check track in each set will be extended 15 positions. The check bits on the extended positions also involve some data-bit values from void positions which will be assumed to have zero binary value.

Each check bit in the eighth track of set A is a vertical parity check over the bits of the same position number m in set A. The mth vertical parity check of set A is given by the equation

$$\sum_{t=0}^{8} A_m(t) = 0 \tag{5.85}$$

Similarly, the mth vertical parity check of set B is given by the equation

$$\sum_{t=0}^{8} B_m(t) = 0 \tag{5.86}$$

5.9.2 Error syndromes

Let $\hat{A}_m(t)$ and $\hat{B}_m(t)$ denote the bit values corresponding to $A_m(t)$ and $B_m(t)$, respectively, as they are read from the tape. These read-back bits may be corrupted by errors. The result of the parity checks of Eqs. (5.83), (5.84), (5.85), and (5.86) applied to the read-back data is called the *syndrome of error*. A nonzero syndrome is a clear indication of the presence of an error.

The mth cross parity check of set A yields the syndrome

$$Sd_m^a = \sum_{t=0}^{7} \hat{A}_{m-t}(t) \oplus \hat{B}_{m-15+t}(t) \tag{5.87}$$

The mth cross parity check of set B yields the syndrome

$$Sd_m^b = \sum_{t=0}^{7} \hat{B}_{m-t}(t) \oplus \hat{A}_{m-15+t}(t) \tag{5.88}$$

The mth vertical check for set A yields the syndrome

$$Sv_m^a = \sum_{t=0}^{8} \hat{A}_m(t) \tag{5.89}$$

The mth vertical check for set B yields the syndrome

$$Sv_m^b = \sum_{t=0}^{8} \hat{B}_m(t) \tag{5.90}$$

The modulo-2 difference between the read $\hat{A}_m(t)$ and the written $A_m(t)$ is called the *error pattern* $e_m^a(t)$ in the mth position of track t of set A.

Therefore, for set A we have

$$e_m^a(t) = \hat{A}_m(t) \oplus A_m(t) \tag{5.91}$$

Similarly, for set B

$$e_m^b(t) = \hat{B}_m(t) \oplus B_m(t) \tag{5.92}$$

Now if we compare the coding equations, Eqs. (5.83), (5.84), (5.85) and (5.86), with the corresponding syndrome equations, Eqs. (5.87), (5.88), (5.89) and (5.90), respectively, and substitute $e_m^a(t)$ and $e_m^b(t)$ from Eqs. (5.91) and (5.92), we get

$$Sd_m^a = \sum_{t=0}^{7} e_{m-t}^a(t) \oplus e_{m-15+t}^b(t) \tag{5.93}$$

$$Sd_m^b = \sum_{t=0}^{7} e_{m-t}^b(t) \oplus e_{m-15+t}^a(t) \tag{5.94}$$

$$Sv_m^a = \sum_{t=0}^{8} e_m^a(t) \tag{5.95}$$

$$Sv_m^b = \sum_{t=0}^{8} e_m^a(t) \tag{5.96}$$

Many different types of errors can be corrected by processing these syndromes. In tapes the predominant errors are track errors caused by large-size defects in the magnetic medium. The erroneous track may be identified by detecting loss of signal, excessive phase error, inadmissible recording pattern, or any other similar external pointer. In the absence of such external pointers, the erroneous track can be identified by processing the syndromes. We will show that any one of the following combinations of track errors can be corrected by processing the syndromes.

1. Up to three known erroneous tracks in one set *and* up to one known erroneous track in the other set

2. Up to two known or one unknown erroneous track in each of the two sets

3. Up to one known *and* one unknown erroneous track in one set, *and* up to one known erroneous track in the other set.

5.9.3 Correction of errors in known erroneous tracks

Errors confined to three known tracks in set A are correctable if set B is error-free or has only one known track in error. The erroneous tracks are

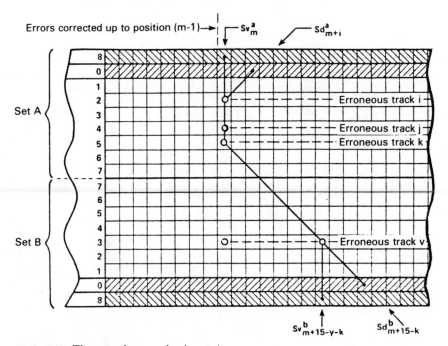

Figure 5.16 Three-track correction in set A.

indicated by track-error pointers i, j, k in set A and y in set B. If y is undefined, then set B is assumed to be error-free.

For convenience in decoding, i is the lowest and k is the highest track index among the erroneous tracks from track number 0 to 7. Track j is the remaining erroneous track so that either $(i < j < k)$ or $(j = 8$ and $i < k)$. Since set B has only one known track in error, the vertical-parity-check syndromes Sv_m^b yield the error patterns for this track. On elimination of the known zero-error patterns corresponding to the error-free tracks, Eq. 5.29 can be rewritten as

$$Sv_m^b = e_m^b(y) \tag{5.97}$$

Assume that all errors are corrected up to byte $m - 1$ and the syndrome equations are adjusted for all corrected error patterns. Then, as shown in Fig. 5.16, the error patterns for the mth position of track i, j, and k of set A can be determined from the syndromes Sd_{m+i}^a, Sd_{m+15-k}^b, and Sv_m^a. We can write the equations for these syndromes from Eq. (5.93), (5.94), and (5.95).

On eliminating the known zero-error patterns corresponding to the error-free tracks and the corrected error patterns up to position $m - 1$ in each track, these equations can be written as

$$Sd_{m+i}^a = e_m^a(i) \tag{5.98}$$

$$Sd_{m+15-k}^b = \begin{cases} e_m^a(k) \oplus e_{m+15-y-k}^b(y) & \text{if } y < 8 \\ e_m^a(k) & \text{if } y = 8 \text{ or set } B \text{ is error-free} \end{cases} \tag{5.99}$$

$$Sv_m^a = e_m^a(i) \oplus e_m^a(j) \oplus e_m^a(k) \tag{5.100}$$

From Eq.(5.97), we have

$$Sv_{m+15-y-k}^b = e_{m+15-y-k}^b(y) \tag{5.101}$$

If $Sv_{m+15-y-k}^b$ in Eq. (5.101) is nonzero for some $0 \le y \le 7$ and y is unknown, then set B must be processed first to identify the unknown track in error.

Equations (5.98) to (5.101) yield the error patterns

$$e_m^a(i) = Sd_{m+i}^a \tag{5.102}$$

$$e_m^a(k) = \begin{cases} Sd_{m+15-k}^b \oplus Sv_{m+15-y-k}^b & \text{if } y < 8 \\ Sd_{m+15-k}^b & \text{if } y = 8 \text{ or set } B \text{ is error-free} \end{cases} \tag{5.103}$$

$$e_m^a(j) = Sv_m^a \oplus e_m^a(i) \oplus e_m^a(k) \tag{5.104}$$

The mth bits in tracks i, j, and k are then corrected, using these error patterns as

$$A_m(i) = \hat{A}_m(i) \oplus e_m^a(i) \tag{5.105}$$

$$A_m(j) = \hat{A}_m(j) \oplus e_m^a(j) \tag{5.106}$$

$$A_m(k) = \hat{A}_m(k) \oplus e_m^a(k) \tag{5.107}$$

Before proceeding for the correction of the next position, we must modify the syndromes affected by these corrections. The modification is shown by an arrow from the previous value of a syndrome (with its modification) to its new value

$$Sd_{m+i}^a \leftarrow Sd_{m+i}^a \oplus e_m^a(i) \tag{5.108}$$

$$Sd_{m+j}^a \leftarrow Sd_{m+j}^a \oplus e_m^a(j) \qquad \text{if } j < 8 \tag{5.109}$$

$$Sd_{m+k}^a \leftarrow Sd_{m+k}^a \oplus e_m^a(k) \tag{5.110}$$

$$Sd_{m+15-i}^b \leftarrow Sd_{m+15-i}^b \oplus e_m^a(i) \tag{5.111}$$

$$Sd_{m+15-j}^b \leftarrow Sd_{m+15-j}^b \oplus e_m^a(j) \qquad \text{if } j < 8 \tag{5.112}$$

$$Sd_{m+15-k}^b \leftarrow Sd_{m+15-k}^b \oplus e_m^a(k) \tag{5.113}$$

Now the decoding procedure can be applied to the next bit position by incrementing the value of m by 1. At the new bit position m, we have met the required condition that all errors up to bit position $m - 1$ are corrected, and the syndromes are adjusted for corrected error patterns. In particular, all error patterns affecting the syndrome value Sd_{m-1}^a are corrected. Consequently, we expect that

$$Sd_{m-1}^a = 0 \tag{5.114}$$

A nonzero value of Sd_{m-1}^a indicates the presence of uncorrected errors. This provides a partial check on the uncorrectable multitrack errors that are beyond the correction capability of the code.

Errors in two known tracks in set A can be corrected if set B has at the most one unknown or two known tracks in error. The erroneous tracks in set A are indicated by track-error pointers i and j, where $i < j$. The error patterns $e_m^a(i)$ and $e_m^a(j)$ for the tracks i and j can be calculated from the local syndromes Sd_{m+i}^a and Sv_m^a as seen from Eqs. (5.102) and (5.104).

Errors confined to only one known track in set A can be corrected by means of only the vertical-parity-check syndrome Sv_m^a of set A. Since the check Sv_m^a ranges over set A only, this correction capability is not affected by the error conditions in set B. The erroneous track in set A is indicated by a track-error pointer j, and the error pattern $e_m^a(j)$ for the known erroneous track j is obtained from Eq. (5.104).

5.9.4 Generation of track-error pointers

5.9.4.1 Pointer to first erroneous track in set A. Errors confined to only one unknown track in set A can be detected and corrected if set B has, at most, one unknown or two known tracks in error. It is assumed that errors in all tracks in set B are corrected up to bit position $m - 1$, and that the syndrome values are adjusted for all corrected error patterns. When all tracks in set A are error-free, the parity-check syndromes Sv_m^a and Sd_{m-i}^a are equal to zero for $0 < i < 7$. When any of these syndromes is found to be nonzero, it is an indication that an error is present in at least one of the tracks in the vicinity—that is, within the next seven bit positions. Assuming that only one erroneous track is affecting the syndromes, the index of the erroneous track can be determined by examining syndromes Sd_{m+7}^a and Sv_m^a as the bit-position value m progresses. The following assertion characterizes the generation of the first track-error pointer in set A:

Assertion 1: Let m_1 and m_2 denote the lowest values of bit positions such that

$$Sd_{m+i}^a \neq 0 \qquad \text{for } m = m_1 \text{ and } i = 7 \tag{5.115}$$

$$Sv_m^a \neq 0 \qquad \text{for } m = m_2 \tag{5.116}$$

Then track q is in error at bit position m_2 and the track index q is given by

$$q = \begin{cases} 7 - (m_2 - m_1) & \text{if } m_1 > m_2 \\ 8 & \text{otherwise} \end{cases} \tag{5.117}$$

The proof of assertion 1 follows from the geometric considerations depicted in Fig. 5.17. In the case when m_2 is greater than or equal to m_1, if the resulting value q is smaller than zero, then the syndromes are affected by two or more unknown erroneous tracks and the errors are uncorrectable. In the special case where m_1 is not captured, even when the bit position value exceeds m_2, the error is in the vertical-parity-check track. The process can be implemented using a running counter with the iterative operation, counting down from the value 7.

Once we have the index value of the erroneous track, the errors can be corrected by applying the procedure for correction of one track with track-error pointer, as discussed previously.

5.9.4.2 Pointer to second erroneous track in set A. Consider the case when set A is being corrected for errors in a known erroneous track, and another unknown track in set A begins to be affected by errors. This second unknown erroneous track can be detected, and both erroneous tracks

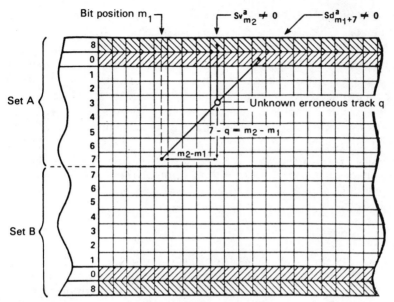

Figure 5.17   Generation of pointer to first erroneous track.

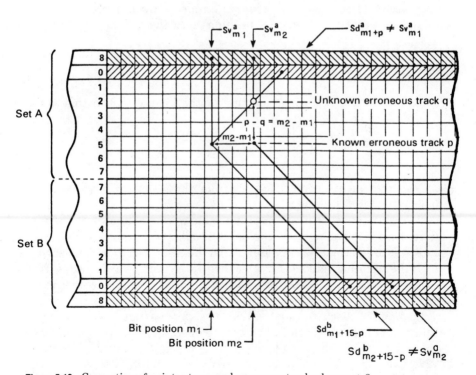

Figure 5.18   Generation of pointer to second erroneous track when $p \neq 8$.

of set A can be corrected provided that set B has, at the most, one known track in error.

For simplicity, first consider the case when tracks 0 to 7 in set B are error-free. Later it will be easy to see how the equations can be modified to include the effect of a known erroneous track in set B. Let p denote the known erroneous track in set A, and assume that, so far, all remaining tracks in set A are error-free. Also assume that all errors are corrected up to bit position $m-1$ and the syndrome values are adjusted for all corrected error patterns.

First consider the case where p is not the vertical-parity track, that is, $p \neq 8$. The error pattern of the mth position of the track p affects the syndromes Sd^b_{m+p}, Sd^b_{m+15-p}, and Sv^a_m. In the absence of errors in any other tracks

$$Sd^a_{m+p} = Sd^b_{m+15-p} = Sv^a_m = e^a_m(p) \tag{5.118}$$

However, as any one of the other tracks begins to be affected by an error, the syndrome relationship of Eq. (5.118) will no longer hold. We can make the following assertion by observing the effect of the new erroneous track on this relationship:

Assertion 2: Let p denote the known erroneous track where $p \neq 8$. Let m_1 and m_2 denote the lowest bit positions such that

$$Sd^a_{m+1} \neq Sv^a_m \qquad \text{for } m = m_1 \text{ and } i = p \tag{5.119}$$

$$Sd^b_{m+15-k} \neq Sv^a_m \qquad \text{for } m = m_2 \text{ and } k = p \tag{5.120}$$

Then the track q is in error starting at the bit position m that is the greater of m_1 and m_2; and q is given by

$$q = \begin{cases} p - (m_2 - m_1) & \text{if } m_2 \neq m_1 \\ 8 & \text{if } m_2 = m_1 \end{cases} \tag{5.121}$$

The proof of assertion 2 follows from the geometric considerations depicted in Fig. 5.18, which considers the case when m_2 is greater than m_1. Note in this case that if the resulting value q is smaller than 0, the syndromes are being affected by two or more unknown erroneous tracks and the errors are uncorrectable. Similarly, in the case when m_2 is smaller than m_1, if the resulting value q is greater than 7, the unknown erroneous track is in set B and will be detected and corrected later by the decoder in set B. In the case when m_2 is equal to m_1, the erroneous track q is in the vertical-parity-check track of set A. The process can be implemented using a running counter with the iterative operation, counting up or down from the value p depending upon the relationship between m_2 and m_1.

Next consider the case where the first unknown erroneous track is the vertical-parity-check track in set A, that is, $p = 8$. The error pattern of the mth position in track p in this case affects the syndrome Sv_m^a only. In the absence of errors in any other track in set A or set B, the cross-parity syndromes must all be zero. However, as any other track in set A begins to be affected by an error, the cross-parity syndromes will no longer be zero as the bit position value m is incremented. We make the following assertion:

Assertion 3: Let p denote the known erroneous track where $p = 8$. Let m_1 and m_2 denote the lowest bit positions such that

$$Sd_{m+i}^a \neq 0 \qquad \text{for } m = m_1 \text{ and } i = 7 \tag{5.122}$$

$$Sd_{m+15-k}^b \neq 0 \qquad \text{for } m = m_2 \text{ and } k = 7 - (m_2 - m_1) \tag{5.123}$$

Then track q is in error beginning at bit position m_2, and the index q is given by

$$q = k \tag{5.124}$$

The proof of assertion 3 follows from the geometric considerations depicted in Fig. 5.19. Note that if the resulting value of q is smaller than zero, the syndromes are affected by two or more unknown erroneous tracks and the errors are uncorrectable. The process can be implemented using a running counter with the iterative operation, counting down from the value 7.

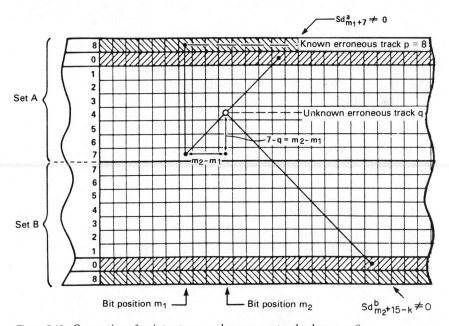

Figure 5.19 Generation of pointer to second erroneous track when $p = 8$.

Now we show the modification for the more general case when set B has, at the most, one track in error. Let y be the erroneous track in set B. The error patterns for this track are all known from the vertical-parity-check syndrome Sv_m^b of set B. If $y \neq 8$, then these error patterns also affect the values of cross-parity-check syndromes Sd_m^b. In order to account for the effect of these error patterns, we use the adjusted value $Sd_m^b \oplus Sv_{m-y}^b$ in place of Sd_m^b for any required value of m. In particular, in assertions 2 and 3, the syndrome Sd_{m+15-k}^b from set B is replaced by a composite syndrome SB where

$$ \text{SB} = \begin{cases} Sd_{m+15-k}^b \oplus Sv_{m+15-y-k}^b & \text{if } y < 8 \\ Sd_{m+15-k}^b & \text{if } y = 8 \text{ or set } B \text{ is error-free} \end{cases} \tag{5.125} $$

If $Sv_{m+15-y-k}^b$ is nonzero for some $0 \leq y \leq 7$ and y is unknown, then set B must be processed first to identify the unknown track in error. The count value $q > 7$ in assertion 2 indicates this condition.

It is interesting to note that the syndromes used in the generation of track error pointers, namely Sd_{m+i}^a, SB, and Sv_m^a, are the same in form as those used in Eqs. (5.102), (5.103), and (5.104) for correction of errors. Thus, with appropriate values for the variables i and k, the same hardware can be used to generate syndromes for both processes.

The coding rules possess a built-in mirror-image symmetry around set A and set B. In particular, the encoding and decoding equations for set B can be obtained from those of set A by substitution of corresponding variables in set B, as shown in Table 5.19.

5.9.5 Implementation of the decoding process

The decoding process consists of two distinct functions: (1) the detection and identification of the erroneous tracks, and (2) the correction of errors in known erroneous tracks. The internal pointer generator identifies the first erroneous track in both sets and subsequently identifies a second erroneous track in one of the two sets. Additional erroneous tracks may be detected and identified by external signals which usually require some form of analog sensing of the playback conditions of the data recorded on the various tracks. This includes detection of loss of read signal, excessive phase error, inadmissible recording code patterns, or any other similar external indicators.

Once an internal or external track-error pointer is generated, it may be

TABLE 5.19 Corresponding Variables in Set A and Set B

Set A	A_m	Sd_m^a	Sv_m^a	SA	i	j	k	e_m^a
Set B	B_m	Sd_m^b	Sv_m^b	SB	x	y	z	e_m^b

kept on for the entire remaining length of the record. Similarly, any track-error pointer may be turned off at an appropriate bit position in a record if the error patterns corresponding to the indicated track turn out to be zero consistently for a significant length of the record, thus confirming that the track is error-free. This allows replacement of the track-error pointer if and when some other track becomes erroneous, particularly in the case of long records.

It may be noted that the errors in a new erroneous track begin to affect the decoding process as much as 15 bit positions ahead of the actual error. Thus the beginning of each new external track-error pointer must be extended 15 bit positions earlier by means of a pointer look-ahead function. For the same reason, in the case of internal generation of track-error pointers, the beginning of two erroneous tracks cannot always be detected successfully if they occur in the vicinity of each other.

The error correction is done at the mth bit position which progresses from the zero value to the last bit position, of value M, in a recurring manner. In order to compute error patterns at the mth bit position, one requires the syndrome values ranging from bit-position m up to $m + 15$. Furthermore, it is necessary that all syndrome values be modified to account for the corrected error patterns prior to position m. The decoding process progresses in synchronism with the incoming read data characters. The actual correction of each received data character is delayed by 15 character positions.

5.10   Error Correction in Magnetic Disk Storage

The earlier disk storage products used either a simple parity check or a cyclic redundancy check (CRC) at the end of the record for the purpose of detection of errors. In these products, rereading was the only means of error recovery. Single-burst-error correction was introduced in disk drives from the year 1971 onward, with a Fire code, with a modified Fire code, and finally with interleaved codes based on Reed-Solomon algebra. Rereading is still an important part of the error-recovery procedure in all these products. The code is designed to provide good error-detection capability, and error correction is limited to only one burst of error in a record which may be as long as the entire track.

5.10.1   Error control in disk products using Fire codes

Fire codes have been used in disk drives introduced in the 1970s. The IBM 3340 and 3350 compatible disk storage products use a Fire code with the following generator polynomial

$$G(x) = (x^{13} + 1)(x^{35} + x^{23} + x^8 + x^2 + 1) \tag{5.126}$$

The degree of $G(x)$ is 48, and the cycle length is $13 \times (2^{35} - 1)$. Thus the code requires 6 check bytes at the end of very long code words. It is capable of correcting one burst of error up to 7 bits long. It can be used to correct any burst of length b bits or less while providing detection of a longer burst of length up to d bits if $b + d - 1 \leq 13$. In a 1973 disk file (IBM 3340), for instance, the parameters b and d are chosen to be 3 and 11, and in a later development (IBM 3350) they are 4 and 10, respectively. The cycle length of $13 \times (2^{35} - 1)$ far exceeds the length of any records in these disk products. Thus any practical record with 6 check bytes at the end forms a much-shortened variable-length code word. The encoding of this code word can be carried out by means of a 48-bit modulo-$G(x)$ shift register, as discussed before. The decoding can be done using the error-trapping technique. This requires two shift registers: one, connected for modulo $(x^{13} + 1)$, determines the pattern of the error burst; the other, connected for modulo $(x^{35} + x^{23} + x^8 + x^2 + 1)$, determines the location of the error burst. Since the cycle length of the second polynomial is very long, backward shifting operations will save time in decoding. This decoding method was also presented earlier in the section on Fire codes.

5.10.2   Modified Fire codes

Some disk products (e.g., IBM 3330) use a modified Fire code (Chien, 1969) where the modification is intended to reduce the number of shifting operations required in the decoding process. The error location, in this code, is determined by means of several short-cycle-length polynomials instead of one long-cycle-length polynomial. The generator polynomial of this code is given by

$$G(x) = G_0(x)G_1(x)G_2(x)G_3(x) \tag{5.127}$$

where
$$G_0(x) = x^{22} + 1 \tag{5.128}$$

$$G_1(x) = x^{11} + x^7 + x^6 + x + 1 \tag{5.129}$$

$$G_2(x) = x^{12} + x^{11} + x^{10} + x^9 + x^8 + x^7 + x^6$$
$$+ x^5 + x^4 + x^3 + x^2 + x + 1 \tag{5.130}$$

$$G_3(x = x^{11} + x^9 + x^7 + x^6 + x^5 + x + 1 \tag{5.131}$$

The degree of $G(x)$ is $22 + 11 + 12 + 11 = 56$. Hence the code requires 7 check bytes with maximum code-word length given by the cycle length of $G(x)$. The cycle lengths of $G_0(x)$, $G_1(x)$, $G_2(x)$, and $G_3(x)$ are 22, 89, 13, and 23, respectively. The cycle length of $G(x)$ is the least common multiple of these numbers, which is the product $22 \times 89 \times 13 \times 23 = 585,442$. The code is used for correcting any single burst of length up to 11 bits. Furthermore, it is claimed that it simultaneously detects any single burst of length up to 22 bits and most of the commonly encountered multiple random errors.

The encoding of this code requires a 56-bit shift register connected according to the composite polynomial $G(x)$, and the encoding process is the bit-by-bit shift-register processing in cyclic codes. The decoding is carried out with four shift registers SR_0, SR_1, SR_2, and SR_3 in which the feedback connections are in accordance with the polynomials $G_0(x)$, $G_1(x)$, $G_2(x)$, and $G_3(x)$, respectively.

In accordance with Eq. (5.32) the error pattern is an 11-bit error burst at position p represented by the polynomial

$$E(x) = x^p E_{11}(x) \tag{5.132}$$

The syndromes of error S_0, S_1, S_2, and S_3 are computed as four separate remainders using the four decoding shift registers. As in Eqs. (5.33) and (5.34), these syndromes can be expressed as

$$S_0(x) = x^p E_{11}(x) \text{ modulo } (x^{22} + 1) \tag{5.133}$$

$$S_1(x) = x^p E_{11}(x) \text{ modulo } G_1(x) \tag{5.134}$$

$$S_2(x) = x^p E_{11}(x) \text{ modulo } G_2(x) \tag{5.135}$$

$$S_3(x) = x^p E_{11}(x) \text{ modulo } G_3(x) \tag{5.136}$$

The decoding operation proceeds as follows: If all four syndromes are zero, then it is concluded that the word contains no error. If some but not all syndromes are zero, then an uncorrectable multiple-burst error is detected which will require rereading of the data. If all four syndromes are nonzero, then they will be processed in their respective shift registers in order to determine the error pattern and location.

The error pattern is trapped as usual by shifting the syndrome S_0 in the SR_0 register until the 11 high-order stages of the 22-stage register contain zeros; thus the burst is trapped in the 11 low-order stages of SR_0. Then the other syndromes are shifted in their respective registers until the contents in each match the error pattern captured in SR_0. Let C_0, C_1, C_2 and C_3 denote the number of shifts required with shift registers SR_0, SR_1, SR_2, and SR_3 in the above process of capturing the error pattern.

Each shift register has a short cycle length (the number of shifts after which it resets to its original content). Thus the error pattern must be captured within the cycle length of each register, otherwise the error is uncorrectable. Thus $C_0 < 22$, $C_1 < 89$, $C_2 < 13$, and $C_3 < 23$. The actual location p of error in the code word satisfies the following simultaneous congruence relations

$$(n-p) \equiv C_0 \text{ modulo } 22 \equiv C_1 \text{ modulo } 89 \equiv C_2 \text{ modulo } 13 \equiv C_3 \text{ modulo } 23$$

The integer $n - p$ may be computed through the Chinese remainder theorem as follows: Find four integers A_0, A_1, A_2, and A_3 such that

$$\left(\frac{n}{22}\right) A_0 + \left(\frac{n}{89}\right) A_1 + \left(\frac{n}{13}\right) A_2 + \left(\frac{n}{23}\right) A_3 = 1 \text{ modulo } n \qquad (5.137)$$

where n is $22 \times 89 \times 13 \times 23$. Then $n - p$ is given by

$$n - p = \left(\frac{n}{22}\right) A_0 C_0 + \left(\frac{n}{89}\right) A_1 C_1 + \left(\frac{n}{13}\right) A_2 C_2$$

$$+ \left(\frac{n}{23}\right) A_3 C_3 \text{ modulo } n \qquad (5.138)$$

The constants A_0, A_1, A_2, and A_3 are predetermined as 17, 11, 7, and 13, respectively. Thus $n - p$ and then p can be easily computed using the above equation.

Thus the decoding process consists of the following steps when the syndromes S_0, S_1, S_2, and S_3 are all nonzero:

Step 1.   Capture error pattern and determine C_0, C_1, C_2, and C_3

Step 2.   Compute $M = 452{,}387\ C_0 + 72{,}358\ C_1 + 315{,}238\ C_2 + 330{,}902\ C_3$

Step 3.   Divide M by 585,442 and find remainder m

Step 4.   Subtract m from 585,442 to obtain position value p

Thus, the fast decoding of Fire code is possible only when a facility for arithmetic operations is readily available. In particular, four multiplications, three additions, one division, and one subtraction are required in order to determine the position value p from the shift counts C_0, C_1, C_2, and C_3 of the shift registers. The number of shifting operations is limited to less than 22 shifts for the error pattern and less than 89 shifts for the error location.

The later disk drives (such as IBM 3370, 3375, and 3380) use byte-oriented processing with single-byte-correcting Reed-Solomon codes and two-way or three-way interleaving. The encoding and decoding processes in these applications are similar to those already described for the various tape products.

References

Adler, R. L., D. Coppersmith, and M. Hassner, "Algorithms for Sliding Block Codes," *IEEE Trans. Inform. Theory*, **IT-29**, 5 (1983).

Brown, D. T., and W. W. Peterson, "Cyclic Codes for Error Detection," *Proc. IRE*, **49**, 228 (1961).

Brown, D. T., and F. F. Sellers, Jr., "Error Correction for IBM 800-bit-per-inch Magnetic Tape," *IBM J. Res. Dev.,* **14,** 384 (1970).

Carasso, M., and H. Nakajima, "What is the Compact Disk Digital Audio System?" *IEEE Int. Conf. on Consum. Electron. Digest of Tech. Papers,* 138 (1982).

Chien, R. T., "Burst-Correction Codes with High-Speed Decoding," *IEEE Trans. Inform. Theory,* **IT-1,** 109 (1969).

Doi, T. T., "Error Correction in the Compact Disk System" *Audio,* **24,** April (1984).

Eggenberger, J., and P. Hodges, "Sequential Encoding and Decoding of Variable Word Length, Fixed Rate Data Codes," U.S. Patent 4,115,768, 1978.

Fire, P., "A Class of Multiple-Error-Correcting Binary Codes for Non-Independent Errors," Sylvania Report RSL-E-2, Sylvania Reconnaissance Systems Lab., Mountain View, Calif., 1959.

Franaszek, P. A., "Sequence-State Methods of Run-Length-Limited Coding," *IBM J. Res. Dev.,* **14,** 376 (1970).

Franaszek, P. A., "Run-Length-Limited Variable Length Coding with Error Propagation Limitation," U.S. Patent 3,689,899, 1972.

Hodges, P., W. J. Schaeuble, and P. L. Shaffer, "Error Correcting System for Serial-by-Byte Data," U.S. Patent 4,185,269, 1980.

Horiguchi, T., and K. Morita, "An Optimization of Modulation Codes in Digital Recording," *IEEE Trans. Magn.,* **MAG-12,** 740 (1976).

Jacoby, G. V., "A New Look-Ahead Code for Increased Data Density," *IEEE Trans. Magn.,* **MAG-13,** 1202 (1977).

Jacoby, G. V., and R. Kost, "Binary Two-Thirds Rate Code with Full Look-Ahead," *IEEE Trans. Magn.,* **MAG-20,** 709 (1984).

MacKintosh, N. D., "The Choice of a Recording Code," *Radio Electron. Eng,* **50,** 177 (1980).

Miller, J. W., U.S. Patent 3,108,261, 1963.

Miller, J. W., U.S. Patent 4,027,335, 1977.

Ouchi, N. K., "Apparatus for Encoding and Decoding Binary Data in a Modified Zero Modulation Data Code," U.S. Patent 3,995,264, 1976.

Patel, A. M., "Zero Modulation Encoding in Magnetic Recording," *IBM J. Res. Dev.,* **19,** 366 (1975).

Patel, A. M., "Error-Recovery Scheme for the IBM 3850 Mass Storage System," *IBM J. Res. Dev.,* **24,** 32 (1980*a*).

Patel, A. M., "Multitrack Error Correction with Cross-Parity-Check Coding." U.S. Patent 4,205,324, 1980*b*.

Patel, A. M., "Improved Encoder and Decoder for a Byte-Oriented $(0, 3)$ § Code," IBM Tech. Disclosure Bull. **28,** 1938 (1985*a*).

Patel, A. M., "Adaptive Cross-Parity Code for High-Density Magnetic Tape Subsystem," *IBM J. Res. Dev.,* **29,** 546 (1985*b*).

Patel, A. M., "On-the-Fly Decoder for Multiple Byte Errors," *IBM J. Res. Dev.,* **30,** 259 (1986).

Patel, A. M., and S. J. Hong, "Optimal Rectangular Code for High-Density Magnetic Tapes," *IBM J. Res. Dev.,* **18,** 579 (1974).

Peterson, W. W., "Encoding and Error-Correction Procedures for the Bose-Chaudhuri Codes," *IEEE Trans. Inform. Theory,* **6,** 459 (1960).

Peterson, W. W., *Error Correcting Codes,* MIT Press, Cambridge, Mass., 1961.

Prusinkiewicz, P., and S. Budkowski, "A Double-Track Error-Correction Code for Magnetic Tape," *IEEE Trans. Computers,* 642 (1976).

Reed, I. S., and G. Solomon, "Polynomial Codes over Certain Finite Fields," *J. Siam,* **8,** 300 (1960).

Shannon, C. E., "A Mathematical Theory of Communication," *Bell Syst. Tech. J.,* **27,** 379 (1948).

Magnetooptical Recording

Dan S. Bloomberg

Xerox Palo Alto Research Center, Palo Alto, California

G. A. Neville Connell

Xerox Palo Alto Research Center, Palo Alto, California

6.1 Overview of Magnetooptical Data Storage

High-density optical recording became a reality in a number of write-once products introduced in the mid-1980s. These drives offer about 2 GB of storage per disk at an areal density of about 4×10^5 b/mm^2 and data rates of about 5 Mb/s. The average access time was about 100 ms, but 20 adjacent tracks that contain about 1 MB of information could be accessed in about 1 ms. The disks themselves were removable. While there continue to be many opportunities for further innovations in this area, there is now keen competition to build systems that add erasability to this list of features. One approach, magnetooptical recording, is the subject of this chapter.

The principle of magnetooptical recording is as follows. A focused laser beam, pulsed to high power for a short time, raises the temperature of a perpendicularly magnetized medium sufficiently for an externally applied magnetic field to reverse the direction of magnetization in the heated region. When the medium returns to a lower temperature for readout, the reverse-magnetized domain persists. Erasure of domains is accomplished by the same thermal process, now aided by an oppositely directed magnetic field. At this fundamental level, there is one important difference between magnetooptical and magnetic recording; that is, the switching rate of the magnetic field may be much lower than the data rate. The latter is determined by the pulse rate of the laser and, therefore, very high-frequency recording should be possible. The price paid is that the write process must first be preceded by a separate erase step when a single magnetooptic layer is used, and unless this is compensated by the disk controller and operating system software, there will be a write delay of one revolution. Alternatively, a complex magnetic structure, consisting of a magnetooptic memory layer supported on a field-controlling magnetic underlayer, may be used to provide a direct overwrite capability (Saito et al., 1987).

Readout of information employs the polar Kerr effect. Linearly polarized light, reflected from a perpendicularly magnetized medium, is rotated to the left or right, according to the direction of magnetization. (Some ellipticity is also introduced into the reflected light, but it may be ignored for now.) By checking the direction of the plane of polarization of the reflected light, magnetization transitions along a recorded track can be read out by the same focused laser beam that was used for recording information. Reading has no influence on the recorded information because the read laser power is low and causes little temperature rise.

The essential features of a magnetooptical recording system are shown in Fig. 6.1a. In this system, collimated linearly polarized light from a GaAs laser passes through a slightly rotated polarizing beam splitter and is focused on a magnetooptical medium. The reflected light is then guided to the detection system by the same polarizing beam splitter. The detection system consists of a second polarizing beam splitter rotated by 45° about the optical axis so that the incident light is split between photodiodes 1 and 2. The detection system itself has three purposes: to sense the magnetic transitions and to provide focusing and tracking signals that maintain the laser spot in precise position relative to the disk surface and data tracks. The performance of each of these features will determine the eventual system performance. Currently, either the laser power and medium sensitivity or the focus and tracking servos impose an upper limit on the disk rotation rate; the signal-to-noise ratio imposes a limit on the minimum bit size; and tracking servo accuracy and read spot size determine the minimum track spacing. It will be seen later that these constraints suggest that, with currently available technology, removable 13-cm-diameter disks will contain 500 MB in systems maintaining a data rate of 5 Mb/s.

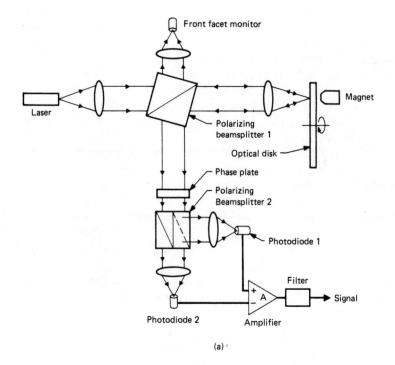

Front facet monitor

Laser

Magnet

Polarizing
beamsplitter 1

Optical disk

Phase plate

Polarizing
Beamsplitter 2

Photodiode 1

Filter

Signal

Photodiode 2   Amplifier

(a)

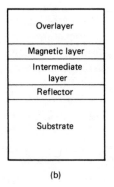

| Overlayer |
| Magnetic layer |
| Intermediate layer |
| Reflector |
| Substrate |

(b)

Figure 6.1 *(a)* Schematic representation of the layout of a magnetooptical recording system; and *(b)* a section of a magnetooptical quadrilayer medium *(after Connell et al., 1983)*.

The first suggestion of thermomagnetic writing (Mayer, 1958) apparently came from the observation that when thin perpendicularly magnetized films of Mn-Bi were heated by the point of a needle, magnetization reversal occurred in the heated region. Called *Curie-point writing*, this phenomenon occurred because the heated region had cooled from above the Curie temperature in the demagnetizing field created by the surrounding sample. A potential benefit of some ferrimagnetic materials for thermomagnetic recording (Chang et al., 1965) is that the rapid decrease of coercivity above the compensation temperature (at which the net magnetization on both sublattices is zero) permits recording to occur with a relatively small temperature rise. This recording process is known as compensation point writing. The writing and erasing processes can therefore, in principle at least, be achieved without thermal degradation of the sample, a problem that plagued much of the later work with Mn-Bi.

The principles for recording were thus established, but practical means for the subsequent readout of the recorded information were absent. The important events that changed this situation were the availability of solid-state lasers, the use of multilayer interference structures to improve the magnetooptical conversivity (Smith, 1965a; 1965b), the understanding and optimization of the system-medium performance (Goldberg, 1967; Mansuripur et al., 1982; Mansuripur and Connell, 1983a; Connell et al., 1983), and the discovery of the amorphous rare-earth transition-metal alloys (Chaudhari et al., 1973; Sunago et al., 1976; Mimura et al., 1976, 1978).

Lasers provide a simple means of delivering to a 1-μm-sized area the thermal energy needed for the writing and erasing processes (Smith, 1967). They therefore offer the potential to record at the densities that are now available in the commercial nonerasable optical recording products. They also make the readout process more amenable to drive integration. Nevertheless, readout with adequate signal-to-noise ratio was, until recently, a major problem. The primary issue is the minuscule size of the polar Kerr effect. When plane-polarized light is normally incident on a magnetooptical medium, a small fraction of the light reflected is polarized at right angles to the incident polarization. We denote the reflectivity of the medium by $R = r_x^2 + r_y^2$, where r_x and r_y are the unconverted (ordinary) and converted (magnetooptical) amplitudes, respectively, and $r_y \ll r_x$. In Sec. 6.3, it will be shown that the signal-to-shot-noise voltage

TABLE 6.1 Comparison of the Magnetooptical Effect in Amorphous and Crystalline Alloys

Material	r_y†
$Tb_{0.2}Fe_{0.8}$	3.0×10^{-3}
Mn-Bi	1.0×10^{-2}
Pt-Mn-Sb	1.3×10^{-2}
$Tb_{0.2}Fe_{0.8}$‡	8.0×10^{-3}

†Data at 633 nm.
‡30-nm layer in a quadrilayer.

ratio in the photodetectors is proportional to r_y, and Table 6.1 shows that the value of r_y in typical amorphous and polycrystalline materials is between 3×10^{-3} and 10^{-2}. Methods to enhance the signal-to-noise ratio by improving signal delivery, material selection, or medium design are essential for the success of the technology.

In early work, an intensity beam splitter was placed in the main beam, and one-half of both the incident laser light and the reflected magneto-optical signal were lost (by reflection and transmission, respectively). Improvement in the signal delivery system was achieved by using instead a polarizing beam splitter, which transmitted essentially all of the incident laser light and reflected all of the magnetooptical signal (Connell et al., 1983). Further improvement was achieved by using a second polarizing beam splitter with a differential detection scheme in the detector module (Goldberg, 1967), instead of a simple analyzer and detector.

Incorporation of the magnetic layer in multilayer interference structures improves the magnetooptical conversivity. These structures couple more incident light into the magnetic layer and couple more of the magnetoptically induced light back out. The quadrilayer structure shown in Fig. 6.1b is one example that will be discussed in more detail later. However, the use of amorphous rare-earth transition-metal alloys has probably been the most important advancement. The alloy composition and multilayer design can be selected to withstand the maximum read power and therefore allow for maximum signal size. This is accomplished by minimizing the writing sensitivity, subject to thermomagnetic recording remaining possible in the system at hand. The method involves the careful adjustment of the Curie and compensation temperatures of the alloy and the heat-flow characteristics of the medium. This is possible only because in amorphous alloys, the Curie and compensation temperatures can be controlled independently and continuously by alloying. Although the magnitude of the polar Kerr effect is smaller in these alloys than in some polycrystalline compounds, as shown in Table 6.1, the magnetic media noise on readout is negligible. This is in contrast with the performance of polycrystalline media in which modulation noise proved to be unacceptable. The difference occurs because the domain walls can be pinned intrinsically in the best amorphous alloys, rather than on grain boundaries or defects as in polycrystalline material. As a result, bit-to-bit fluctuations in the location of the transitions are minimal in the amorphous media. In well-designed systems, shot-noise–limited performance can be approached.

There have also been important advances in noise suppression by both optical and electronic means. The differential detection scheme in Fig. 6.1a reduces medium noise associated with substrate and medium inhomogeneities (Goldberg, 1967), while laser noise can be reduced to acceptable levels by rf drive techniques (Ohishi et al., 1984). Finally, the substrate quality itself has improved through the earlier development of write-once commercial products.

Several review articles are recommended for further information on magnetooptical recording. There is an excellent introduction to the physics of magnetooptical effects (Freiser, 1968), and there are several historical reviews with expositions of thermomagnetic processes, optical properties, and the micromagnetics of magnetooptical recording (Smith, 1967; Hunt, 1969; Dekker, 1976). There are also two recent reviews: a comparison between magnetooptical and magnetic recording, as they are envisioned to develop during the 1980s (Bell, 1983) and a summary of the status of magnetooptical technology with rare-earth transition-metal films (Kryder, 1985). For an introduction to the technology of read-only and nonerasable optical memories, which have many features in common with erasable magnetooptical systems, there is a set of review articles on digital audio (Carasso et al., 1982; Heemskerk and Immink, 1982; Hoeve et al., 1982) and a book on optical-disk systems (Bouwhuis et al., 1985).

The primary technical problems in the construction of an erasable magnetooptical system have historically been, and continue to be, in the development of a suitable storage medium. For this reason, the physical properties of the magnetooptical materials (Sec. 6.2) and the enhancement and optimization of the optical and thermal response of structures fabricated from these materials (Sec. 6.3) occupy a central position in this chapter. Section 6.2 covers the magnetic, optical, and electronic properties of the rare-earth transition-metal films, including film preparation and chemical stability. Section 6.3 describes the use of multilayered structures both to enhance the magnetooptical signal and to protect the magnetooptical film. Additionally, some problems associated with fabricating and preformatting the substrate are discussed. The other components of the magnetooptical system—the optics, laser, detectors, servos, and magnet—are described in Sec. 6.4. System performance is given in Sec. 6.5 in terms of the various noise sources, signal-to-noise ratio, modulation transfer function, areal data density and data rate, data encoding, error rate, and error correction. Finally, Sec. 6.6 describes the position of magnetooptical memories in the evolution of both digital recording systems and digital computers.

6.2 Physical Characteristics of Magnetooptical Materials

6.2.1 Basic properties

The alloys currently of most interest for magnetooptical recording are based on the amorphous $(Tb, Gd)_x (Fe, Co)_{1-x}$ system with $0.2 \leq x \leq$

0.3. While the actual structural topology of these materials is unknown (Cargill, 1975; Rhyne, 1977), measurements on $TbFe_2$, $GdFe_2$, and $SmFe_2$ (D'Antonio et al., 1982) show some distinct differences between the coordination of the rare-earth atoms in the amorphous and corresponding crystalline Laves phase. For example, the tetrahedral Tb-Tb coordination of the crystal is approximately doubled in the amorphous phase, an effect that is accompanied by a 10 percent increase in the Tb-Tb distance. In contrast, the Fe-Fe coordination shells and distances are not very much changed. There are therefore both qualitative and quantitative differences between many properties of the amorphous and crystalline phases. This is exemplified by the diverse trends for the Curie temperatures of amorphous alloys and crystalline compounds (Heiman et al., 1975). Understanding the structural origin of these differences is of paramount importance when devising high-performance magnetooptical media.

As far as the magnetic properties are concerned, the key intrinsic structural feature of the amorphous phase is that no two atomic sites are equivalent (Coey, 1978). Three effects can result: a distribution in magnitude of the atomic moments, a distribution in the magnitude and sign of the exchange interaction, and the possibility of local anisotropy dominating the magnetic behavior in alloys containing highly anisotropic rare-earth ions. Some important cases are sketched in Fig. 6.2. In the simplest case, represented by Gd-Co, the local anisotropy is zero at the rare-earth site because Gd is an S-state ion. Thus, because the Co subnetwork is strongly ferromagnetic and the coupling between the Gd and Co spin angular momenta is weakly antiferromagnetic, simple ferrimagnetic behavior occurs. In the two other cases, the random anisotropy at the non-S-state rare-earth site competes with exchange to determine the orientation of the rare-earth moment. Moreover, according to Hund's rule, $J = L + S$

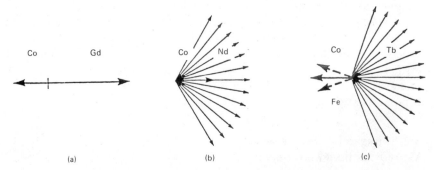

(a)   (b)   (c)

Figure 6.2 Competition between exchange and anisotropy in determining magnetism in amorphous rare-earth transition-metal materials. *(a)* S-state ion (Gd); *(b)* non-S-state ion (light rare earth); *(c)* non-S-state ion (heavy rare earth). In *(a)* there is no local anisotropy. In *(b)* and *(c)*, a large local anisotropy results in either ferromagnetic (for light rare-earth elements) or ferrimagnetic (for heavy rare-earth elements) alignment. Sperimagnetic distribution of resulting moments is also shown *(after Coey, 1978).*

for the heavy elements and $J = L - S$ for the light elements. Thus, while both Nd-Co and Tb-Co are sperimagnetic (Coey, 1978), only the latter will show ferrimagnetic-like behavior. Figure 6.2 also shows one final complication, namely the scatter in the orientation of the iron moments in Tb-Fe alloys that arises either through the distribution in the exchange interaction between iron atoms (see Sec. 6.2.3) or through the effect of local anisotropy at the rare-earth sites on the iron subnetwork.

The resulting magnetic behavior of the heavy rare-earth alloys is indicated in Fig. 6.3 (Connell et al., 1982). With a suitable choice of composition, the saturation magnetization M_s will be close to zero over a wide temperature range and exactly zero at the compensation temperature. In contrast, the magnetization of each subnetwork is large at temperatures reasonably below the Curie temperature. For example, low-temperature magnetization measurements suggest that the magnetic moment per Fe atom approaches two Bohr magnetons at compositions of interest, while the magnetic moment per rare-earth atom, after accounting for sperimag-

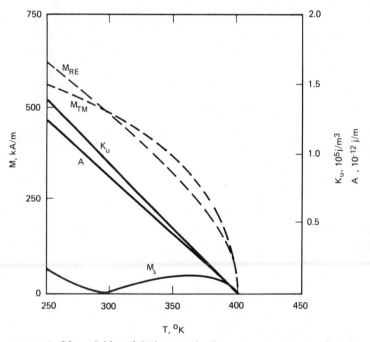

Figure 6.3 Mean-field model of magnetization versus temperature for a ferrimagnetic alloy with compensation point near room temperature. M_{RE} and M_{TM} are the magnetization of the rare-earth and transition-metal sublattices, respectively, and M_s is the net magnetization. The temperature dependencies of the exchange stiffness A and anisotropy energy density K_u are also shown *(after Mansuripur and Connell, 1984)*.

netic scatter, takes on its normal ionic value (Rhyne, 1977). The magnetooptical effect, which is found to depend almost entirely on the magnetic moment of the transition metal (Argyle et al., 1974; Connell and Allen, 1983), has the important property of being finite and smoothly varying with temperature at the compensation temperature.

There is also a key extrinsic feature in the structure of these alloys that arises from the deposition process itself. Namely, a uniaxial anisotropy develops in the growth direction that ranges from columnar microstructure in poor material to more subtle pair-ordering effects (Mizoguchi and Cargill, 1979; van Dover et al., 1985) in the best material. The latter establish a uniaxial anisotropy in the averaged local order around each atomic site. Alloys containing non-S-state rare-earth ions, whose moments are directed by the local order through single-ion anisotropy, thus have a mechanism not present in S-state alloys for developing a large macroscopic uniaxial magnetic anisotropy ($K_u > 0$) in the growth direction. (This mechanism could also operate through the local anisotropy of transition metal ions and could be responsible for the weak magnetic anisotropy in S-state alloys.) The temperature variation of K_u in these alloys is sketched in Fig. 6.3. As mentioned earlier, the exact orientation of each non-S-state rare-earth moment is set by the competition between single-ion anisotropy and exchange, and this leads to a scatter of the moments about the growth direction from site to site. The ensuing fluctuations can stabilize domains at submicrometer dimensions by providing an intrinsic mechanism for the pinning of walls (Gambino et al., 1974). This important interplay between structural and magnetic anisotropy, which occurs in alloys containing non-S-state ions, is absent in the S-state systems, such as Gd-Co.

The final key structural feature that affects many aspects of the media performance is the flexibility inherent in the amorphous structure itself. For example, the Curie temperature can be increased linearly by substituting Co for Fe (Tsunashima et al., 1982; Connell and Allen, 1983) or Gd for Tb (Imamura et al., 1985) in Tb-Fe alloys. Or the compensation temperature can be varied by changing the alloy composition (Chaudhari et al., 1973; Rhyne, 1977). Other elements that may be added, such as Al, Cr, and Pt, have very little effect on the magnetic properties but significantly improve the corrosion resistance (Imamura et al., 1985).

Along with these advantages of the amorphous phase, there is the possibility that crystallization may limit its performance and life in a recording system. While there are currently no general arguments to provide a priori rules for stable alloy design, empirical evidence has shown that crystallization is not likely to occur either at the high temperatures present during the recording and erasure cycles or at the lower ambient temperatures present in a drive or storage environment (Luborsky et al., 1985).

6.2.2 Preparation

Rare-earth transition-metal alloys can be made in the amorphous state by a large variety of vacuum-deposition processes. In the time since their discovery, thermal (Biesterbos et al., 1979) and electron-beam (Hasegawa and Taylor, 1975; Hansen and Urner-Wille, 1979) evaporation, rf-diode sputtering (Chaudhari et al., 1973; Mimura et al., 1976), magnetron sputtering (van Dover et al., 1986), and ion-beam sputtering (Harper and Gambino, 1979) have all been used successfully to make amorphous specimens. Some of the properties of the resulting material are influenced by the character of the deposition process itself rather than by the alloy composition (Heitmann et al., 1985), including, in some cases, the ability of the material to maintain perpendicular domains (Sakurai and Onishi, 1983) and to resist corrosion (Imamura et al., 1985). Of the various evaporation and sputtering methods available, magnetron sputtering, with its ability to coat thermally fragile substrates, is most favored for disk production. The remainder of this section therefore is addressed to the problems encountered when sputtering these materials.

Various forms of sputtering targets have been used. In early work, mosaic targets (Mimura et al., 1978) were often employed. In this case, the composition of the alloy is controlled by selecting the number and location of rare-earth platelets on a transition-metal base target. Small samples with a fixed or uniformly varying composition are readily obtained for study (Connell and Allen, 1983), an approach that has been particularly valuable for developing the ternary and quaternary alloys. Scaled-up production can follow two routes: the use of alloy or sintered powder targets that produce the required alloy film directly, or the use of multiple targets whose flux is mixed at the substrate.

Whichever method is selected, the problem of film contamination, particularly through rare-earth oxidation, must be faced both at the stage of target fabrication and particularly during film growth. Since adequate, multicomponent alloys are now becoming available from a number of target manufacturers, only the prevention of oxidation and contamination are significant barriers for successful production. Three important techniques have been used: a high vacuum with base pressures of 10^{-7} torr or less (Biesterbos et al., 1979); high-purity sputtering gases, often stripped of oxygen just prior to entering the chamber (Connell et al., 1982); and a gettering geometry in the system design (Theuerer et al., 1969). One example of the latter employs dual magnetron sputtering targets to prepare Tb-Fe alloys (van Dover et al., 1986). In this case, the magnetron guns and substrates are surrounded by a liquid nitrogen–cooled copper Meissner trap to getter both residual gases in the system and impurities introduced with the sputtering gas. The effective pressure of reactive gases inside the trap during sputtering is estimated to be in the vicinity of 5×10^{-10} torr. For production, a similar method using either an ambient or water-cooled trap is sufficient.

It is also necessary to control the composition of the alloys to a few tenths of an atomic percent over the entire surface of the disk to ensure uniform performance. This in fact is one of the most difficult requirements to meet under production conditions, requiring both system stability and reproducible target material. Full servo controls on the gas pressure, gas flow, and target voltage are needed, and stringent quality control on target material must be exercised.

The final essential component of a magnetooptical medium, required both for environmental stability and optical performance, is provided by dielectric overlayer and underlayer materials. These materials are again frequently deposited by rf-magnetron sputtering, and often reactively sputtered. The material chosen must prevent chemical reactions with the rare-earth transition-metal alloy from destroying the magnetooptical performance. It must therefore be a diffusion barrier to oxygen and water vapor and must itself not react with either rare-earth or transition metals. The search for such materials has been empirical, and to date some attractive candidates appear to be SiO (Nagao et al., 1986), AlN (Deguchi et al., 1984), and TiO_2 (Fukunishi et al., 1985). More information is provided on this in Sec. 6.2.6.

6.2.3 Magnetic properties

This section will present the phenomenology of the magnetic properties of the amorphous rare-earth transition-metal alloys. The major topics are the dependence of the magnetization, exchange, anisotropy, and coercivity on alloy composition and temperature and the dependence of the Curie and compensation temperatures on alloy composition. The relationship between the magnetooptical properties and these magnetic properties will be described in Sec. 6.2.4.

The typical behavior of the magnetization is shown in Fig. 6.4 for three Fe-based alloys (Mimura et al., 1978). In each case, the composition is chosen such that a compensation temperature exists. Under these conditions, the magnetization is less than 100 kA/m (100 emu/cm^3) at any temperature of interest for system operation, and each alloy, when prepared in thin-film form, will have only a small demagnetizing field to compete with perpendicular anisotropy (Chaudhari et al., 1973). The condition $K_{\text{eff}} = K_u - \frac{1}{2}\mu_0 M_s^2 > 0$ ($K_{\text{eff}} = K_u - 2\pi M_s^2 > 0$), required to obtain perpendicular magnetization, can therefore be readily achieved in each case over a useful temperature range. Figure 6.5 demonstrates, however, that this feature is not without cost. Namely, the compensation temperature varies rapidly with composition, at a rate of about 50°K/at %, implying

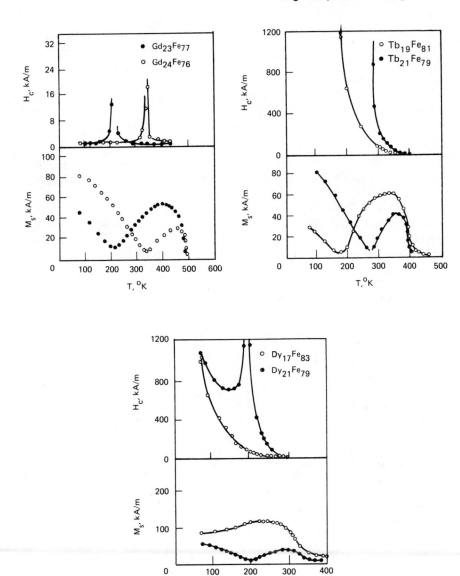

Figure 6.4 Magnetization and coercivity of various rare-earth-Fe films versus temperature. Each film shows a compensation temperature characterized by zero net moment and infinite coercivity *(after Mimura et al., 1978)*.

that compositional and batch-to-batch uniformity must be high to provide reliable system performance. The degree of uniformity required depends on a complicated interplay of system read and write parameters.

Figure 6.5 shows that the Curie temperature is much less sensitive to alloy composition and, to a first approximation, may be regarded as constant over the range of perpendicular anisotropy. However, it can be altered deliberately to advantage in ternary and quaternary alloys by substituting for the base rare-earth and transition-metal elements. Figure 6.6 shows the result of substituting transition metals T at levels up to about 10 at % in $Tb_{0.2}(Fe_{1-x}T_x)_{0.8}$ alloys. There is some uncertainty in the exact values of dT_c/dx because of difficulties in measuring x, but a value of about 10°K/at % Co seems reasonable from the published data and the corresponding result for crystalline $Fe_{1-x}T$ alloys. Figure 6.7 shows the equivalent results of substituting heavy rare earths R for Tb in $(Tb_{1-x}R_x)_{0.2}Fe_{0.8}$ and $(Tb_{1-x}R_x)_{0.2}(Fe_{0.9}Co_{0.1})_{0.8}$ alloys (Uchiyama, 1986). In this case, the Curie temperature is found empirically to vary linearly

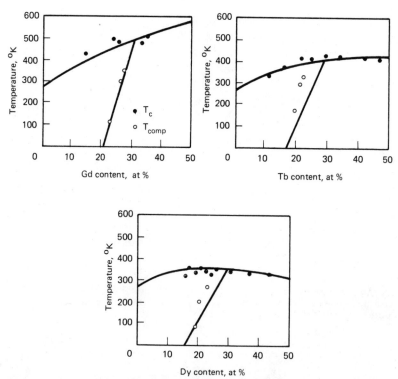

Figure 6.5 Variation of Curie and compensation temperatures for the films of Fig. 6.4, versus rare-earth content *(after Mimura et al., 1978)*.

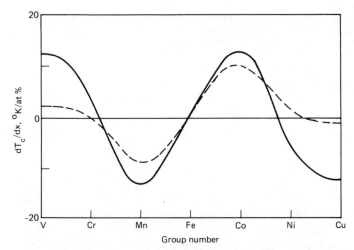

Figure 6.6 Variation of Curie temperature with substitution of small amounts of various transition metals in amorphous $Tb_{0.2}(Fe_{1-x}T_x)_{0.8}$ (dashed) and in crystalline $Fe_{1-x}T_x$ (solid line) materials.

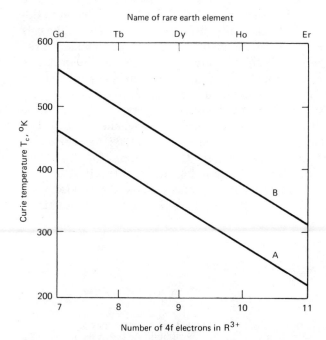

Figure 6.7 Variation of Curie temperature with substitution of various rare earths for Tb in (curve A) $(Tb_{1-x}R_x)_{0.2}Fe_{0.8}$ and (curve B) $(Tb_{1-x}R_x)_{0.2}(Fe_{0.9}Co_{0.1})_{0.8}$ where x is determined from the average number of $4f$ electrons in the alloy *(after Uchiyama, 1986)*.

TABLE 6.2 The Magnetic Moment of the Transition Metal and Exchange Integrals
for Amorphous Gd-Co, Gd-Fe, and Tb-Fe Alloys

Material	TM moment, m_B	Exchange integral, J		
		J_{TM-TM}	J_{RE-TM}	J_{RE-RE}
Gd-Co	1.2	25×10^{-22}	-2.5×10^{-22}	0.3×10^{-22}
Gd-Fe	1.9	$\approx 10 \times 10^{-22}$	-1.7×10^{-22}	0.3×10^{-22}
Tb-Fe	1.9	$\approx 10 \times 10^{-22}$	-1.0×10^{-22}	0.3×10^{-22}

with the average number of $4f$ electrons for the full range of x, when R is
Gd, Tb, Dy, Ho, and Er. (To the accuracy of the measurements, the Curie
temperature could possibly vary linearly with the deGennes factor rather
than the number of $4f$ electrons.) There have been many attempts to
derive mean-field parameters from these results (Gangulee and Kobliska,
1978). While there are numerous intricacies involved in obtaining reliable
parameters, the values for the magnetic moment and exchange integrals
in Table 6.2 express the observed trends.

There are several features to note. First, the magnetic moment of the
transition metal is independent of the rare-earth element. In contrast, the
transition metal moment decreases rapidly on going from Fe-based to Co-
based alloys (Masui et al., 1984). While this parallels the trend in the corresponding crystalline alloys, the moments themselves are smaller in the
amorphous phase. There are both qualitative similarities and quantitative
differences in these results that relate to the electronic density of states
of the two phases, a topic that will be pursued further in Sec. 6.2.4.

Second, while J_{TM-TM} dominates the other exchange forces in all the
alloys, its value in Fe-based alloys is considerably smaller than in Co-
based alloys. One explanation (Taylor and Gangulee, 1982; Mansuripur et
al., 1985) is based on the Bethe curve for the variation of J_{TM-TM} with
normalized interatomic distance, shown in Fig. 6.8 (Bozorth, 1951). Owing
to the reduction with separation for distances below the Goldschmidt
radius and the range of interatomic distances that occur in these amor-

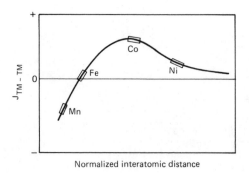

Figure 6.8 Bethe curve for the
dependence of the exchange
integral J_{TM-TM} on the normalized interatomic spacing r. The
boxes indicate the expected variation in spacing for each amorphous metal. A negative
exchange integral results in antiferromagnetic ordering *(after
Taylor and Gangulee, 1982).*

phous alloys, the average $J_{\text{Fe-Fe}}$ experiences a large downward displacement from its value in crystalline Fe. Indeed, it is conceivable that a fraction of the iron subnetwork might be antiferromagnetically coupled. It follows that a simple two-subnetwork mean-field theory cannot be expected to provide a detailed description of the Fe-based alloys. In contrast, $J_{\text{Co-Co}}$ is not similarly sensitive to interatomic separation, and neither downward displacement of $J_{\text{Co-Co}}$ nor significant deviations from the mean-field theory are to be expected.

Finally, the mean-field parameters themselves provide an estimate of the exchange stiffness $A(T)$. For nearest-neighbor exchange, it has been shown (Gangulee and Kobliska, 1978) that

$$A(T) = \tfrac{1}{6}N \sum_{i=1}^{n} B_i(T)X_i \sum_{j=1}^{n} B_j(T)J_{ij}Z_{ij}r_{ij}^2 \tag{6.1}$$

where $B_i(T)$ = Brillouin function that represents the total angular momentum of atom i

X_i = atomic fraction of atom i

r_{ij}, J_{ij} = interatomic distance and exchange interaction between nearest neighbor atoms i and j

Z_{ij} = number of j atoms that are nearest neighbors of atom i

N = total number of atoms per unit volume

Figure 6.3 shows the result for the simple two-network mean-field model for a Tb-Fe alloy. The exact quantitative value must be used with caution, but the order of magnitude of $A(T)$ and its temperature dependence are vital to both the understanding of domain stability and wall width in these materials.

Figure 6.4 also shows how the coercivity varies with temperature in Fe-based alloys. The Gd alloys are qualitatively different from the others in that the coercivity is highly peaked around the compensation temperature but falls precipitously to values less than 4 kA/m [50 Oe] over most of the temperature range. In principle, this material is suited for compensation-point writing, but its small coercive energy density $(\mu_0/2)M_sH_c \sim 10^2$ J/m^3 ($M_sH_c \sim 10^3$ erg/cm^3) precludes the formation of stable submicrometer domains (Kryder et al., 1983). The coercivity of both the Tb and Dy alloys has a different character. While still peaked around the compensation temperature, the coercivity falls much more slowly, and the coercive energy density [$\sim 10^4$ J/m^3 ($\sim 10^5$ erg/cm^3)] is several orders of magnitude larger. Submicrometer-sized domains are therefore stable throughout the system operation range.

A clue to the origin of this behavior is provided by Fig. 6.9, in which the anisotropy constant K_u is plotted versus rare-earth content x in $(\text{Gd}_{1-x}R_x)_{0.3}\,\text{Co}_{0.7}$ alloys (Sato et al., 1985). Upon substitution of non-S-state ions, changes are induced in K_u, the magnitude and direction of which are proportional to the magnitude and sign of the local anisotropy effects of the rare-earth ion in corresponding crystalline alloys. For Tb-substitution in particular, the result is an increase of K_u by over an order of magnitude at the point at which the Gd has been completely replaced. The conclusion is that the magnetic anisotropy of Tb-based alloys is dominated by local forces at each Tb site that arise from pair ordering of Fe atoms in the plane of the film (Nishihara et al., 1978). On average, these forces align each $4f$ electron cloud in the plane of the film, with its electron spin perpendicular to the plane, in analogy with the crystal-field effects that occur in the corresponding crystalline alloys. This mechanism is crucial for obtaining materials with large magnetic anisotropy and, as a consequence, with small wall widths that allow high-density recording.

Numerous mechanisms have been proposed for the origin of the alignment forces. Of these, pair ordering (Mizoguchi and Cargill, 1979) and internal planar stress due to substrate constraint (Takagi et al., 1979) are most significant in good-quality material. Since both are subject to variation and control by adjustments to the deposition process, the deposition conditions are generally chosen to minimize the internal planar stress near room temperature. The value of K_u in stress-free Tb-Fe alloys made by magnetron sputtering has a peak anisotropy of about 3×10^5 J/m³ [3×10^6 erg/cm³] (van Dover et al., 1985), although variations in K_u by a factor of about 2 to 4 can be induced by modifying the degree of resputtering during growth (Okamine et al., 1985). Conveniently, the peak values occur across the region of most interest for recording media, namely, between 15 and 25 at % Tb.

Armed with this information for the Tb-based alloys, it is reasonable to write (Alben et al., 1978)

$$K_u(T) \simeq N W_{\text{Tb}} \, X_{\text{Tb}} \, B_{\text{Tb}}^2(T) \tag{6.2}$$

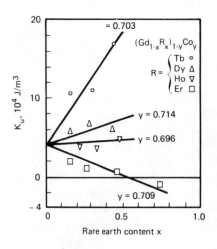

Figure 6.9 Variation of anisotropy constant K_u in $(Gd_{1-x}R_x)_{1-y}Co_y$ versus amount x of substituted rare earth, for R = (Tb, Dy, Ho, and Er) and for alloys with $y \simeq 0.7$ *(after R. Sato et al., 1985)*.

where W_{Tb} is a phenomenological parameter that expresses the strength of the single-ion anisotropy at each Tb site, and the other parameters are taken from the mean-field theory. While W_{Tb} must depend implicitly on the degree of pair ordering, and therefore on alloy composition, the expression itself is quite different qualitatively and quantitatively from that for S-state ions (Mansuripur et al., 1985), where dipolar interactions are assumed to lead to the anisotropy (Gangulee and Kobliska, 1978; Mizoguchi and Cargill, 1979). A typical result is shown in Fig. 6.3 for the simple two-network model for a Tb-Fe alloy. Again the precise values must be used with caution, but the order of magnitude and temperature dependence are vital for the studies of domain recording and stability.

There is as yet no detailed argument for the relationship between magnetic anisotropy and the wall coercivity in these non-S-state rare-earth alloys. However, since the exact orientation of each non-S-state rare-earth moment is set by the competition between single-ion anisotropy and exchange (Harris et al., 1978), there will be a scatter of the moments about the easy axis from site to site. The ensuing fluctuations can stabilize domains because the wall energy can be minimized by deforming the wall so that it passes through regions of lower-than-average anisotropy (Alben et al., 1978). Movement of the wall away from this optimized location can then be achieved only by thermal activation over a magnetic field-dependent energy barrier. The correlation length of the fluctuations will be comparable with the wall width, unlike the situation in crystalline materials where wall pinning is infrequent and defect related. The result is an intrinsic mechanism for the wall coercivity, in that it is a feature of the natural fluctuations of the amorphous state of non-S-state alloys and does not occur in Gd-based alloys.

Intrinsic coercivities of this type are not observed as a static coercive force (Egami, 1973). Rather, the process should be treated statistically as wall diffusion that is accelerated both by temperature and applied field. Studies of the characteristic time for magnetization reversal at various temperatures versus inverse magnetic field for a $Tb_{0.21}Fe_{0.79}$ alloy have been made (Connell and Allen, 1981). The conclusion is that any finite field is sufficient to cause wall movement, but for well-chosen medium and system parameters, data recorded in micrometer-sized domains will not be endangered in 10 years of exposure to magnetic fields and temperatures that might be expected to occur under normal operating conditions.

The initial stage of the magnetization reversal process in a uniformly magnetized film appears to occur by the growth of a small number of nuclei that form soon after the reversal field is applied (Connell and Allen, 1981). The number of these nuclei varies inversely with the size of the magnetizing field (Ohashi et al., 1979), but the origin and number of nuclei in well-magnetized films present interesting but unanswered questions (Harris, 1980). For magnetooptical recording, however, the nucleation process during field-driven magnetization reversal is not of great concern. Instead domain nucleation is achieved thermomagnetically by cooling the medium through the Curie temperature in a reverse field. The ensuing stability is then determined by the wall coercivity described earlier.

Table 6.3 summarizes the properties of the alloys that have been discussed. The polar Kerr rotation θ_K has been measured at 633 nm through the supporting glass substrate, and a reflectivity of about 0.45 is typical of all the alloys. The relation between θ_K and the magnetooptical component r_y is given in Sec. 6.3.4. The Tb-Fe-Co alloys are presently favored for their combination of high magnetic anisotropy, high coercive-energy density, and optimized Curie temperature. The addition of a fourth element to improve corrosion resistance will be discussed in Sec. 6.2.6.

6.2.4 Magnetooptical effect and electronic properties

Many empirical efforts have been made to increase the magnetooptical effect of amorphous rare-earth transition-metal alloys. These efforts have been retarded by the lack of any workable model for the relation between the electronic structure and the optical properties of highly disordered metals. Recently, ideas that were originally proposed for the optical properties of amorphous semiconductors (Hindley, 1970) have been extended (Connell and Bloomberg, 1985; Connell, 1986) to develop a model that, as a minimum, provides useful guidelines for materials engineering. It recognizes that the usual free-electron model of optical conductivity fails when the structural disorder is large, because the uncertainty in the wavevector becomes comparable to the wavevector itself. Under these conditions, the optical absorption can be described by a convolution of filled

TABLE 6.3 Magnetic Properties of Various Amorphous Rare-Earth Transition-Metal Alloys

Material	Curie temp. T_c, °K	Compensation temp. T_{comp}, °K	Coercive energy density $M_s H_c$, 10^4 j/m^3	Anisotropy constant K_u, 10^4 j/m^3	Kerr[†] rotation θ_K, deg.
$Tb_{23}Fe_{77}$	400	300	4	30	0.23
$Tb_{21}Co_{79}$	· · ·	300	4	14	0.33
$Gd_{26}Fe_{74}$	480	300	0.02	2.5	0.29
$Gd_{21}Co_{79}$	· · ·	300	0.02	1	0.33
$Tb_{22}Fe_{66}Co_{12}$	~500	<300	8	10	0.38
$Gd_{22}Tb_4Fe_{74}$	450	300	1.5	4	0.30
$Gd_{16}Tb_6Co_{78}$	· · ·	280	1.1	4	0.32

[†]Measured second surface at 633 nm.
SOURCE: After Heitmann et al., 1986.

and empty states that are separated by the photon energy, together with a slowly energy-dependent matrix element. The magnetooptical properties then follow in the same manner from the difference between the optical absorption of left and right circularly polarized light (Bennett and Stern, 1965).

In this approach, the frequency dependence of the conductivity, $\sigma^{xx} = \sigma_1^{xx} + j\sigma_2^{xx}$, can be described solely in terms of transitions between filled and empty states separated by energy ω and occurring with a dipole matrix element $|x|^2$. When spin-flip contributions that occur through spin-orbit interaction are treated as negligible in comparison with spin-conserving transitions, the absorptive component of the conductivity, σ_1^{xx}, becomes a sum of separate contributions from spin-up and spin-down electron states. Thus the contribution from spin-up states is

$$\sigma_{1,u}^{xx}(\omega) \sim \omega^2 |x|^2 \int_{E_f - \omega}^{E_f} \frac{N_u(E_f - E)N_u(E_f + \omega - E)\, dE}{\omega} \tag{6.3}$$

The contribution from spin-down states is formally identical but, because of exchange splitting in magnetic materials, the two contributions must be evaluated separately. It has been found empirically that the dipole matrix element for the classical oscillator provides a good estimate of $|x|^2$ in amorphous Ge and Si (Jackson et al., 1985). Therefore, with this approximation,

$$\sigma_{1,u}^{xx}(\omega) \sim \frac{1}{\omega^2 + \Delta^2} \int_{E_f - \omega}^{E_f} \frac{N_u(E_f - E)N_u(E_f + \omega - E)\, dE}{\omega} \tag{6.4}$$

where Δ is a fitting parameter of the order of the bandwidth. The frequency dependence of σ^{xx} can therefore be derived directly from the electronic density of states. From an expansion of the density of states near E_f, it can be shown that the frequency dependence of the conductivity is Drude-like, but the parameter formerly interpreted as a relaxation time is now proportional to the logarithmic derivative of the density of states at the Fermi energy (Connell and Bloomberg, 1985).

The polar Kerr effect is determined by the off-diagonal conductivity, $\sigma^{xy} = \sigma_1^{xy} + j\sigma_2^{xy}$, and results from the difference between the absorption of left and right circularly polarized light. When spin-flip processes are again neglected, the off-diagonal absorptive component of the conductivity, σ_2^{xy}, is the sum of separate contributions from spin-up and spin-down electron states. For spin-up states it is

$$\sigma_{2,u}^{xy}(\omega) \sim \omega^2(|r_-|^2 - |r_+|^2) \int_{E_f - \omega}^{E_f} \frac{N_u(E_f - E)N_u(E_f + \omega - E)\, dE}{\omega} \tag{6.5}$$

where $|r_-^2|$ and $|r_+^2|$ are the dipole matrix elements for left and right circularly polarized light, respectively. Making the harmonic oscillator approximation for the matrix elements as before

$$\sigma_{2,u}^{xy}(\omega) \sim \frac{\omega}{(\omega^2 + \Delta^2)^2} H_u \int_{E_{f-\omega}}^{E_f} \frac{N_u(E_f - E)N_u(E_f + \omega - E)\, dE}{\omega} \quad (6.6)$$

where H_u is an effective field seen by the spin-up electrons as the result of the spin-orbit interaction and is of opposite sign for transitions between spin-down states. Therefore, above the Curie temperature, where the spin polarization is zero, $\sigma_2^{xy} = 0$. On the other hand, σ_2^{xy} depends on both the spin-orbit interaction and the spin polarization at low temperature, and specific information about the density of states is needed for a discussion of its behavior.

Theoretical work (Malozemoff et al., 1983) provides important insights into the development of the electronic density of states in amorphous transition metal alloys and, in particular, into the role that is played by the rare earth in developing the magnetism in amorphous 12-coordinated iron. Hybridization between the rare-earth and Fe d states leads to a strong depression of the density of states at the top of the Fe d band and an accompanying accumulation of states at lower energy, sufficient to satisfy the Stoner criterion. A clear gap also separates the Fe-dominated low-lying band from the higher-lying rare-earth band. The location of the Fermi level and the net moment per Fe atom are found to be very sensitive to the Fe-Fe distance. However, the minority spin density is always much greater than the majority spin density at the Fermi energy, and optical effects for photon energies less than 2 to 3 eV should be dominated by transitions within the minority states. Experimental photoemission and inverse photoemission measurements on amorphous $Tb_{0.25}Fe_{0.75}$ (Connell et al., 1984) appear to confirm these views.

Figure 6.10 summarizes the results. The parts of the density of states represented by solid lines are derived directly from the experimental measurements, while those represented by broken lines are estimated using the hybridization scheme proposed. These data provide the basis for the comparison of the measured and calculated magnetooptical properties of Tb-Fe–based alloys. The measured conductivity $\sigma_1^{xx}(\omega)$ is nearly frequency-independent between 0 and 3 eV, with a value of about 5×10^5 $(\Omega \cdot m)^{-1}$ (5×10^{15} sec^{-1}). This is consistent with the conductivity calculated from Eq. (6.4) using $\Delta = 10$ eV. The contributions to the conductivity from transitions within the nearly full Fe majority spin band can be neglected. The calculation shows that the Fe d states dominate the conductivity behavior below 3 eV, although transitions between Tb d and f states begin to occur at about 1.5 eV and provide a significant contribution by 3 eV.

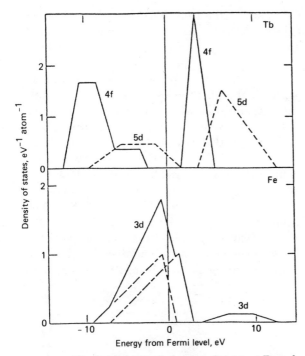

Figure 6.10 The local electronic density of states at Fe and Tb sites versus energy relative to the Fermi level. Solid lines are derived from experimental measurements; dashed lines for Tb are estimates for the hybridized bonding and antibonding states; dashed lines for Fe are estimates for the spin-up and spin-down bands *(after Connell, 1986).*

The experimental diagonal and off-diagonal elements of the dielectric tensor, shown in Fig. 6.11, suggest that

$$\frac{\varepsilon_1^{xy}(\omega)}{\varepsilon_2^{xx}(\omega)} = \frac{\sigma_2^{xy}(\omega)}{\sigma_1^{xx}(\omega)} \simeq \text{const} \tag{6.7}$$

to within a factor of 2 for the range $1 \leq \hbar\omega \leq 3$ eV. This is qualitatively similar to the simple prediction of Eqs. (6.4) and (6.6) for $\Delta \simeq 10$ eV and $H_u = \text{constant}$. Quantitative agreement will require better approximations for both the matrix element and density of states used in Eqs. (6.4) and (6.6), and the inclusion of the contribution of the Tb d-to-f transitions to $\sigma_2^{xy}(\omega)$ in the range $2 \leq \hbar\omega \leq 3$ eV. Nevertheless, there is an indication that contributions to $\sigma_2^{xy}(\omega)$ from d-to-f transitions in light rare earths that occur near 1 eV should enhance r_y.

There are a number of major points that come out of this. First, the Fe d states dominate the magnetooptical properties, as they do the conductivity at energies below 3 eV, explaining the long-known relation between the polar Kerr and extraordinary Hall effects. The similarity of the temperature dependences of the magnetization of the Fe subnetwork and the polar Kerr effect is also rationalized.

Second, variations of $\sigma_2^{xy}(\omega)$ at low temperatures in $Tb_x(Fe_{1-y}T_y)_{1-x}$ alloys, where T is a transition metal that replaces Fe in small amounts $y \leq 0.1$, should occur largely through changes within the Fe d states rather than through changes in the spin-orbit interaction. For example, the measured dependence of σ_2^{xy} at low temperatures on Co, Mn, V, or Cu substitution can all be explained in this way (Connell et al., 1982). The increase that is found at 300°K with Co substitution then derives almost entirely from the increase in Curie temperature. Figure 6.12, which shows the polar Kerr rotation at 300°K versus Curie temperature for numerous rare-earth Fe-Co alloys, confirms this viewpoint (Uchiyama, 1986). While this is extremely significant for magnetooptical recording, there is no intrinsic enhancement of σ_2^{xy}. Figure 6.6 shows that some substitutions cause very little change of the Curie temperature, and there is only a small reduction of the magnetooptical conversivity (Smith, 1965a). Therefore,

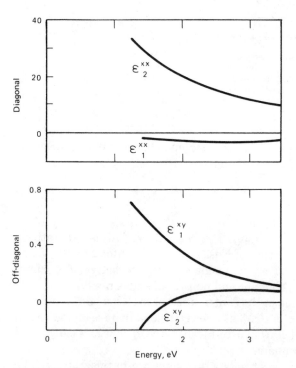

Figure 6.11 Experimental diagonal and off-diagonal elements of the complex dielectric tensor $\varepsilon = \varepsilon_1 + j\varepsilon_2$ versus photon energy *(after Connell, 1986).*

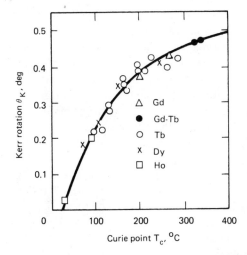

Figure 6.12 Kerr rotation angle θ_K versus Curie temperature for various rare-earth-substituted Fe-Co alloy films *(after Uchiyama, 1986)*.

these substitutions can improve the oxidation or corrosion resistance of the alloy, the price paid in magnetooptical performance is small.

The third and final point is that, while Tb d-to-f transitions do not contribute to σ_2^{xy} at energies of interest for magnetooptical recording, the results obtained suggest that enhancements of $\sigma_2^{xy}(\omega)$ over its value in Tb-Fe should be possible in $(Tb_{1-y}R_y)_x Fe_{1-x}$ alloys when R is a suitable light rare earth. Photoemission and inverse photoemission data for Tb, Gd, Sm, and Nd metals (Lang et al., 1981) suggest that both Nd and Sm substitution should produce useful enhancements in the low-temperature value of σ_2^{xy}, because transitions between their d and f states are possible at energies between 1 and 2 eV. There should be no such contribution upon substituting Gd.

Figure 6.13 shows the change in the low temperature value of the polar Kerr rotation, measured at 2 eV, versus Nd and Sm substitution. As expected $\Delta\theta_K(0)$ is positive, with Sm substitution being especially effective. This is consistent with work on Nd and Pr alloys with iron and cobalt (Gambino and McGuire, 1986). The maximum improvement obtained, however, still leaves the value of r_y much below those shown in Table 6.1 for crystalline Mn-Bi and Pt-Mn-Sb. Figure 6.13 also shows that $\Delta\theta_K(0)$ for Gd substitution is small.

6.2.5 Domain formation and stability

The process of thermomagnetic recording involves the formation of a reverse-magnetized domain in the region of the hot spot produced by a focused laser beam and the ensuing growth or contraction of this domain as the heating and subsequent cooling progress. The approach used to assess the stability of the domain at any point during the recording process follows the analysis established for magnetic bubble domains (Thiele, 1970), in which a cylindrical domain is assumed to be already present in the film, and its wall is subject in general to a contracting or expanding radial force. In bubble materials, a domain is stable when this force is exactly zero. In magnetooptical materials, the condition for domain stability is less stringent, because a force on the domain wall must overcome the large wall coercivity of these materials for domain collapse or expansion to occur (Huth, 1974).

This principle is most readily used to obtain the conditions for domain stability when the temperature distribution in the film is uniform (Kryder et al., 1983). There are four contributions to the energy of formation of the domain (Mansuripur and Connell, 1984): the external field energy, the anisotropy energy, the exchange energy, and the self (demagnetizing) energy. The force on the domain wall may then be calculated by differentiating the total energy E_t with respect to the domain radius. The min-

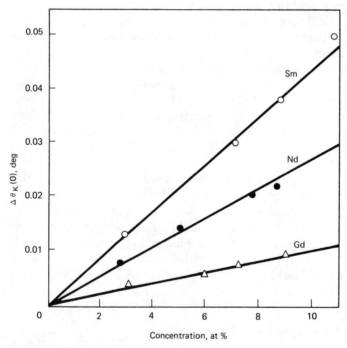

Figure 6.13 Measured change of the low temperature value of the polar Kerr rotation of $Tb_{0.2}(Fe_{1-y}R_y)_{0.8}$ alloys versus concentration of substituted rare-earth element R, where R is Gd, Nd, or Sm. *(after Connell, 1986).*

imum stable domain diameter, obtained when this force is exactly balanced by the coercive force, is

$$d_0 = \frac{4\,(AK_u)^{1/2}}{\mu_0 M_s(H_c - H)} + \frac{M_s L_f}{2(H_c - H)}\,Q\!\left(\frac{r}{L_f}\right) \tag{6.8}$$

where H_c = coercivity
$\quad r$ = domain radius
$\quad L_f$ = film thickness
$\quad Q$ = complicated function of r/L_f

Typical amorphous rare-earth transition-metal materials that are used for magnetooptical recording have their compensation point near room temperature; thus $M_s \approx 0$ and the applied recording fields are much less than H_c. Consequently, a good estimate of the minimum domain size is

$$d_0 \simeq \frac{4(AK_u)^{1/2}}{\mu_0 M_s H_c} \tag{6.9}$$

The magnitude of d_0 can be obtained from the experimental data in Table 6.3 and the mean field value for A in Eq. (6.1). Typical values at room temperature are $\mu_0 M_s H_c = 10^4$ J/m^3 [$M_s H_c = 10^5$ erg/cm^3], $K_u \approx 5 \times 10^5$ J/m^3 [5×10^6 erg/cm^3] and $A \approx 10^{-12}$ J/m [10^{-7} erg/cm]. The minimum domain size d_0 is therefore about 50 nm at room temperature. For small excursions about room temperature, the coercive energy density decreases slowly with temperature, indicating that variations of d_0 will be controlled by the temperature dependence of the wall energy. Since this decreases with temperature, as shown by the mean field results in Fig. 6.3, domains stable at room temperature will remain stable over a reasonable system operating range. The currently available magnetooptical media can therefore be pushed to much higher recording densities when the optical-head technology is improved.

Equation (6.9) provides an important design rule for the preferred magnetic characteristics of the magnetooptical media. At first examination, it might appear that the requirement is for materials with low wall energy

TABLE 6.4 Thermal and Optical Properties of Materials Used in Media Engineering

Layer	Complex refractive index	Specific heat J/cm^3/°K	Thermal conductivity, J/cm/°K/s	Diffusivity, cm^2/s
Substrate	1.46	1.7	0.002	0.0012
Aluminum	$2 + 7.1i$	2.7	2.4	0.89
Quartz	1.5	2.0	0.015	0.0075
Tb-Fe	$3.6 + 3.7i$	2.6	0.16	0.062

and high coercive-energy density. However, these are mutually exclusive for the following reason: in the non-S-state rare-earth alloys, the wall coercivity derives from fluctuations in the anisotropy constant, as explained earlier, and these fluctuations are greater in materials with larger K_u. It is therefore reasonable to write

$$H_c \simeq \chi \frac{K_u}{\mu_0 M_s} \tag{6.10}$$

where $\chi \approx 10^{-2}$ is the ratio of the average fluctuation in K_u to the mean value of K_u. It follows that

$$d_0 \simeq \frac{1}{\chi} \left(\frac{A}{K_u} \right)^{1/2} = \frac{\delta_w}{\chi} \tag{6.11}$$

where δ_w is the domain-wall width parameter. The minimum domain diameter thus scales with the wall width.

The appropriate design rule therefore is to engineer materials with small exchange stiffness and large magnetic anisotropy. Since the magnitude of A is almost entirely determined by J_{TM-TM} and J_{TM-RE} in Eq. (6.1), it follows from the earlier discussion of the mean-field data that the first requirement, small exchange stiffness, is best met in Fe-based alloys. The second requirement, large magnetic anisotropy, requires the presence of a non-S-state rare earth that has a large single-ion anisotropy term W_{RE} in Eq. (6.2), and this is best met in Tb-based alloys. It therefore appears that Tb-Fe–based alloys provide the best prospect for high density recording.

The behavior of the spin system under conditions of thermomagnetic writing can be investigated only by intensive computer simulations (Mansuripur and Connell, 1984). There are no simple expressions for the contributions to the total energy because the radial changes in the magnetic film temperature during the recording process make each individual energy contribution a complicated integral of radius and time. Nevertheless, the evolution of the domain energetics can be followed, and considerable insight into the process can be gained from numerical results.

As an example, consider the simulated recording of a quadrilayer medium (see Fig. 6.1b) consisting of a 50-nm-thick aluminum layer, an 80-nm-thick quartz intermediate layer, a 15-nm-thick magnetic layer, and a 120-nm-thick quartz overlayer. (The thermal and optical parameters used in the calculation are given in Table 6.4.) The procedure is as follows: First, calculations are made of the temperature profiles that develop in the magnetic material when a 5-mW, 50-ns pulse from a Ga-As laser is focused to a Gaussian spot of 1-μm diameter at the e^{-1} point. These are shown in Fig. 6.14 for different radii and times. Second, the total magnetic

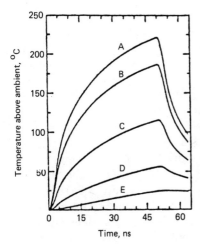

Figure 6.14 Increase in temperature within the magnetic film of a quadrilayer (described in the text) versus time, generated by a 5-mW, 50-ns pulse from a beam with full-width-at-half-maximum $W = 0.83$ μm. Curves are for different radii: (curve A) 0, (curve B) 250 nm, (curve C) 500 nm, (curve D) 750 nm, and (curve E) 1000 nm *(after Mansuripur and Connell, 1984)*.

energy E_t for domain formation is calculated versus domain radius at different times during the heating and cooling cycle, using the magnetic properties represented by Fig. 6.3 and Eq. (6.10) and assuming an applied field of 32 kA/m [400 Oe]. (These properties are typical of good magnetooptical materials and system operating conditions.) The results are shown in Fig. 6.15. At $r = 0$, before the laser is switched on, the total energy is an increasing function of radius, and therefore no reversed domain of any radius can be formed. At $t = 8$ nS, there is a minimum of energy around 500 nm, and the coercivity at $r = 0$ is less than the applied field. Domain nucleation therefore occurs, and the domain expands until the force on the wall is balanced by the wall coercivity. This condition is met when

$$\left(\frac{\partial E_t}{\partial r}\right)_{r_c} = 4\pi\mu_0 r_c L_f M_s(r_c) H_c(r_c) \tag{6.12}$$

which will occur at some radius r_c less than 500 nm. At $t = 50$ nS, the end of the heating period, points at $r \leq 520$ nm are above the Curie temperature, and the minimum of energy has moved to $r = 1200$ nm. The domain wall itself, however, will not progress beyond $r = 1000$ nm because of the effect of the wall coercivity.

The calculations also demonstrate the relative importance of the different contributions to the total energy. In this example, the self-energy is much smaller than the other terms. This is a feature of the magnetic characteristics and medium design chosen. Were the magnetic film greater than 100 nm thick or had it a much lower compensation temperature, the self-energy could no longer be ignored. For system optimization, such a situation might be an advantage since recording could then be achieved with lower applied fields. The disadvantage is that erasure would require relatively greater applied fields to overcome the larger demagnetizing field.

The total energy curves for the cooling period can also be calculated. It is found that depending on the position of the wall at a given instant, the forces on the domain wall can be contractive or expansive. Just after the end of the laser pulse, when coercivities are low and the wall is between the center of the beam and the radius of minimum energy, the forces tend to be expansive and prevent the written spot from collapsing. By the time the temperature drops back to room temperature, the coercivities have increased drastically, and despite the presence of contractive forces, there is no longer any danger of domain collapse.

Similar considerations apply to the erasure process. It is found that if the demagnetizing energy is negligible, it is possible to erase the recorded spots without an external field or in the presence of only a small reverse field by raising the local temperature until the coercivity drops below the contractive forces that act on the domain wall. If the demagnetizing energy is large, however, an appreciable external field in the reverse direc-

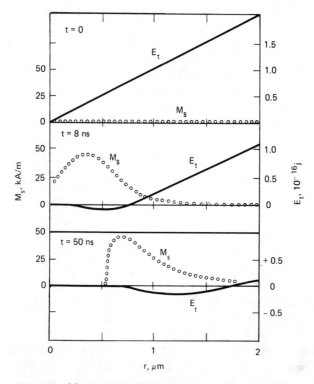

Figure 6.15 Magnetization M_s and net magnetic energy E_t produced on formation of a domain in the quadrilayer of Fig. 6.14 versus domain radius r. The magnetization M_s is given by the dotted curve. Three graphs are shown: at $t = 0$ (just before heating), and at times $t = 8$ and 50 ns during heating *(after Mansuripur and Connell, 1984)*.

tion together with local heating is needed to achieve erasure. The erasure process in this case is likely to be a combination of two processes: nucleation of a new domain within the recorded spot and the destabilization and contraction of the recorded spot itself. If the reverse field is not large enough, the process could lead to partial erasure and the creation of ring-like domains.

6.2.6 Chemical stability and aging

The stability and aging characteristics of the rare-earth transition-metal alloys have received much attention (Luborsky et al., 1985). Two independent and equally important issues have been addressed: first, the intrinsic stability of the magnetic and magnetooptical properties, and second, the stability of the magnetic information after thermomagnetic write-erase cycling. The purpose of this section is to distill some general principles from the large body of experimental work and, in so doing, expose the areas of concern.

Many studies agree that the principal mechanism of degradation of rare-earth transition-metal alloys is oxidation, particularly by OH^- species. In general, the rare-earth element has a stronger affinity for oxygen than the transition metal and tends to oxidize preferentially (Allen and Connell, 1982). The complex oxidation profile that results when unprotected Tb-Fe films are annealed in air at 200°C is shown in Fig. 6.16 (van Dover et al., 1986). There is first a rapid formation of an outer Fe_2O_3 layer on a Tb_2O_3 layer. Below these layers a mixed Tb-deficient alloy and Tb-suboxide layer then grows at the expense of the underlying unoxidized region. The process continues until the whole film is consumed.

Several approaches have been taken to overcome this potentially catastrophic failure mechanism. First, attempts have been made to reduce the oxidation rate by selecting the deposition parameters that minimize the microstructure of the film; voids, residual gas, and columnar structure all present a larger reaction surface by which oxidation can progress. As an example of what is possible, the perpendicular anisotropy of smooth, dense, magnetron-sputtered Tb-Fe-Co films changes very little on air exposure at 300°K, in contrast to similar data taken on gas-containing diode-sputtered samples (Hong et al., 1985). This result should not necessarily be taken as indicating a limitation on the diode method but rather the significant effects of microstructure on the corrosion properties.

Second, while Co-containing alloys are chemically more stable than Fe-containing alloys (Ichihara et al., 1985), the addition of a component chosen to passivate the surface of the alloy upon its first exposure to air is necessary in both cases for the best stability. It is found that the addition of small amounts of Cr to Tb-Fe alloys is most effective in this regard. Recently, it has been shown that a few atomic % of Pt, Al, or Ti also greatly improve the corrosion resistance and suppress the generation and growth of pinholes (Imamura et al., 1985). Furthermore, the addition of Pt slightly increases the polar Kerr rotation. The quaternary alloys of Tb-Fe-Co-Pt, therefore, appear to offer the best total performance characteristics.

Finally, there has been much work done to find effective protection layers between which to sandwich magnetooptical media to prevent both the

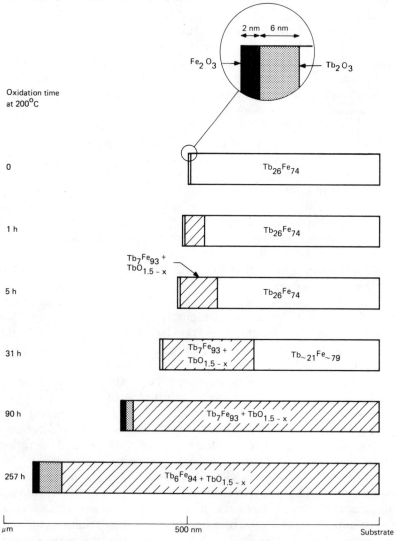

Figure 6.16 Oxidation profile in an unprotected Tb-Fe film, annealed in air at 200°C *(after van Dover et al., 1986).*

direct exposure of the medium to air and the occurrence of substrate-medium interfacial reactions. The latter is a particular problem when using plastic substrates. For example, Gd-Tb-Fe films deteriorate rapidly both on polymethyl methacrylate (PMMA) substrates (M. Sato et al., 1985) and in humid conditions on 2P-coated glass substrates (Hartmann et al., 1984; Jacobs et al., 1984). As one response to this, techniques for directly formatting glass substrates have been developed (Deguchi et al., 1984), but most of the current effort is to find suitable dielectric protection layers. Of those examined to date, it appears that SiO (Nagao et al., 1986), AlN (Deguchi et al., 1984), ZnS (Togami et al., 1985), Al_2O_3 (Bernstein and Gueugnon, 1985), and TiO_2 (Fukunishi et al., 1985) have been used with some success. It is apparent, however, that much effort will be required in this area to find an optimum solution.

The preceding focus on the macroscopic magnetic and chemical properties provides an indication of both the intrinsic stability of the magnetooptical material and the stability issues in medium design. It is now possible to understand how these observed limitations affect the stored data integrity. For a multielement alloy, it was concluded that the rate-limiting factor in the life of the data is the change of K_u that occurs with thermal annealing (Luborsky et al., 1985). This is not surprising given that the magnetic anisotropy derives almost entirely from the single-ion anisotropy of non-S-state rare earths, and the rare-earth constituents are oxidized in preference to the transition metals. Then from Arrhenius plots, it is projected that the shelf life of recorded data at 20°C is 3000 years and the number of write-erase cycles is 2×10^{11}. Actual real-time experimental results put lower bounds of greater than 5 years and 10^6 cycles on the same parameters.

While these estimates are encouraging, they completely ignore extrinsic effects, such as the existence and growth of microscopic defects that can destroy the integrity of the recorded information. These are best studied by bit error rate measurements. Figure 6.17 shows bit error rate data on various magnetooptical media that have been exposed at 55°C and 60 percent RH for different times (Nagao et al., 1986). For Tb-Fe-Co–based films with an SiO overcoat, there is no apparent increase in error rate in 1000 hours of exposure. Error maps produced during other accelerated aging experiments (Freese et al., 1985) show that observed increases in bit error rate are caused primarily by the growth of preexisting defects, rather than the nucleation of new defects. These results demonstrate that the medium design itself and the cleanliness of the preparation conditions, rather than the alloy composition, can determine the data life. Since thin films are more susceptible to pinhole formation and local crystallization during growth than thick films, the trade-off to be made between the ultimate magnetooptical performance available from a quadrilayer structure (Sec. 6.3.1) and the data integrity is a major issue for study. Finally, estimates of the aging factor for the accelerated test data in Fig. 6.17 suggest that data written on films protected by SiO will not experience degradation for many years under normal operating conditions.

6.3   Medium Design and Optimization

There are many properties of magnetooptical media that must be considered in the design of practical films for storage and retrieval. The basic design goals are to maximize the signal-to-noise ratio, achieve archival stability and high write-erase cyclability, achieve a low medium modulation noise with small bit shift and low defect (error) rates, and work in a reflectivity range that is sufficient for reliable servo operation but not high enough to cause problems with laser noise.

In this section, some of these problems are described, and the methods that have been devised to solve them are explored. These involve the nature and preparation of the substrate; the material properties and dimensions of the magnetooptical and adjacent films; and the optical, thermal, and protection characteristics provided by the medium design. As expected, multiple trade-offs occur. For example, the use of second-surface recording potentially gives greater protection to the magnetic

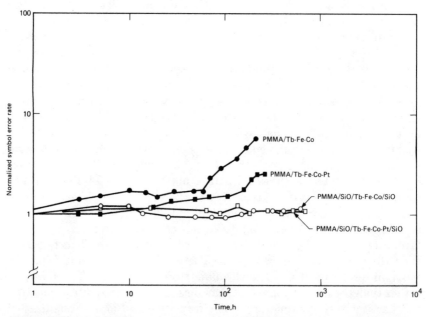

Figure 6.17   Symbol (8-bit) error rate on magnetooptical disk versus time, with accelerated aging at 55°C and 60% relative humidity. Results are shown for ternary and quaternary Tb-Fe alloys, with and without SiO overcoat and undercoat protection. The absence of degradation in films with SiO protection is notable *(after Nagao et al., 1986).*

layer and reduces errors due to surface contaminants. However, the substrate and optics birefringence must be limited, and it is more difficult to achieve a low reflectivity. As another example, the use of a quadrilayer structure for interference enhancement of the readout signal allows a greater choice of materials for the protection layers but at the cost of fabrication complexity and expense.

The first step in media design is to calculate the magnetooptical response of multilayer media. As shown in Fig. 6.18, normally incident plane-polarized light, when reflected from a perpendicularly magnetized specimen, becomes in general elliptically polarized with magnetooptical (polar Kerr) amplitude r_y and unrotated amplitude r_x. Because the magnetooptical signal is weak, $r_y \ll r_x$ and

$$r_x = \sqrt{R} \tag{6.13}$$

where R is the reflectivity. For the present purpose, it is sufficient to understand that in the shot noise limit

$$\text{SNR} \sim \sqrt{P_i}r_y \tag{6.14}$$

where P_i is the power incident on the medium during readout. The optical goal for a multilayer design therefore is to increase r_y through interference

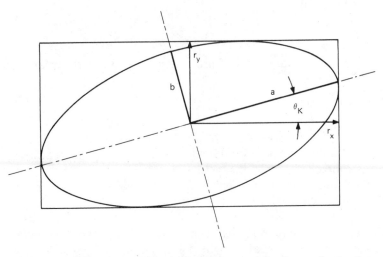

Figure 6.18 Schematic of elliptical polarization resulting from reflection from a magnetooptical medium of incident linearly polarized (along x axis) light. The magnetooptical component r_y is much smaller than the unrotated component r_x. The Kerr rotation angle is θ_K, and the ellipticity is $\eta_K = \pm b/a$, where a and b are the major and minor ellipse axes.

enhancement without decreasing R so much that servo control is made difficult. Typical results of this step, using the techniques described in Sec. 6.3.1, are given in Sec. 6.3.2.

The second step is to select those designs that can withstand large incident laser power during readout, without any degradation of r_y occurring through an increase in the temperature of the magnetic film. One of these will provide the maximum signal-to-noise ratio. The physical basis of this thermal engineering process is described in Sec. 6.3.3, and the different roles of the various thermal time constants are pointed out. In Sec. 6.3.4, two figures of merit for media performance in the shot noise limit are derived, and several media designs are compared. The final step is to make the trade-off between maximum signal performance and data and system integrity. The decisions that must be taken and the areas of concern are reviewed in Sec. 6.3.5.

6.3.1   Magnetooptics of multilayer media

A general method for calculating the magnetooptical response of any multilayer media has been described (Smith, 1965a, 1965b). The polar Kerr effect, however, can be treated in a much simpler way. The starting point is the dielectric tensor that describes the magnetooptical response of a perpendicularly magnetized medium (Landau and Lifshitz, 1958):

$$\varepsilon = \begin{bmatrix} \varepsilon^{xx} & \varepsilon^{xy} & 0 \\ -\varepsilon^{xy} & \varepsilon^{xx} & 0 \\ 0 & 0 & \varepsilon^{zz} \end{bmatrix} \qquad \mathbf{M}_s \parallel z \qquad (6.15)$$

It follows that the normal modes of propagation are circularly polarized with refractive indices

$$n_{\pm} = \sqrt{(\varepsilon^{xx} \pm j e^{xy})} \qquad (6.16)$$

The polar Kerr effect arises because of the difference between the optical constants for these right and left circularly polarized modes (Jackson, 1967) and is controlled by the magnetic dependence of the off-diagonal component ε^{xy} of the dielectric tensor. The diagonal component ε^{xx} is independent of magnetic effects to the first order.

To calculate the polar Kerr rotation θ_K and ellipticity η_K of the elliptically reflected light in Fig. 6.18, let $\mathbf{r}_+$ and $\mathbf{r}_-$ be the external complex Fresnel coefficients that correspond to the amplitude and phase of the reflected circular modes of propagation (Jackson, 1967). Then defining Ψ and Δ from

$$\mathbf{r}_- / \mathbf{r}_+ = \tan \Psi \exp j\Delta \qquad (6.17)$$

allows θ_K and η_K to be expressed as

$$\theta_K = \frac{\Delta}{2} \tag{6.18}$$

and

$$\tan \eta_K = \frac{1 - \tan \Psi}{1 + \tan \Psi} \tag{6.19}$$

It is therefore sufficient to calculate $\mathbf{r}_+$ and $\mathbf{r}_-$ independently, using the optical constants and thicknesses of the layers that constitute the multi-layer structure, and then combine them in Eq. (6.17). The values of $\mathbf{r}_+$ and $\mathbf{r}_-$ also give the reflectivity, since

$$R = \tfrac{1}{2}(r_+^2 + r_-^2) \tag{6.20}$$

where $r_+ = |\mathbf{r}_+|$ and $r_- = |\mathbf{r}_-|$. The complex Fresnel coefficients themselves can be calculated by a number of simple iterative schemes that were originally proposed for the calculation of the Fresnel coefficients for plane polarized light (Heavens, 1965). They apply equally to this case when the refractive indices for circular polarization replace those for plane polarization in the formalism.

For completeness, it should be noted that the polar Kerr rotation and ellipticity can also be expressed in terms of the external complex Fresnel coefficients for linearly polarized light $\mathbf{r}_x$ and $\mathbf{r}_y$ since

$$\mathbf{r}_x = \tfrac{1}{2}(\mathbf{r}_+ + \mathbf{r}_-) \tag{6.21}$$

and

$$\mathbf{r}_y = \tfrac{1}{2}(\mathbf{r}_+ - \mathbf{r}_-) \tag{6.22}$$

The reflectivity amplitudes used elsewhere in this chapter are then

$$r_x = |\mathbf{r}_x| \tag{6.23}$$

and

$$r_y = |\mathbf{r}_y| \tag{6.24}$$

In terms of the polar Kerr rotation and ellipticity, these are

$$r_x^2 = R(\sin^2 \theta_K \sin^2 \eta_K + \cos^2 \theta_K \cos^2 \eta_K) \tag{6.25}$$

and

$$r_y^2 = R(\sin^2 \theta_K \cos^2 \eta_K + \cos^2 \theta_K \sin^2 \eta_K) \tag{6.26}$$

where the reflectivity expressed in these terms is

$$R = r_x^2 + r_y^2 \tag{6.27}$$

6.3.2 Interference enhancement

The tools of the previous section can now be used to evaluate the effectiveness of multilayer interference structures for increasing the magnetooptical component r_y. The design goals are twofold: first, to provide an antireflective coating to couple the incident radiation into the medium, and second, to provide interference conditions that will capture all of the converted magnetooptical radiation and emit it in phase from the medium with the normally reflected light. In the following, the type of medium (e.g., bilayer, trilayer, quadrilayer) is defined from the number of optically active layers deposited on the substrate. In a completed assembly, it is conceivable that other layers would be added to provide further protection, but it is assumed that these would not affect the optical performance.

Bilayer, trilayer, and quadrilayer media can be constructed for both first-surface and second-surface recording. In Figs. 6.19 and 6.20, the bilayer and quadrilayer cases are shown. In both cases, the dielectric overlayer must act both as an antireflection layer and as a protection layer against chemical attack from either the atmosphere or substrate. A trilayer medium for magnetooptical recording differs from a quadrilayer medium by the removal of either the intermediate dielectric layer or the metallic reflecting layer, such that the magnetic layer remains protected by under- and overlayers in either situation. This contrasts with the

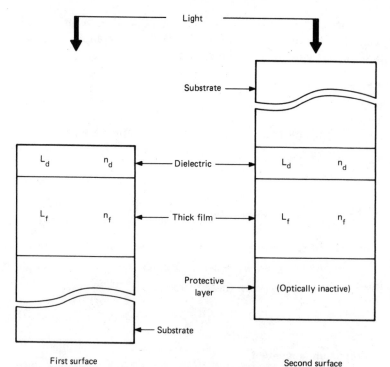

Figure 6.19 First- and second-surface bilayer media. Layer thickness and index of refraction are L and n, respectively. The magnetooptical film, labeled by L_f and n_f, is sufficiently thick that no light penetrates.

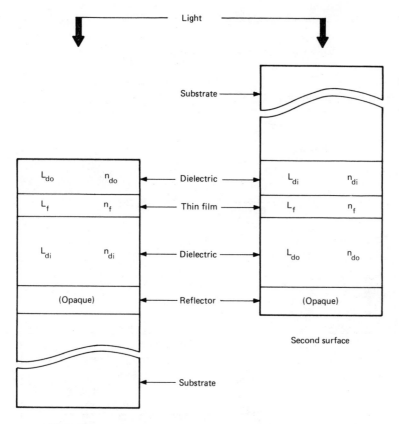

Figure 6.20 First- and second-surface quadrilayer media with two dielectric layers: an intermediate layer near the substrate and an outer layer deposited on top of the thin magnetooptical film.

accepted definition of a trilayer in nonerasable recording (Bell and Spong, 1978), where the ablative recording material is the uppermost, unprotected layer.

To establish a baseline, Figs. 6.21 and 6.22 show the reflectivity, rotation, ellipticity, and r_y of bare Tb-Fe: first, as an optically thick film versus wavelength, and second, as a thin film on a glass substrate versus film thickness for a wavelength of 840 nm. The components of the dielectric tensor from which these results are derived (Connell, 1983) were already presented in Fig. 6.11. There are several points to notice. First, the optical and magnetooptical parameters are all weakly dependent on wavelength, and Eq. (6.14) indicates that only a small decrease of signal-to-noise ratio will occur with decreasing wavelength. Therefore, the media will remain useful as solid-state lasers with high-output power become available at

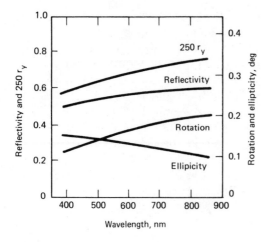

Figure 6.21 Reflectivity, rotation, ellipticity, and $250\ r_y$ from a thick bare Tb-Fe film versus wavelength *(after Allen and Connell, 1982).*

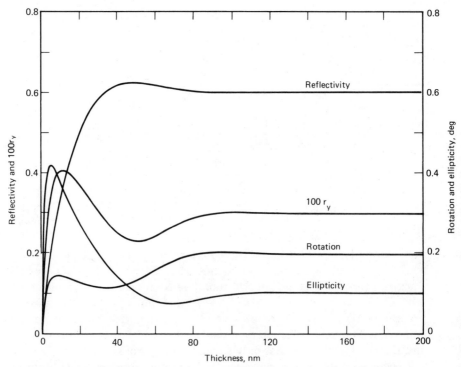

Figure 6.22 Reflectivity, rotation, ellipticity, and $100\ r_y$ from a thin bare Tb-Fe film versus film thickness for a fixed wavelength $\lambda = 840$ nm.

shorter wavelengths. Second, the optical and magnetooptical parameters all vary rapidly with thickness below about 80 nm. In particular, r_y rises strongly to a peak near 10 nm, and Eq. (6.14) shows that this leads to an increase of signal-to-noise ratio. The object of the multilayer designs is to make further improvements of this kind and provide protection to the magnetic film at the same time. A simple argument can be made for the degree of improvement that is possible (Connell et al., 1980; Connell, 1982). Consider a thick magnetooptical film of reflectivity R that is illuminated with plane-polarized light. From the light that penetrates the surface of incidence, a perpendicularly polarized component is created by the magnetooptical interaction. The intensity of this converted radiation, which is emitted from the surface of incidence I_s is then (Loudon, 1965)

$$I_s \backsim \frac{(1 - R)^2}{2\alpha} \tag{6.28}$$

where α is the absorption coefficient of the magnetic film.

This equation also holds for an overcoated thick magnetic film. In this case, a proper choice of the thickness and refractive index of the overcoating material will cause the reflectivity of the specimen to be reduced considerably. The maximum intensity enhancement of the bilayer signal over that from the bare thick sample is therefore

$$G_b \simeq \frac{1}{(1 - R)^2} \tag{6.29}$$

Now consider the quadrilayer configuration shown in Fig. 6.23. For very thin magnetic films ($\alpha L_f \ll 1$), the light reflected from the sample can be nearly cancelled by light that has suffered at least one reflection from the back reflector, when both the overlayer thickness is about one-half wavelength and the intermediate layer thickness is about one-quarter wavelength. As a result, most of the light is absorbed by the magnetic film. Furthermore, the converted radiation suffers relatively little reabsorption in the film, and the interference conditions that have been set up to create total absorption of the incident beam are exactly those required to maximize the exiting radiation I_q by in-phase addition. (The difference between the behavior of R and I_q stems from a 180° phase shift experienced by R at the top surface of the film, that is absent for I_q.) It follows that the intensity of converted radiation I_q emitted by the ideal quadrilayer film from its surface of incidence is

$$I_q \backsim 4L_f \tag{6.30}$$

where the factor of 4 results from the assumption that the converted radiation is emitted equally in the forward and backward directions, and the two components exit the film in phase. The intensity enhancement of the quadrilayer signal over that from the bare thick film is then

$$G_q \sim \frac{8\alpha L_f}{(1 - R)^2} \tag{6.31}$$

This holds for $\alpha L_f \leq 0.2$, beyond which an exact calculation indicates that G_q rapidly approaches a maximum value. Consequently, using $R \approx 0.6$, $\alpha \approx 10^8 \mathrm{m}^{-1}$ and $L_f \approx 3\mathrm{nm}$, it is estimated that $G_b \leq 8$ and $G_q \simeq 15$.

Figure 6.24 shows exact calculations of the reflectivity, rotation, ellipticity, and r_y of a first-surface bilayer versus overcoat thickness. The refractive index of the overlayer n_0 was chosen to be 2.0, a value typical of several of the dielectric materials currently used and close to the condition

$$n_0 = (n_i |\mathbf{n}|)^{1/2} \tag{6.32}$$

at which a minimum reflectivity occurs at the antireflection thickness. In this equation, n_i and $\mathbf{n}$ ($= n + jk$) are the refractive indices of the incident medium and the magnetic material. It follows that r_y and therefore the signal-to-noise ratio in Eq. (6.14) reach a maximum value at an overlayer thickness of about 85 nm. This is close to the quarter-wave thickness of $\lambda/4n_0 = 105$ nm but is different because of the large absorptive component in $\mathbf{n}$. From the maximum value of r_y and that of the bare thick film in Fig. 6.22, it is seen that $G_b \approx 4.9$, in reasonable agreement with

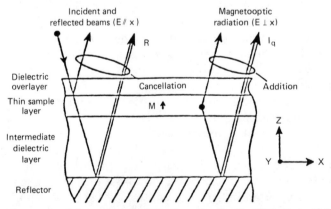

Figure 6.23 Schematic representation of reflection from a quadrilayer. The reflected component R consists of the superposition of a beam reflected from the magnetooptical film and all beams having at least one bounce from the reflector. Similarly the rotated component I_q consists of the back emitted radiation from the magnetooptical film plus radiation having at least one bounce from the reflector (after Connell and Allen, 1981).

the earlier estimate. It should also be noted that the rotation is peaked at the maximum of r_y, having a value of 1.0°, while the ellipticity drops to near 0.1°. The ellipticity therefore makes a negligible contribution to r_y in Eq. (6.26), which can be of some significant value for efficient system design.

Finally, Fig. 6.25 shows the reflectivity and r_y versus magnetic film thickness for numerous second-surface quadrilayer designs. The dielectric overlayer and intermediate layer again have refractive indices of 2.0, but now their thicknesses have been allowed to vary between 150 and 350 nm at each magnetic layer thickness, so that r_y is maximized. While the value of the refractive index is no longer of such great consequence for establishing a large r_y, in this second-surface configuration it is still necessary for n_0 to meet the condition (6.32) to obtain low reflectivities. It is then found that the maximum of r_y is broad, being close to 9×10^{-3} for thicknesses between 5 and 20 nm. Its value together with that for the thick film data in Fig. 6.22 indicates that $G_q = 10.2$. This is again in reasonable agreement with an earlier estimate. The breadth of this maximum provides wide latitude in which to optimize other features: the protection

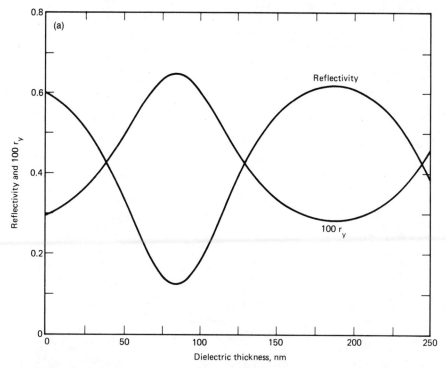

Figure 6.24 Optical parameters at $\lambda = 840$ nm from a first-surface Tb-Fe bilayer versus overcoat thickness. (a) Reflectivity and 100 r_y (b) ellipticity and rotation. The Tb-Fe layer thickness is 200 nm, and the overcoat dielectric has an index of refraction $n_0 = 2.0$.

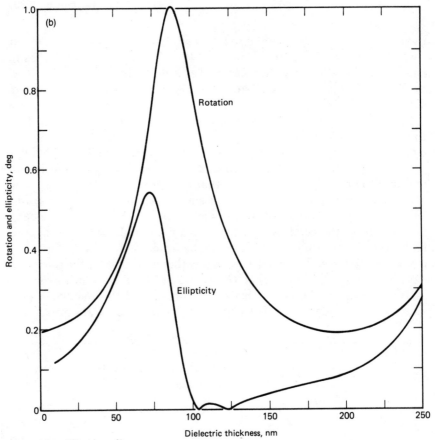

Figure 6.24 (*Continued*)

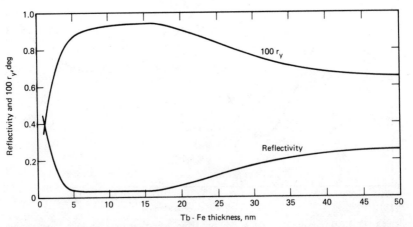

Figure 6.25 Reflectivity and $100\, r_y$ from a second-surface quadrilayer versus Tb-Fe thickness. The thicknesses of dielectric overlayer ($n_0 = 2.0$) and intermediate layer ($n_i = 2.0$) have been allowed to vary to maximize r_y at each value of Tb-Fe thickness.

provided by the dielectric stacks, the integrity of the media, the balance between rotation and ellipticity in r_y, and the thermal response of the medium. This flexibility is of great importance in the final stages of system engineering.

6.3.3 Thermal engineering

From the media designs available with high magnetooptical component r_y, it is now necessary to select those which can withstand large readout powers such that the signal-to-noise ratio is maximized. There are two approaches that can be taken. First, the Curie point can be raised by alloying, such as by substituting Co for Fe in iron-based alloys (see Sec. 6.2.3). Second, and of concern here, the stack can be arranged to increase heat flow out of the magnetic film, typically into an underlying conducting layer, thus reducing the temperature rise for a given incident optical power. This method is particularly useful for recording at high density in that the lateral heat spreading is reduced and very small domains can be written given sufficient laser power (Mansuripur and Connell, 1983b).

In this thermal engineering problem, the maximum temperatures reached during recording or erasing are low enough that radiation can be ignored. There are then three characteristic thermal time constants for the magnetooptical media that determine the dynamic response and equilibrium conditions (Corcoran and Ferrier, 1977). First, for a magnetic film of thickness L_f, thermal conductivity κ, specific heat (per unit volume) C_f, and diffusivity

$$D_f = \frac{\kappa}{C_f} \tag{6.33}$$

the time constant τ_{eq} that determines when a uniform temperature is established through the thickness of the film is given by

$$\tau_{eq} \simeq \frac{L_f^2}{D_f} \tag{6.34}$$

With $L_f \simeq 10^{-5}$ cm and $D_f \simeq 10^{-1}$ cm^2/s from Table 6.4, $\tau_{eq} \simeq 10^{-9}$ s. This is much shorter than any practical laser pulse time, and as a result the temperature through the thickness of the magnetic layer may be treated as uniform. Second, there is the penetration diffusion time constant τ_d, which determines the characteristic time for loss of energy to a conducting layer through an intervening dielectric layer. This can be approximated using a one-dimensional analysis (Kivits et al., 1981) in which the adiabatically heated volume is proportional to the diffusion length

$$L_{dif}(t) = \sqrt{D_d t} \tag{6.35}$$

where D_d is the diffusivity of the adjacent dielectric layers. Then as time proceeds, a thickness L_{dif} of each adjacent layer is heated in addition to the magnetic film, and the fraction of energy deposited into the magnetic film is approximately

$$\zeta = \frac{C_f L_f}{C_f L_f + 2 C_d L_{dif}} \tag{6.36}$$

where C_d is the specific heat per unit volume of the adjacent layers. For a laser pulse time of 10^{-7} s, a bilayer with the thermal parameters in Table 6.4 and $L_f \simeq 10^{-5}$ cm has $\zeta \simeq 0.2$. There is therefore a great gain in energy efficiency to be made by shortening the laser pulse time to the order of 10^{-8} s. This provides other advantages for system operation, which are discussed in Sec. 6.5.2. In a quadrilayer, one of the dielectric layers, of thickness L_d, separates the magnetic film from a conducting layer, so that at a time

$$\tau_d \simeq \frac{L_d^2}{D_d} \tag{6.37}$$

the further rate of heating is considerably slowed. The laser therefore should apply sufficient power to heat these layers close to the required temperature in a time τ_d, because for times longer than τ_d, the temperature will not rise appreciably further. For typical quadrilayer designs, $L_d \simeq 10^{-5}$ cm, and from Table 6.4, $D_d \simeq 10^{-2}$ cm^2/s, giving $\tau_d \simeq 10^{-8}$ s. Therefore, in both bilayer and quadrilayer designs, the same limit on laser pulse time is obtained.

Finally, there is a characteristic lateral diffusion time within the film, found from a two-dimensional solution to the diffusion equation (Carslaw and Jaeger, 1959),

$$\tau_{lat} \simeq \frac{r_b^2}{4 D_f} = \frac{0.09 \, W^2}{D_f} \tag{6.38}$$

where r_b and W are the e^{-1} intensity radius and full-width-at-half-maximum of the Gaussian beam, respectively. After the pulse is turned off, the temperature of the medium at a radius greater than $r_c \simeq r_b$ from the beam center continues to rise because of lateral heat diffusion and can cause domain spreading. This must be inhibited, for example, by thermal diffusion into the substrate, if the minimum size of the written bit is to be limited by the optics.

Figure 6.14 shows an exact calculation for the increase of temperature of a magnetic layer in a quadrilayer structure versus time at different radii (Mansuripur and Connell, 1983b). Over two-thirds of the temperature rise occurs in the first 15 ns, and thereafter only a small rate of increase is achieved by the incident radiation. The magnetic media could therefore be heated to the same final temperature by a 15 ns pulse of 50 percent more power than was used in the original 50-ns pulse of the calculation. This confirms the simple arguments that follow from Eq. (6.37).

These exact calculations also provide a detailed appreciation of the balance between axial and radial heat flow and the effects of the motion of the media during writing on the result (Mansuripur and Connell, 1983c). The calculations show that films with poor heat sinking exhibit lateral spread, both in the cross-track and down-track directions, whereas the heated area for films with good heat sinking approaches the shape of the writing beam. The central temperature increase also differs considerably in the two cases. The conclusion is that structures designed to withstand the greatest incident laser power during readout and requiring the highest write power provide the potential for the highest-density recording.

In selecting the media design, therefore, from the group of high r_y candidates, the system designer should first consider the least thermally sensitive design, because of its highest signal-to-noise ratio in Eq. (6.14) and ability to achieve the highest recording densities. If lack of writing sensitivity cannot be compensated with a larger applied magnetic field, then the search for the appropriate media structure can be systematically pursued by increasing the thermal sensitivity to obtain a structure that provides the best compromise between the writing and reading constraints of the magnetooptical system.

6.3.4 Magnetooptical signal and figures of merit

It is now necessary to establish figures of merit for the medium, and two such figures of merit are derived in this section. Consider the simplest detection scheme, a single analyzer oriented at an angle α to the unrotated direction r_x, as shown in Fig. 6.26. Assume that all of r_y and a fraction ξ of r_x is reflected toward the detectors by the first polarizing beam splitter (PBS) in Figure 6.1a. (This is approximately correct if ξ is small, but the detailed operation of the polarizing beam splitter is given in Sec. 6.4.1.) After compensating for the phase difference between r_x and r_y (i.e., removing the ellipticity), the polarization is linear and makes an angle $\pm \theta_s$ for the two directions of magnetization in the film, where

$$\theta_s = \tan^{-1} \frac{r_y}{\xi r_x} \simeq \frac{r_y}{\xi r_x} \ll 1 \tag{6.39}$$

Because θ_s is small, the optical power P transmitted through the analyzer is approximately

$$P \simeq P_i R \xi^2 \cos^2 \alpha \tag{6.40}$$

where P_i is the incident power on the medium, and $R \simeq r_x^2$.

The magnetooptical signal P_s is the change in optical power transmitted through the analyzer to the detector when the magnetization state is reversed:

$$P_s = P_i R \xi^2 [\cos^2 (\alpha - \theta_s) - \cos^2 (\alpha + \theta_s)] \tag{6.41}$$
$$= P_i R \xi^2 \sin 2\alpha \sin 2\theta_s$$

This is maximized by placing the analyzer at $\alpha = 45°$, so that

$$P_s = P_i R \xi^2 \sin 2\theta_s \simeq 2P_i R \xi^2 \theta_s \simeq 2P_i \xi r_x r_y \tag{6.42}$$

and the magnetooptical signal is proportional to the product of P_i and r_y and to a fraction of the unrotated component ξr_x that must be coherently added to r_y.

The first figure of merit K_r that represents the rotating power of the film is defined by (see Fig. 6.18)

$$K_r \equiv r_y \simeq \theta_K r_x \simeq \theta_K \sqrt{R} \tag{6.43}$$

Table 6.1 gives values of 3.0×10^{-3} for K_r for a Tb-Fe bilayer and 8.0×10^{-3} for a Tb-Fe quadrilayer.

Because K_r does not take into account the increase in signal with incident read power P_i, a second figure of merit K_p is required that incorporates P_i and the limit on P_i set by the reduction of r_y at elevated temperatures. The shot-noise power on the detector is proportional to $\sqrt{P_i}$, and

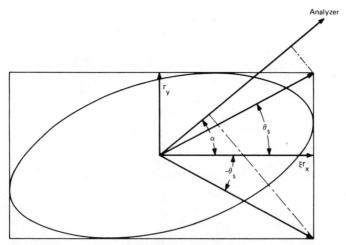

Figure 6.26 Ordinary r_x and magnetooptical r_y components, after attenuation of r_x by the factor ξ in the polarizing beam splitter and compensation to bring them into phase. The resulting linear polarization can have two directions, represented by angles θ_s and $-\theta_s$, depending on the magnetization direction. Maximum projections on an analyzer at angle α are also shown for the two magnetization directions.

under the best circumstances, the signal is shot-noise limited. Hence, K_p is defined to be proportional to the signal to shot-noise ratio

$$K_p \equiv K_r \sqrt{P_i} = r_y \sqrt{P_i} \sim \frac{P_s}{P_{\text{shot}}} \tag{6.44}$$

Although K_p is not dimensionless, it provides a better measure of the obtainable signal-to-noise ratio than K_r under shot-noise limited conditions. Both figures of merit are of significant value at the early stages of media design.

6.3.5 Disk integration and design compromises

There is a continuing effort, through the use of new materials, to (1) improve the optical, mechanical, and chemical properties of the substrate to ensure the integrity of the written data; (2) reduce the cost of the substrate, preformatting, and disk assembly; and (3) improve the packaging of the disk. These goals often compete with each other and with the media design itself. In this section, some of these conflicts will be addressed. We will look at construction of the multilayer stack, choice of substrate, substrate preformatting, and packaging of nonerasable and magnetooptical disks.

In comparison with a bilayer, the quadrilayer stack has an improved signal-to-noise ratio from optical and thermal construction, and it has improved flexibility in the choice of dielectric layers. This comes at a price: a more expensive and complex medium construction and, because the magnetic film must be thin in the quadrilayer, potentially a larger defect density.

Metal substrates are not practical for magnetooptical recording because the large inhomogeneous magnetic field applied during erase and write cycles would cause unacceptable drag on the disk. Of the possible nonconducting substrates, glass provides the smoothest and most inert surface on which to deposit magnetooptical media. Medium noise and defects increase with surface roughness, and it is critical to prevent chemical attack on the magnetic film from the substrate. A glass substrate with etched preformatted information would require the least protection, but it represents the most difficult preformatting process (Deguchi et al., 1984). Further, when glass is coated with the photopolymer 2P process (van Rijsewijk et al., 1982; Kloosterboer et al., 1982), it loses the advantage of chemical neutrality, and undercoats must be used to protect the magnetooptical film from the 2P layer (Hartmann et al., 1984). Plastic substrates engender a variety of problems, such as increased surface roughness, lack of rigidity (especially at elevated temperatures), birefringence, and retention of water. The advantages relative to glass are that plastic substrates are less fragile and less expensive. Additionally, 2P preformatting technology has already been highly developed for digital audio disks, and in the future the preformatting may be molded during the substrate manufacturing process itself.

Birefringence can cause serious problems in second-surface recording systems (Treves and Bloomberg, 1986a). With uniaxial substrate birefringence typical of PMMA, the substrate principal birefringence axis will rotate with the disk, causing a low-frequency oscillation (four times the rotation rate) in both the sum and difference channels of a magnetooptical system. With additional birefringence in the system optics, the variation has in general a twofold rotational symmetry. The substrate birefringence effects add on each pass through the substrate, and for a one-way phase lag of 5°, the modulation of the difference channel signal is of the same order of magnitude as the magnetooptical signal itself. This modulation increases as the square of the birefringence and degrades system performance primarily through an amplification of laser and medium noise due to a large unbalance in the difference channel. The constraint of a 5° one-way phase lag is equivalent to a substrate birefringence

$$\frac{\Delta n}{n} \le 10^{-5} \tag{6.45}$$

There remains much activity in the search for suitable transparent and inexpensive substrate materials.

Preformatted grooves are favored for nonerasable disks because they can increase contrast by interference effects, help confine the written (ablated) marks, and improve track-seeking performance by providing a continuous-tracking servo signal. Grooved disks, however, are disadvantageous for magnetooptical media. They can potentially lower medium quality because of the difficulty in coating over groove edges without losing the integrity of the protection layer. Medium noise is increased due to groove nonuniformities and the magnetooptical signal can be reduced through complex interference effects at the groove edges. Loss of signal-to-noise ratio is much more critical with magnetooptical media than with nonerasable media because the magnetooptical signal is relatively small. Some compromises in media design will have to be made in order to use both types of media on the same drive.

Nonerasable media are typically packaged into a sandwich composed of two disks with data on facing surfaces, separated by ring spacers at the hub and outer edges. Second-surface recording is used on each disk. This configuration has two disadvantages. First, the data surfaces are not well protected from gases and ions inside the air space. Second, the total package is relatively thick (about 3.5 mm). A preferable configuration for magnetooptical media is to cement the two disks together with the films facing each other, again using second-surface recording. The advantages are maximum protection of the data layer and a more compact packaging arrangement. As yet, this construction cannot be used with ablative non-erasable media because the constraint that the underlayer places on material movement pushes the recording threshold to unacceptably high energies. Nonerasable phase change or dye media would not suffer from this limitation and would provide a better match with magnetooptical media.

6.4 Components and System Design

Erasable magnetooptical systems differ from nonerasable write-once systems in several respects that make design of the former more difficult. The signal is smaller by more than one order of magnitude when using the Kerr effect for signal readout, and a large amount of light is necessarily reflected back into the laser. Also, a magnet is required in magnetooptical systems for writing and erasure. In this section, the operation of the optical, mechanical, electrical, and magnetic components are discussed, and those system design issues that relate directly to the functioning of these components are addressed.

Figure 6.1a shows the optical path for a typical magnetooptical system. The laser light is initially polarized in the x direction (perpendicular to the figure) and is collimated and made of circular cross section with a Gaussian intensity distribution by a lens and two prisms (not shown). After transmission through a slightly rotated polarizing beam splitter, the light is focused by the objective lens onto the medium, achieving a theoretical minimum spot size with a full-width-at-half-maximum intensity

$$W = \frac{0.56\,\lambda}{N_A} \tag{6.46}$$

where λ is the wavelength, and N_A, the numerical aperture of the objective lens of diameter d_l and focal length f, is defined by

$$N_A = \sin\,(\tan^{-1} d_l/2f) = \sin\,\theta_h. \tag{6.47}$$

The half-angle of the lens, as viewed from the focal plane, is θ_h. The depth of focus δ for Gaussian optics varies inversely with the square of the numerical aperture

$$\delta = \frac{\pm\lambda}{2N_A^2} \tag{6.48}$$

Thus, whereas use of a large numerical aperture reduces the spot size, the objective lens is expensive and the reduction in depth of field makes focus servoing much more difficult. A reasonable compromise is to use an objective with $N_A \simeq 0.5$. For $\lambda = 0.82 \ \mu m$, the spot size is then 0.9 μm and the depth of focus is 1.6 μm.

The light returning from the magnetooptical medium is incident on the polarizing beam splitter in the main beam, which is constructed to reflect almost all of the r_y component and a fraction ξ of the r_x component. Such a beam splitter can be visualized as a perfect one that transmits all of r_x and reflects all of r_y, but is rotated about the x axis by an angle $\phi = \sin^{-1} \xi$, as shown in Fig. 6.27. This rotation causes the fraction ξ of r_x and $\sqrt{(1 - \xi^2)}$ of r_y to be reflected by the beam splitter. The magnetooptical signal is proportional to the product of x and y amplitudes reflected to the detectors. Consequently, when using a rotated beam splitter, the magnetooptical signal from a single detector [Eq. (6.42)] is more accurately given by

$$P_s = 2P_i\xi r_x \sqrt{(1 - \xi^2)}r_y = 2P_i r_x r_y \sin \phi \cos \phi \tag{6.49}$$

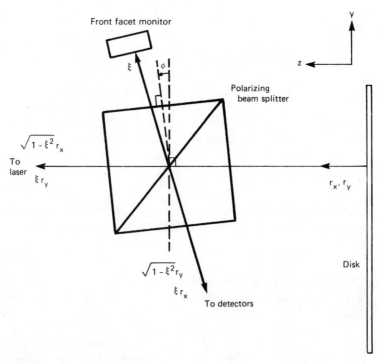

Figure 6.27 Operation of a polarizing beam splitter in the main beam, showing amplitude r_x and r_y upon reflection from the disk. The relation between ξ and the physical rotation angle ϕ of the polarizing beam splitter is $\xi = \sin \phi$. A front surface monitor is conveniently placed as shown.

But for a small beam splitter rotation ϕ, essentially all of the magneto-optical signal r_y goes to the detector, so that Eq. (6.42) is a satisfactory approximation. The beam splitter therefore optimizes the detection of the magnetooptical signal.

The rotation of the polarizing beam splitter also provides a useful side benefit: No light is directly reflected back to the laser from either its front or back face, and laser mode instabilities that could so result are prevented. In contrast, most of the light reflected from the medium returns to the laser, and noise control techniques must be used (see Sec. 6.4.2). In nonerasable or read-only systems, such light feedback is almost entirely eliminated by inserting a quarter waveplate between the beam splitter and the medium at 45° to the incident polarization direction. The polarization of the incident beam is therefore converted from linear to circular, and upon reflection the handedness is reversed. Traversing the quarter waveplate again, the light becomes polarized perpendicular to the incident direction and is reflected by the polarizing beam splitter toward the detector. Unfortunately, this scheme does not work for a magnetooptical system, because the light incident on the medium must be linearly polarized, and no analogous practical method to rotate the polarization has yet been devised. (A Faraday rotator can be used, but it is necessarily large and expensive.)

The light reflected into the detection module is linearized by a phase plate that brings r_y into phase with r_x. A second polarizing beam splitter then divides the light into two components polarized along orthogonal directions for differential detection. Thus, upon reversal of the film magnetization, the phase of r_y changes by 180°, and the change in the difference signal (i.e., the magnetooptical signal) is $2P_s$. The differential scheme therefore doubles the signal and, at the same time, eliminates the common mode noise.

The mass of the moving components should be minimized to achieve fast track-seeking performance. This is accomplished by splitting the optical system into two parts: a stationary laser, magnet, and detector and a moving objective lens. Focus is maintained with a small actuator attached to the objective lens. This is necessary because the optical path length between the lens and the magnetic film is not constant because of variations in lens-disk spacing and, for second-surface recording, variations in substrate thickness as well. In the latter case, the lens-disk spacing averages about 1.5 mm, and the substrate thickness is typically 1.2 mm. The objective lens can be mounted on either a moving carriage or a rotating arm, but the linear motion of the latter simplifies the design if the optical system is split.

Coarse and fine tracking positioners are also required. Coarse tracking may be accomplished by either using a stationary actuator attached by bands to a moving objective carriage or allowing a coil on the carriage to move between stationary permanent magnets. Fine tracking can be accomplished by either attaching a small radial actuator to the objective lens or using a mirror mounted on a high-speed galvanometer. The fine-tracking actuator or galvanometer typically adjusts the radial tracking position by ± 30 μm. Thus, with a track spacing of 1.5 μm, about 40 tracks are accessible with an access time of a few milliseconds. Use of a fine-tracking positioner can thus mimic the performance of a cylinder in a multihead multidisk magnetic recording system. The necessity for a separate fine-tracking positioner will be discussed in Sec. 6.4.3.

6.4.1 Semiconductor laser operation and collimation

Semiconductor diode lasers can now be made and packaged in small, low-power, efficient, and inexpensive units. Whereas the efficiency of gas lasers is about 10^{-3}, efficiencies near 50 percent have been achieved for semiconductor lasers. Because of their small size, however, diode lasers exhibit a relatively large amount of beam spread, which requires optical correction. They also typically excite several widely spaced modes, and exhibit mode instabilities when pumped with reflected light. These instabilities in turn create intensity modulations (laser noise), which, if not prevented, can significantly degrade the signal.

The beam emitted from the laser has a range of divergence angles. It must both be collimated and given a circular cross section, and the resulting beam must be directed along a given axis. For this purpose, a collimating lens followed by two prisms is used. The lens is arranged to collimate the rays. However, because of the different divergence angles parallel and perpendicular to the plane of the laser, the beam is elliptical in cross section after collimation. The purpose of the prism pair is to provide an anamorphic correction for this ellipticity without rotating the optic axis.

The optical power from a semiconductor laser is zero below a threshold current of about 50 mA and rises linearly with current thereafter. For a typical 820-nm infrared laser, the gain envelope is positive for about 15 modes. The mode separation is determined by the length of the lasing cavity and is typically about 0.4 nm. Redistribution of the energy between different modes results in laser noise, a variation in output light intensity. This variation is strongly affected by light reflected back into the laser from optical components or the disk. The physics of the effect is not completely understood. In the absence of reflected light, the laser usually oscillates in a single longitudinal mode (Stubkjaer and Small, 1984). Figure 6.28 shows the variation of laser noise with the logarithm of light intensity reflected back into the laser (Biesterbos et al., 1983). Very low levels (less than 0.1 percent) of reflected light cause the laser to oscillate noisily in a single mode. The noise is a maximum for a reflected light level of about 0.3 percent, where two modes share the energy. For reflected light levels above this, many modes are typically excited, and the noise monotonically decreases with light level. Above 10 percent and especially at higher laser power, it is observed that the laser can operate between two metastable states, of which one typically has large non-Gaussian noise spikes.

Three approaches must be taken to reduce laser noise. First, direct reflections from optical elements back into the laser should be minimized. This light is highly coherent and can stimulate rapid fluctuations in both the coherency and intensity of the light (Arimoto and Ojima, 1984). Second, stray light that reaches the detectors after reflection from optical elements should be minimized to reduce coherent interference effects. Finally, superposition of a high-frequency (300 to 700 MHz) current on the (essentially) dc current driving the laser has been shown to reduce the laser noise to an acceptable level (Ohishi et al., 1984; Kanada, 1985).

In nonerasable systems, a diode near the back facet of the laser can be used to monitor laser power. This is useful for two reasons: it may be desired to control the write and read power at different levels for the inner and outer tracks, and a gradual increase in injection current is necessary to compensate for a decrease in laser power with aging. This arrangement is not satisfactory for magnetooptics, because the dc output from front and back facets of the laser varies oppositely with light feedback. Therefore, it is preferable to use a front facet monitor to maintain constant or desired output. The signal can be obtained from a small amount of the laser light that is diverted from the main beam by the polarizing beam splitter (see Fig. 6.27), and a feedback loop is closed through a several-megahertz servo to adjust the laser driving current.

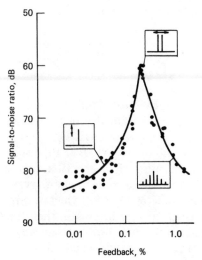

Figure 6.28 Variation of laser noise with light reflected back into the laser. Three types of excited mode structure are indicated. The signal-to-noise ratio is measured in a 10-kHz bandwidth *(after Biesterbos et al., 1983).*

Use of feedback to correct for the slow decrease in light output due to aging cannot continue indefinitely. The rate of aging depends on the operating temperature through an Arrhenius relation with an activation energy of about 0.8 eV. The useful life is reduced by a factor of 10 for each 25°C temperature increase (Tsunoda et al., 1983). Commercially available diode lasers give stable operation for 10^4 h at a continuous output power of 15 to 30 mW.

6.4.2 Detection methods

In the best detection method, the signal is derived differentially from a detector module containing a polarizing beam splitter and two photodiode detectors (Fig. 6.1a). This method has two advantages over the single-analyzer scheme described in Sec. 6.3.4. First, with a single analyzer, half the signal is lost, whereas the polarizing beam splitter divides the signal essentially without loss. Second, differential detection allows common mode rejection of additive noise, and if the detection threshold is correctly placed, it can also reduce the zero-crossing shift from multiplicative (e.g., laser) noise.

Using Eq. (6.40) with $\alpha = 45°$, the optical power on each detector is

$$P = 0.5\, P_i R \xi^2 \tag{6.50}$$

As in Eq. (6.42), when the magnetic state of the storage medium is reversed, the power on each detector changes by

$$P_s = 2P_i r_y \xi r_x = 2P_i K_r \xi \sqrt{R} \tag{6.51}$$

and the peak-to-peak differentially detected signal is twice this. Reduction of the component of r_x that is reflected by the beam splitter in the main beam improves the contrast

$$\frac{P_s}{P} = \frac{4K_r}{\xi \sqrt{R}} \tag{6.52}$$

The contrast is readily measured by dividing the peak-to-peak signal by the average power that reaches the detectors. Knowledge of ξ and the reflectivity then allows a determination of the figure of merit K_r.

The detectors can be either avalanche photodiodes (APD) or PIN photodiodes. An analysis of the signal and noise from APDs and PIN diodes is given in Secs. 6.5.1 and 6.5.2. Here, a qualitative comparison is given. Avalanche photodiodes have a built-in gain factor $G \simeq 20$, which is typically sufficient for the shot noise from the source to be greater than the electronic noise of the detector. The signal-to-noise ratio is then shot-noise limited, and the results of Sec. 6.3.4 show that it depends only on P_i

and r_y. Because APDs can handle only a small dynamic range of light power and require a large contrast, $\xi = \sin \phi$ should be small, and a beam splitter with $\xi^2 = 0.1$ is a good choice. The primary disadvantages of APDs are that they are expensive and require high voltages.

PIN diodes are inexpensive and can handle a higher light level. There are two reasons for increasing the light on the photodetectors. First, without the added gain of the APD, the signal is lower and the output noise is typically dominated by the electronics of the amplifier. Thus for PIN diodes, the signal-to-noise ratio (SNR) is proportional to the signal itself

$$\text{SNR} \sim P_i r_x r_y \sin \phi \cos \phi \tag{6.53}$$

For constant laser power, the power P_i incident on the disk is proportional to $1 - \xi^2 = \cos^2 \phi$ of the incident laser power. Maximization of Eq. (6.53) then occurs for $\phi = 30°$, whence a polarizing beam splitter with $\xi^2 = \sin^2 \phi = 0.25$ is more appropriate. A second advantage is that by increasing the light level on the detectors, the focus and tracking signals will be larger. There are two disadvantages incurred with increasing ξ: first, the contrast is reduced, and the system will be more sensitive to alignment error and to laser noise; second, less laser power will arrive at the disk for writing.

6.4.3 Focus and tracking servos

Focus and tracking servos are required in all optical recording systems to compensate for the variation in optical path length and radial runout. For a 13-cm disk, the optical path length will typically vary by up to ± 200 μm as the disk rotates, and a focus servo is therefore required to maintain focus within about ± 1 μm that is set by the depth of field. Likewise, because the disks are removable, a typical radial runout of ± 20 μm must be reduced by a tracking servo to ± 0.1 μm. The extremely high track density additionally requires that tracking servo information be embedded in the disk. The servos can derive information from the disk either continuously or at specified times. In either case, this information is preformatted on the disk substrate during fabrication. For example, a continuous tracking signal can be derived from grooves between tracks. Alternatively, a sample and hold servo can receive updates periodically, at equally spaced intervals around the track, from a preformatted header that has focus, timing, track-following, and sector identification data. In both cases, the focus and tracking methods must provide feedback signals that are linear in the error.

Several focus methods have been used, and the obscuration, knife-edge, biprism, and astigmatic methods are illustrated here (Bouwhuis and Braat, 1983). They all rely on the divergence of the reflected rays increasing (decreasing) near the detector as the disk approaches (moves away from) the objective lens. This change in divergence is then translated into a change in differential output from two or more photodiodes. The better techniques combine a large slope of the error signal near focus with an acquisition range that is 100 times the depth of focus.

In the knife-edge or Foucault approach (Kocher, 1983), a one-sided obstruction is placed at the nominal focal point of the detector lens, as in Fig. 6.29, and the focus-error signal is derived differentially from the split diode. A disadvantage with this method is the loss of half the signal by the obstruction.

A biprism method which retains the full signal and requires two pairs of diode detectors is shown in Fig. 6.30. A double wedge splits the beam in two parts, each of which is detected by a split diode. The focus error signal is derived from $(S_1 + S_4) - (S_2 + S_3)$.

In the astigmatic method (Cohen et al., 1984), a cylindrical lens is placed within the detector module. If the beam is out of focus on the disk, its image on the quadrant photodetector will be elliptical, with the major axis orientation dependent on the out-of-focus direction, as shown in Fig.

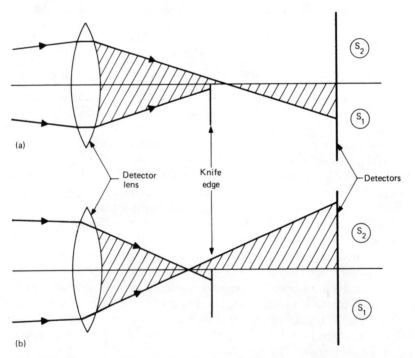

Figure 6.29 Focus signal from knife edge. The knife edge is placed at the nominal focal plane of the detector lens. When the disk is *(a)* too close to, or *(b)* too far from the objective lens, the differential signal $S_1 - S_2 > 0$ and $S_1 - S_2 < 0$, respectively *(after Bouwhuis and Braat, 1983).*

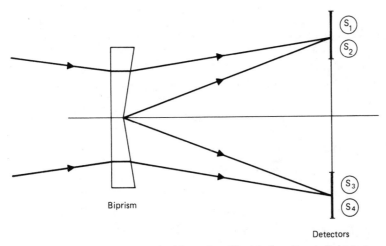

Figure 6.30   Focus signal from double wedge. The biprism directs light to two split photodiodes. As the disk moves out of focus, the beam-splitting angle changes, and the focus error signal is $S_1 + S_4 - (S_2 + S_3)$ *(after Bouwhuis and Braat, 1983).*

6.31. The error signal can then be constructed from the differential output $S_1 + S_3 - (S_2 + S_4)$.

All tracking-servo methods work by diffracting light out of the optical path to the detector. In a sample and hold system, alternate off-track pits are used with precise timing to develop an error signal. The pit depth is $\lambda/4$ to diffract light as efficiently as possible. When the beam is in the center of the track, the light reflected to the detector is the same for each pit. When the beam is off-center, the light reflected to the detector differs for each pit, and an error signal can be developed. This is fed back into the track-following servo.

For continuous far-field tracking, an error signal is derived from pre-formatted grooves. The objective lens must then capture the zeroth- and first-order diffraction beams, shown in Fig. 6.32 imaged in the far-field on a detector, that these grooves produce. The tracking signal is developed from a split detector, which covers the zeroth-order beam and part of the first-order beam, as shown. When the read beam is centered on or between grooves, the first-order diffraction beams are symmetrical, and the differential signal is zero. However, when the beam is not centered on or between grooves, the two first-order beams are no longer identical, and they will interfere differently in the region of overlap with the zeroth-order beam. The differential signal from the split detector can then be used to center the beam on the track. It has been shown that rectangular phase grooves with a depth of about $\lambda/8$ provide an adequate tracking signal (Jipson and Williams, 1983), and optimal readout is obtained from grooves that are half the width of the track pitch (Pasman et al., 1985).

Among other suggested tracking methods, the three-spot approach (Kryder, 1985) has been used effectively in read-only systems like the digital audio disk, but the use of laser power is too inefficient for writeable systems. It has also been shown that the far-field diffraction pattern from the data itself (a phase grating) can be used to develop both focus and

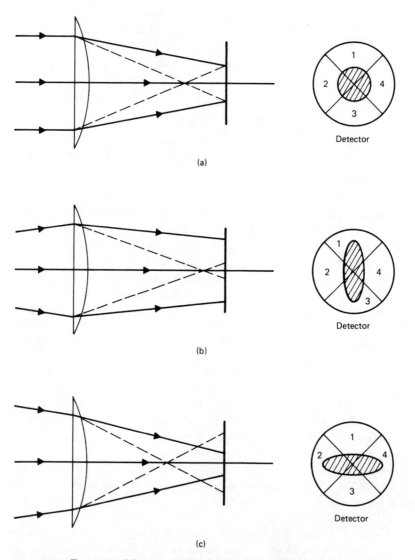

(a)

(b)

(c)

Figure 6.31 Focus signal from astigmatic lens. *(a)* The quadrant detector is placed equidistant between focal planes for the two orthogonal axes, when the beam is focused on the disk. Out-of-focus conditions are shown in *(b)* and *(c)*. Dashed rays are in the perpendicular plane *(after Bouwhuis and Braat, 1983).*

tracking servo signals (Braat and Bouwhuis, 1978), but this too is probably restricted to a read-only system.

In summary, all of the tracking techniques have advantages and disadvantages. The sample and hold technique works well with short sectors of a few hundred bits, but for larger sectors it is necessary to use updates which are more frequent than once per sector. This method therefore is implemented at the cost of formatted data density. The continuous far-field tracking technique provides the ability to count tracks during track seeking, which can shorten access time. Moreover, for some write-once systems, the data written in the grooves are physically contained by the groove walls, thus enhancing readback uniformity and reducing crosstalk from adjacent tracks. In contrast, for magnetooptical media, data in the grooves can have significant additional amplitude modulation (medium noise) due to groove irregularities, as well as loss of signal due to interference. If the data are written on wide "land" areas between grooves, it is difficult to acquire enough tracking signal. It would appear, therefore, that sample and hold techniques for tracking might be better suited to magnetooptical recording at present data densities.

In write-once systems, the data, focus, and tracking signals can all be generated from a quadrant PIN diode, by taking appropriate output combinations (Bell, 1983). In a magnetooptical system, there are many options. One simple approach is to generate the servo signals in separate arms of the data detection module. For example, astigmatic focusing can be implemented in one arm by placing a cylindrical lens in front of a quadrant detector. Far-field tracking can then be implemented in the other arm by using a split detector placed beyond the focal plane of the

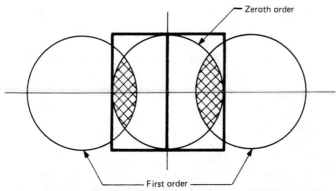

Figure 6.32 Diffraction pattern produced at the detector by the groove structure on the disk. The detector (heavy outline) is split and covers the zeroth order diffraction beam. Interference of zeroth- and first-order beams occurs at the overlap (crosshatched) *(after Braat and Bouwhuis, 1978).*

lens. Another method is to use a separate optical arm for the servo signals (Imamura et al., 1985).

The necessity for a separate fine-tracking actuator can be seen from the following considerations. Suppose it is desired to operate the disk at a 50-Hz rotation rate. To prevent instability, the closed-loop gain of the servo is constrained to have a slope of approximately -1 near the unity gain point. (Away from this point, the slope can be increased to -2.) With this constraint, a bandwidth (unity-gain frequency) of about 2 kHz is required for a closed-loop track servo to follow a 20-μm runout with an error of not more than 0.1 μm. To avoid positive feedback, no poorly damped mechanical resonances should exist below about ten times the closed-loop unity-gain frequency (20 kHz). A small galvanometer or transverse lens actuator can be designed without sharp resonances below 20 kHz, but this is not possible for a 10- to 50-gm carriage and optical head assembly.

It is beyond the scope of this chapter to give details of the large number of lens focus and tracking actuators that can be built into the optical head. The most common approach is to combine focus and tracking into two orthogonal actuators, typically using a voice-coil arrangement for focus motion. For radial track-following motion, there are several possibilities. The optical system can be mounted off-center with respect to a rotated shaft (Okada et al., 1980); a voice coil or variable reluctance structure can be used to translate the lens (Bouwhuis et al., 1985), or the beam can be shifted by a mirror on a stationary galvanometer which is not on the moving head.

6.4.4 Magnet design

As described in the introduction, erasure and writing in currently practical systems occur in two passes with an oppositely directed magnetic field. That this will remain a requirement of magnetooptical recording can be seen from an order-of-magnitude calculation for the time required to switch a several-hundred-oersted field from an air coil. The time constant

$$\tau_m = \frac{L}{R} \tag{6.54}$$

of a coil with inductance L and resistance R must be comparable with or less than the inverse of the data rate. No specific assumptions are made about the coil shape, other than it has n turns of average radius r, and the n turns have a total cross-sectional area $A \simeq r^2$. The coil inductance is

$$L = q\mu_0 r n^2 \tag{6.55}$$

where q is a factor on the order of unity that depends on the coil shape. The coil resistance is

$$R \simeq \frac{\kappa 2\pi rn}{A/n} \simeq \frac{2\pi\kappa n^2}{r} \tag{6.56}$$

where the resistivity of copper is $\kappa = 1.7 \times 10^{-8}\ \Omega\cdot\text{m}$. For a given cross-sectional area, the time constant is thus independent of n

$$\tau_m = \frac{q\mu_0 r^2}{2\pi\kappa} \tag{6.57}$$

The coil radius and number of turns are determined by requirements on the field and coil resistance. When energized with a current i, the magnetic field near the coil center is

$$H \simeq \frac{ni}{2r} \tag{6.58}$$

and the coil resistance is designed to match the power supply. For example, suppose the power supply can deliver 1 A at 5 V. Using $i = 1$ A, $R = 5\ \Omega$, and $H = 40$ kA/m [500 Oe] in Eqs. (6.56), (6.58), and (6.57) gives

$$n \simeq 600 \text{ turns} \tag{6.59}$$

$$r \simeq 7 \times 10^{-3} \text{ m} \tag{6.60}$$

$$\tau_m \simeq 10^{-3} \text{ s} \tag{6.61}$$

This time constant is four orders of magnitude too great to operate at a data rate of 10 Mb/s. The problem with the air coil is that it is too large: the inductance is about 1 mH. Some improvement can be achieved by designing a smaller coil, but the coil radius cannot be made less than a few millimeters, which is the spacing between the coil and the medium. This relatively large separation prevents a low-inductance structure from focusing the magnetic field on the small region where thermomagnetic writing is occurring. It is possible to use a one-sided disk, with a flying magnetic recording head on the side opposite to the optics, to build a system which uses the magnetic field to switch the heated medium at the data rate. However, mechanical complications, including alignment and co-tracking of the flying head with the laser beam, make this an unattractive alternative. Therefore, should direct overwrite become an important system requirement, it must be accomplished through medium engineering, such as the use of multiple magnetic layers to obviate the need for a separate erase cycle (Saito et al., 1987).

The requirements on a practical (two-pass system) magnet are that it (1) generates a field of several hundred oersteds at the medium, (2) switches direction in a few milliseconds (a fraction of a disk rotation period), (3) does not interfere with the tracking and focus actuators on the optical head, (4) causes no problems with heat dissipation, (5) does not substantially increase the seek time, and (6) fits into the packaging constraints, which are particularly severe in small drives.

These requirements can be met by several designs. An electromagnet can be placed on either the front (optics) or the back side of the disk. An electromagnet on the front side must be designed around the objective lens, but its size should be minimized for maximum field generation from the ampere-turns. A greater magnetic efficiency can be achieved by winding a coil around a soft-iron pole on the back side, and then shimming with soft iron to reduce both the magnetic circuit reluctance and the demagnetization near the pole tip. Removal of joule heating is simpler for this back-side geometry, because the magnet can be thermally tied to a structure not directly attached to the optical path. Alternatively, a high-energy permanent magnet on the back side can be physically rotated to reverse the field direction. With either front or back (electro-) magnets, it is best if the magnet covers all useful tracks on the medium and hence does not move during track seeking. The substrate must be nonconducting because of eddy-current drag induced in conductors that move in an inhomogeneous magnetic field.

The power dissipated by the magnet depends on the coil shape and ampere-turns. If no portion of the electromagnet saturates during writing or erasure, the field is proportional to the ampere-turns. The power is then given by

$$P = \frac{\kappa c}{FA} \left(\frac{H}{h}\right)^2 \tag{6.62}$$

where c = average length per turn in the coil
F = coil filling factor (typically $F \simeq 0.6$)
h = efficiency (in A/m of field per ampere-turn of current)

Because the power varies directly as the square of the field to be attained and inversely as the square of the coil efficiency, the requirement on the former should be minimized and the efficiency should be maximized, subject to packaging constraints. Efficiencies of 60 A/m per ampere-turn and time constants of a few milliseconds can be achieved in low-power back-side designs.

6.5 System Performance Characteristics

System performance is ultimately characterized by data rate and by corrected bit error rate as a function of track density and linear bit density. Internal performance measures that are crucial for design are the signal-to-noise ratio and the uncorrected bit error rate. Adequate signal-to-noise ratio is required to guarantee that the error-correction circuitry is not overwhelmed by soft errors due to Gaussian noise, and an uncorrected bit error rate of less than 10^{-4} is needed to allow a corrected bit error rate of better than 10^{-12}. Consequently, the emphasis in this section is on the signal and noise in magnetooptical systems and on encoding and error-correction techniques designed to maximize areal bit density while maintaining the required corrected error rate.

The magnetooptical signal depends on the medium, the optics, the size and spacing of the written marks, the detectors and electronics, and the signal processing for removing intersymbol interference. The noise originates in the electronics, the quantized nature of light, the intensity modulations of the laser output, and the medium. Some of the noise components, like the electronic and shot noise, have a Gaussian amplitude distribution. Others, like the laser noise and the noise due to previously written data, are non-Gaussian, with negligible tails, and act to narrow the timing window. The hard medium errors are also non-Gaussian, but they have a large probability in the tail region and cannot be treated analytically. All of these noise sources ultimately cause errors in the interpretation of the digital message. For example, when the message is encoded by the presence or absence of a pulse at a specified time, noise can cause errors by shifting the signal amplitude across the detection threshold. Or when the message is encoded by zero crossings, the noise can move the zero-crossing position out of the timing window.

The magnetooptical signal and the shot, electronic, and thermal Gaussian noise sources in APD and PIN detectors are presented in Sec. 6.5.1. The signal-to-noise ratio in typical magnetooptical systems and medium noise are then discussed in Sec. 6.5.2. Additional loss of signal-to-noise ratio, due to the modulation transfer function of the optical system and the loss of timing window through run-length-limited encoding, is described in Secs. 6.5.3 and 6.5.4, respectively. Finally, error-correction methods are briefly described in Sec. 6.5.5.

6.5.1 Magnetooptical signal and noise

The photodetector converts optical power (watts) to electrical current, with a quantum efficiency $\eta \simeq 0.5$ A/W

$$i = \eta P \tag{6.63}$$

An avalanche photodiode (APD) has a built-in current gain G, so that in general, the detector current is

$$i = G\eta P \tag{6.64}$$

A transimpedance amplifier can be used to convert the current generated by an APD or a PIN diode into a voltage signal. The amplifier holds the diode collector at ground and forces all current through the feedback resistor R, producing an output voltage

$$V = GR\eta P \tag{6.65}$$

A 10 kΩ resistor allows operation up to 20 MHz, a reasonable bandwidth for the passage of the signals through the electronics.

The peak-to-peak differentially detected magnetooptical signal on each diode is given by Eq. (6.49)

$$P_s = 2P_i r_x r_y \sin \phi \cos \phi \tag{6.66}$$

and the average power on each detector is

$$P = 0.5P_i(r_x^2 \sin^2 \phi + r_y^2 \cos^2 \phi) \simeq 0.5P_i r_x^2 \sin^2 \phi \tag{6.67}$$

The rms shot noise current from each detector is then (Horowitz and Hill, 1980)

$$i_{\text{shot}} = G\sqrt{2e\eta\beta P} \simeq G\sqrt{e\eta\beta P_i}\, r_x \sin \phi \tag{6.68}$$

where $e = 1.6 \times 10^{-19}$ C, and β is the amplifier bandwidth. There will also be some electronic and thermal noise current in the amplifiers, denoted by

$$i_n = \sqrt{i_{\text{ne}}^2 + i_{\text{nt}}^2} \tag{6.69}$$

A good amplifier has an equivalent noise input current of about 3 pA/$\sqrt{\text{Hz}}$. Thus,

$$i_{\text{ne}} = 3 \times 10^{-12} \sqrt{\beta} = 0.013 \ \mu\text{A} \tag{6.70}$$

with the bandwidth $\beta = 20$ MHz. The rms thermal noise current from the feedback resistor is (Horowitz and Hill, 1980)

$$i_{\text{nt}} = \sqrt{4kT\beta/R} = 0.006 \ \mu\text{A} \tag{6.71}$$

for $k = 1.38 \times 10^{-23}$ J/$^\circ$K, $T = 300^\circ$K, and a nominal value of $R = 10$ kΩ.

Electrical power out of the amplifiers is proportional to the square of these currents (and also to the square of the input optical power). Because the three noise contributions are uncorrelated, the total noise current is the sum of squares of individual currents. From Eqs. (6.64), (6.66), and (6.68), the signal-power-to-noise-power ratio is (Mansuripur et al., 1982)

$$\text{SNR} = \frac{4G^2\eta^2 P_i^2 r_x^2 r_y^2 \sin^2 \phi \cos^2 \phi}{e\eta\beta G^2 P_i r_x^2 \sin^2 \phi + i_n^2} \tag{6.72}$$

Due to the high internal gain of APDs, the shot-noise dominates the electronic and thermal noise, and

$$\text{SNR} = \frac{4\eta P_i}{e\beta}\, r_y^2 \cos^2 \phi \tag{6.73}$$

Note that the gain G does not appear in this equation; the system is limited by the signal-to-shot-noise ratio. However, with PIN diodes ($G = 1$), the electronic and thermal noise together are typically larger than the shot noise, and

$$\text{SNR} \simeq \frac{4\eta^2}{i_n^2}\, P_i^2 r_x^2 r_y^2 \sin^2 \phi \cos^2 \phi \tag{6.74}$$

6.5.2 Signal-to-noise ratio in magnetooptical systems

It is useful to represent the detector currents by an equivalent input optical power, the noise equivalent power (NEP). This is found, using Eq. (6.64), by dividing the detector current by $G\eta$. For illustration, the optical power and noise equivalent powers for practical systems constructed with APD and PIN diode detectors are given in Table 6.5. Both systems use a magnetooptical medium with parameters $R = 0.2$, $r_x = \sqrt{R} = 0.45$, and $r_y = 5 \times 10^{-3}$. For APD detectors, the power on each detector should not exceed about 10 μW, and we choose $P_i = 2$ mW, $\xi^2 = 0.05$, and $G = 20$. For PIN diodes, the parameters $P_i = 2$ mW, $\xi^2 = 0.25$, and $G = 1$ are more appropriate. It is seen that the contrast P_s/P for the PIN system is quite low, and any increase obtained by reducing ξ is accompanied by a loss of signal.

TABLE 6.5 Power and Noise-Equivalent Powers on APD and PIN Detectors

Quantity	Formula	APD	PIN
P_s (peak-to-peak)	$2P_i r_x r_y \sin \phi \cos \phi$	2.0 μW	4.0 μW
P (average)	$0.5 P_i r_x^2 \sin^2 \phi$	10 μW	50 μW
Contrast	P_s/P	0.20	0.08
NEP_{shot} (rms)	$\sqrt{(e\beta P/\eta)}\, r_x \sin \phi$	0.011 μW	0.025 μW
NEP_e (rms)	$3 \times 10^{-12}\, \sqrt{\beta}/\eta G$	0.001 μW	0.027 μW
NEP_t (rms)	$\sqrt{(4kT\beta/R)}/\eta G$	0.001 μW	0.012 μW
$\text{NEP}_{\text{total}}$ (rms)	$\sqrt{\begin{array}{l}(\text{NEP}_{\text{shot}})^2 + \\ (\text{NEP}_e)^2 + (\text{NEP}_t)^2\end{array}}$	0.011 μW	0.039 μW

The signal-to-noise ratio can be calculated from the shot, thermal, and electronics noise equivalent powers, using

$$\text{SNR} = 10 \log \frac{P_s^2}{\text{NEP}_{\text{shot}}^2 + \text{NEP}_t^2 + \text{NEP}_e^2} \quad \text{dB} \tag{6.75}$$

and the results are given in Table 6.6 for both the APD and PIN detector systems. The APD will actually perform a few dB worse than shown because of extra noise generated with the gain (Mansuripur et al., 1982). The two systems show similar performance, because the extra electronic noise of the PIN is partly compensated by the larger magnetooptical signal that it can detect, because of its greater dynamic range.

The signal-to-noise ratios calculated above are for a single detector with the signal given by the peak-to-peak change when the magnetization reverses in a saturated region of magnetooptical film. The signal-to-noise ratios from data in a system with differential detection differ from this in several ways, as is also shown in Table 6.6. The corrections to the signal-to-noise ratio are +3 dB from using two detectors, −9 dB for converting the signal from peak-to-peak to rms, about −6 dB for loss of readback signal due to the modulation transfer function (see Sec. 6.5.3), and various effective losses in signal-to-noise ratio due to narrower timing margins with RLL encoding (see Sec. 6.5.4).

Because the PIN detector system is not shot-noise limited at 2 mW incident read power, an increase in power delivered to the disk causes a proportional increase in signal-to-noise ratio. This trend is being followed, with the availability of higher power lasers to write on media designed to be less thermally sensitive. In a reasonably designed system, a 25 mW

TABLE 6.6   Representative Signal-to-Noise Ratio in Magnetooptical Systems with APD and PIN Detectors.

System configuration	Correction to SNR, dB	SNR, dB APD	SNR, dB PIN
One-detector†	0	45	40
One-detector‡	−9	36	31
Two-detectors‡	+3	39	34
At moderate data density‡	−6	33	28
With enhanced NRZ [≈0.9]§	0	33	28
With 1,3 code [1.0]	−3	30	25
With 1,7 code [1.33]	−3	30	25
With 2,7 code [1.5]	−6	27	(22)

†Ratio (signal, peak-to-peak)/(total noise-equivalent-power, rms), from Table 6.5.
‡rms/rms.
§The figures in brackets are the changes in data density achieved with RLL encoding.

laser can deliver about 12 mW to the disk during writing. With a 4:1 ratio of write-read power, this allows 3 mW of read power to be used.

The various noise sources are shown in Fig. 6.33, in which the signal power and the various noise equivalent powers in a 20-MHz bandwidth are plotted as a function of read power. With increasing read power, the shot-noise contribution overtakes the fixed electronic noise, and for higher read powers where the PIN response is shot-noise limited, there is no signal-to-noise ratio advantage to using APD detectors rather than PIN diodes. Above some value of read power, the medium noise overtakes the

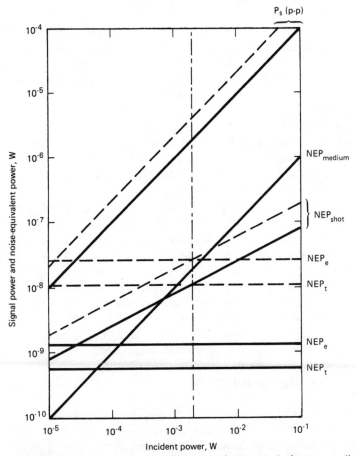

Figure 6.33 A log-log plot of signal power and noise-equivalent power (in a 20-MHz bandwidth) versus incident power for the PIN (dashed line) and APD (full line) systems described in the text. The signal power is peak-to-peak, and all noise-equivalent powers are rms. The vertical line is the assumed operating point $P_i = 2$ mW of Tables 6.5 and 6.6. For the APD system with P_i near 2 mW, the total noise-equivalent power (neglecting medium noise) is given by NEP_{shot}.

other noise sources and provides the ultimate limitation. The medium noise line given in Fig. 6.33 is from the best available film deposited on a glass substrate, and has a signal-to-noise ratio of about 42 dB in a 20-MHz bandwidth. Thus, at the nominal power $P_i = 2$ mW at the disk, the medium noise is only slightly greater than the shot noise. From these curves, the typical dependence of the signal-to-noise ratio on read power can be derived. At very low read powers where thermal and electronic noise are dominant, the ratio is proportional to the incident power; at intermediate powers where shot noise is dominant, the ratio is proportional to the square root of the read power; and at higher powers where medium noise limits the performance, the ratio becomes independent of power.

In addition to electronic and shot noise, two other sources of noise, the laser and the medium, adversely affect system performance. Laser noise has been discussed in Sec. 6.4.1. When rf driving is used on the laser, and back-reflection to the laser and extra internal reflection to the photodetectors are minimized, the noise from commercial diode lasers can be reduced to acceptable levels.

Medium noise is usually multiplicative (proportional to the incident power) and has several sources. First, there are microscopic substrate inperfections, including surface roughness and wobble in the preformatted grooves. Flat glass substrates offer the lowest variation in medium reflectivity. If preformatted grooves are used to generate a continuous tracking signal, medium noise is reduced if the magnetooptical data are written on a wide "land" area rather than in the grooves. Second, there are magnetooptical film imperfections, including voids resulting from contaminants and other process problems, that result in hard repeatable errors. These hard errors must be corrected through redundancy of information (see Sec. 6.5.5). Third, there is modulation noise due to a scatter in the shape and position of the recorded transitions. For stable systems in which focus and tracking fluctuations can be ignored, modulation noise is related to the domain-wall pinning mechanisms in the thermomagnetic process described in Sec. 6.2.5. It is very small in amorphous Tb-transition-metal alloys, often significant in ablative nonerasable materials, and usually unacceptably large in polycrystalline magnetooptical materials such as Mn-Bi.

Those aspects of medium noise exclusive of hard errors are most easily characterized experimentally from a power spectrum—the readback spectrum of a single-frequency low-density pattern on an optimized system with shot-noise limited detectors. The carrier-to-noise ratio, defined as the signal-to-noise ratio in a 30 kHz bandwidth from such a data pattern, is commonly used to compare performance from different systems. Without medium noise, the carrier-to-noise ratio from a shot-noise limited system is about 61 dB (add 28 dB for bandwidth correction to the 33-dB value in Table 6.6). The modulation component of the medium noise can be found by comparing the noise when the medium is saturated (erased) with the noise after the pattern is written. The signal is a series of spikes at multiples of the fundamental, and the modulation noise is seen as an elevation of the noise baseline after writing the low-frequency pattern. Figure 6.34 shows typical results from a spectrum analyzer with 200-kHz bandwidth, for a 2.5 μm bit length 101010 . . . pattern recorded at 5 Mb/s with 10 mW incident power (50 percent duty cycle) on a preerased area. There is no evidence of modulation noise, but the carrier-to-noise ratio (add 8 dB for bandwidth correction) is about 55 dB, which is slightly below the shot-noise limit of about 61 dB mentioned earlier. Therefore, while the measured carrier-to-noise ratio compares favorably with current nonerasable recording systems, further improvements will be possible when better substrate materials become available. It should be noted that hard errors due to medium defects have little rms noise power, and their presence is not detectable by spectrum analysis.

As the read power (and temperature) increases, the magnetooptical component r_y decreases. Consequently, the signal is not strictly proportional to P_i, and the carrier-to-noise ratio, when limited by medium noise, will decrease at higher read power. For example, a broad maximum in the carrier-to-noise ratio of 52 dB was obtained near $P_i \simeq 1$ mW on a Tb-Fe-Co disk deposited on a plastic substrate (Aoki et al., 1985).

The signal-to-noise ratio required for a reliable system may be estimated in the following way: Considering only noise fluctuations with a

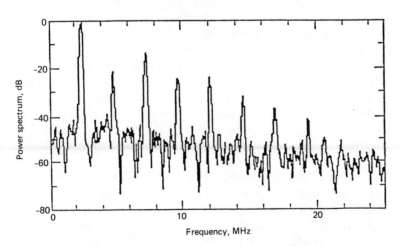

Figure 6.34 Readback power spectrum in a 200-kHz bandwidth for data written at 2.5 MHz. (The carrier-to-noise ratio in a 30-kHz bandwidth is obtained by adding 8 dB.) The signal-to-noise ratio is limited by the medium noise. In this case the shot noise power is about 6 dB below the medium noise *(after Bloomberg and Connell, 1985)*.

Gaussian amplitude distribution, an rms ratio of 17 dB has a probability of 10^{-12} for a noise fluctuation to exceed the signal, and a probability of 10^{-4} corresponds to 12 dB. A soft-error rate below 10^{-12} is clearly satisfactory, whereas an error rate of 10^{-4} would not be sufficiently correctable. To guarantee a worst-case signal-to-noise ratio greater than 17 dB (after corrections such as in Table 6.6), margins must be built in to anticipate signal loss through (1) reflection, absorption, and diffraction in the optics; (2) variations in medium quality; and (3) inadequate performance of various components (laser, detectors, electronics). Considering that halving the signal-to-noise ratio corresponds to a drop of 6 dB, the nominal corrected value should be at least 25 dB. Table 6.6 indicates that this is presently achievable with both APD and PIN detectors at moderate bit densities and with an enhanced NRZ code that does not reduce the timing window. Performance of PIN detectors with codes that reduce the timing window may be marginal at high density.

6.5.3   Areal bit density and data rate

6.5.3.1   Modulation transfer function.

The magnetooptical signal given in Eq. (6.66) assumes that the entire read beam illuminates regions that are uniformly magnetized in both the written and erased states. However, for actual written data, the mark size is smaller than the Gaussian beam. Consequently, the maximum signal amplitude is found to decrease as the written bits get smaller. The signal decreases rapidly below a characteristic bit length that depends on the wavelength, numerical aperture of the objective lens, and shape of the written bits.

This is best summarized by the modulation transfer function S_{MTF} that gives the zero-to-peak amplitude of the output for a repeated pattern of magnetization reversals. For convenience, all such outputs are normalized to half the change in output when a uniformly magnetized region is reversed. As will be shown later, S_{MTF} depends on three properties of the recorded pattern: the repeat distance (recorded "wavelength"), the mark width, and the actual placement of the upward- and downward-going transitions within the repeated pattern.

At the very high data density for which a full recorded wavelength is incorporated within W, the full-width-at-half-maximum of the optical read beam, the modulation transfer function is small. In fact, at such short recorded wavelengths, the signal is limited by diffraction, and for Gaussian optics, the transfer function from a coherent source vanishes for recorded wavelengths smaller than (Goodman, 1968)

$$\lambda_{\mathrm{cutoff}} = \frac{0.5\,\lambda}{N_A} \simeq W \tag{6.76}$$

As expected from the geometrical (as opposed to diffraction) argument given above, the cutoff wavelength is comparable to W, given in Eq. (6.46). For typical values of $\lambda = 0.83$ μm and $N_A = 0.5$, $\lambda_{\text{cutoff}} = 0.83$ μm. Therefore, the average distance between flux reversals at cutoff is only 0.415 μm and is far smaller than any useful bit size.

In a proper treatment of the magnetooptical signal, the film acts as a 180° phase grating on the small signal component r_y, and the signal is determined from the light coherently diffracted into the objective lens. In a simpler geometrical model, the signal is assumed to be proportional to a convolution of the reversed magnetization in the written pattern with the Gaussian read spot intensity

$$I(r) = I(0)e^{2.77 \ (r/W)^2} \tag{6.77}$$

and results from this model are in good agreement with data from magnetooptical systems (Bloomberg and Connell, 1985). When calculating the normalized signal $S(t)$, it is not necessary to subtract the convolution of $I(r)$ with the unreversed background, because this would merely double the dynamic range and cause a level shift.

The convolution can also be performed as a superposition of responses from properly positioned $+ \rightarrow -$ and $- \rightarrow +$ transitions. The response from one such transition is shown in Fig. 6.35, with the level adjusted so that $S(-x) = -S(x)$. The normalized signal from such an isolated transition has a long linear portion and attains a maximum value S_{trans}. The

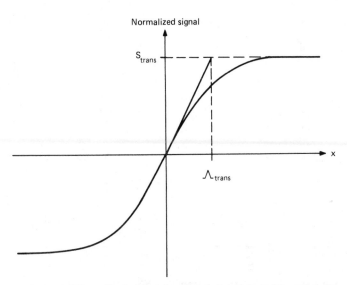

Figure 6.35 Normalized signal from an isolated transition, showing definition of S_{trans} and Λ_{trans}.

tangent at the zero crossing equals this maximum at a characteristic distance Λ_{trans} from the zero crossing, given by

$$\Lambda_{\text{trans}} \equiv \frac{S_{\text{trans}}}{dS/dx} \simeq 1.1\,W = \frac{0.62\,\lambda}{N_A} \tag{6.78}$$

where dS/dx is the slope at the zero crossing (Bouwhuis et al., 1985). The maximum signal S_{trans} depends on the ratio w/W of the written mark width to the read beam full-width-at-half-maximum, as shown in Fig. 6.36. Several points should be noted about Λ_{trans}. First, it gives the approximate location of the knee of S_{MTF}: the modulation transfer function drops precipitously when the distance between transitions is reduced below Λ_{trans}. Second, Λ_{trans} is determined entirely by W. Third, it is essentially independent of either the written mark width w or the relative curvature of the written mark and the read beam.

The modulation transfer function is then derived by superposition of a pattern of isolated transitions, each with a magnitude given by $S_{\text{trans}}(w/W)$, that repeats every two transitions. The location of the transitions

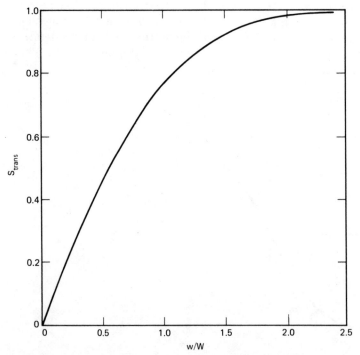

Figure 6.36 Maximum amplitude of an isolated transition S_{trans} (as a fraction of saturation amplitude), as a function of the ratio of mark width w to the read beam full-width-at-half-maximum W.

depends on the mark shape and the average distance between transitions ("bit length" b). For a typical bit length $b \simeq 1.5W$, the waveform thus composed is nearly sinusoidal, with magnitude S_{MTF}.

In addition to spatial frequency and the ratio w/W, the S_{MTF} for actual written patterns depends on the duty cycle of the write laser. Consider the simplest encoding method, NRZ level detection, in which a region of reversed magnetization (mark) is written in the data cell for a 1 and is not written for a 0. The mark shape is then determined by the write power, size of the write spot, and the pulse time. It ranges between two extremes; namely, from nearly circular marks for pulse times much less than the inverse data rate to elongated marks, with rounded ends which protrude from data = 1 cells into adjacent data = 0 cells, for the maximum pulse time (equal to the inverse data rate).

Two examples of written bits are given in Fig. 6.37 for 50 percent and 5 percent duty cycles. The parameters in the model are the bit length b, the skid length l, the written mark width w, and the full-width-at-half-maximum of the read beam W. At high bit density, the modulation transfer function falls off most rapidly for the 50 percent duty cycle written marks because the space between the marks is sharply reduced. This is seen in Fig. 6.38, where a 5 Mb/s 101010 ... pattern was written with 5 mW. Both b and l are 0.92 μm, and W is 0.75 μm. It is seen that the (peak-to-peak) signal amplitude is 0.30 of saturation. Furthermore, the encroachment of the marks into the 0 bit cells produces a loss or "sag" equal to 0.26 of the saturation signal. The data can be fit to the model with one adjustable parameter, $w = 0.5$ μm. At other write powers, equally good fits are obtained with different values of write power.

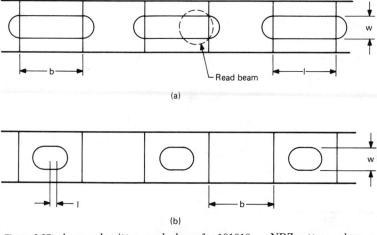

Figure 6.37 Assumed written mark shape for 101010 ... NRZ pattern, where w is the mark width, b is the bit length, and l is the skid length. *(a)* $l = b$ (50 percent duty cycle); *(b)* $l = b/10$ (5 percent duty cycle).

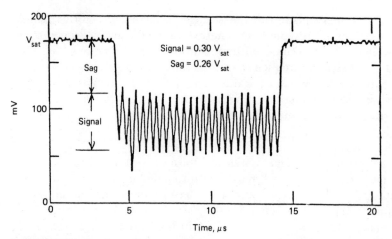

Figure 6.38 Output from 101010 ... pattern written at high density with $b = l = 0.92$ μm, $W = 0.75$ μm, and data-rate $D = 5$ Mb/s. Because of the large sag, the modulation transfer function for this pattern is only 0.30 *(after Bloomberg and Connell, 1985).*

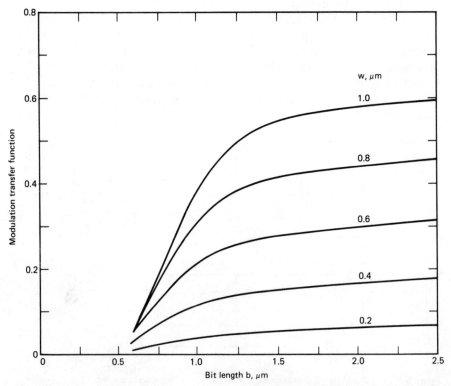

Figure 6.39 Computed modulation transfer function versus bit length b, for $l = b/10$, $W = 1.0$ μm, and various mark widths. The theoretical cutoff is at $b = 0.45$ μm.

The modulation transfer function is constructed by taking such 101010 ... data patterns at different bit lengths. Calculated transfer functions are shown in Fig. 6.39 for a series of mark widths and for $l/b = 0.1$ and $W = 1.0$ μm. The written mark width will have the least sensitivity to laser power if the mark extends to the maximum slope of the Gaussian write beam, which occurs at

$$w = 0.85W \tag{6.79}$$

Therefore, for a practical case of $b = 1.5$ μm and $w = 0.8$ μm, the signal is reduced to about 0.41 of the saturated value (nearly -8-dB loss).

Two modulation transfer functions are shown in Fig. 6.40, with $W = 0.75$ μm. Curves A and B show the calculated fractional signal versus bit

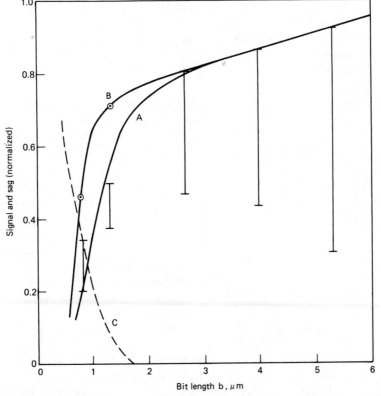

Figure 6.40 Comparison of calculated modulation transfer function (solid lines) with observed values versus bit length. Data are read from an NRZ 101010 ... pattern with $W = 0.75$ μm. The vertical lines show the range of measured values of the transfer function for different write powers for $l = b$, and their envelope should be compared with curve A. The circles are measured for $l = b/10$ and fall on the theoretical curve B. Curve C is normalized sag for $l = b$ (after Bloomberg and Connell, 1985).

length b, maximized with respect to mark width, for $l = b/10$ and $l = b$, respectively. The vertical lines show the range of experimental values obtained using different write powers and $l = b$. The two data points are the measured signal for $l = b/10$, and the dashed curve C is the observed sag for $l = b$. Note that the sag rises steeply as the bit length is decreased below the knee of the transfer function curve.

Figure 6.41 compares theory (solid lines) and experiment for signal and sag versus write power at the disk. When the bit length is 0.92 μm and $l = b$ (50 percent duty cycle), the signal attains a maximum over a narrow range of write power, and then rapidly decreases because of sag as the written marks become larger. In contrast, when $l = b/10$ (5 percent duty cycle), the sag remains small and the signal increases monotonically with write power. Essentially the same power is required to write the bits with either 200-ns pulses or 20-ns pulses, indicating that the thermal time constant τ_d for heat to flow into the adjacent dielectric layers is short compared with 200 ns. This is a demonstration of the argument presented in Sec. 6.3.3 on thermal engineering. The maximum signal amplitude is obtained when the transitions in a 101010 . . . NRZ pattern are equally spaced, so that the mark and space are of equal length. This criterion can be approached at high density by use of very short duration pulses.

As noted earlier, there is an optimum mark width w for writing information. Figure 6.42 shows how w varies as a function of the incident laser power and the disk speed. In the figure, w is inferred by comparing the signal amplitude from 101010 . . . NRZ data with the prediction of the readback model. The dashed curves are smooth fits to the experimental points, and w is given in units of W ($= 0.75$ μm). At low linear velocities, the optimum mark width is readily achieved, but at higher velocities the required power rises rapidly and may not be attainable.

Thus, choice of the mark width depends on several factors. When sag is minimized through short pulses, the signal increases with the mark width, as shown indirectly in Fig. 6.41. It is feasible therefore to approach the criterion given by Eq. (6.79), subject to medium sensitivity and available laser power. When long pulses are used, the signal at high data densities may be reduced by sag before the condition in Eq. (6.79) is reached. The mark width is also constrained by erasure requirements. That is, the erase width must be sufficiently larger than the written mark width to allow for worst-case tracking errors during writing and erasure. This is best achieved through higher laser power during erasure, rather than higher bias field, so that if laser power is limited, the written mark width may be less than the value given by Eq. (6.79).

In summary, media can be designed so that extremely short pulses can be used to write data. The knee of the modulation transfer function sets a practical upper limit on useful flux-reversal densities. For example, from Fig. 6.40, the -6-dB point for the short pulse transfer function at $W =$

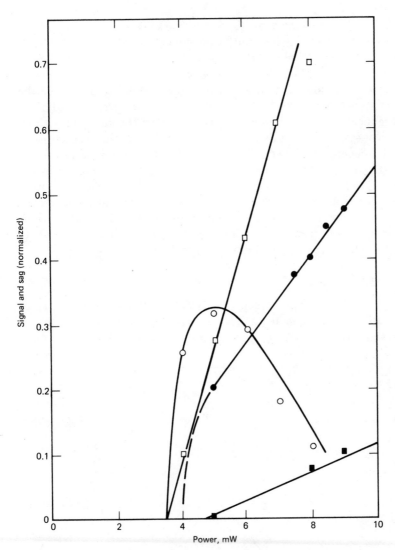

Figure 6.41 Modulation transfer function (circles) and normalized sag (squares) for NRZ 101010 ... pattern with $b = 0.92$ μm, for $l = b$ (open symbols) and $l = b/10$ (filled symbols). Theoretical predictions are given by solid lines; measured values are given by symbols. Read beam $W = 0.75$ μm *(after Bloomberg and Connell, 1985).*

0.75 μm occurs at $b \simeq 1.0$ μm. For an NRZ code, this corresponds to a linear data bit density $D_l = 10^3$ bits per millimeter. In Sec. 6.5.5, the possibility of further increasing this linear data density without decreasing the minimum distance between transitions through run-length-limited encoding will be discussed.

6.5.3.2 Limitations on track density. When marks are present on adjacent tracks, the tail of the Gaussian read beam will always generate cross-track interference. Figure 6.43 shows the crosstalk that occurs in the carrier-to-noise ratio as the read beam is moved between two adjacent tracks separated by $t = 2.0W$ and written with 2.5- and 3.5-MHz signals. When the beam is centered on one track, the carrier-to-noise ratio from the adjacent track is reduced by more than 40 dB. With perfect tracking and with an ideal optical system, crosstalk from adjacent tracks is thus negligible at this track spacing. However, in a practical system, tracking errors are about ± 0.1 μm, effectively moving the tracks together by 0.2 μm in the worst case. A second major contribution to crosstalk is the expansion of the spot size because of focus error. Additionally, crosstalk is increased somewhat with spatial frequency, owing to diffraction effects (Bouwhuis et al., 1985). Therefore, a conservative value for the minimum track spacing t_{min}, for which crosstalk will not be a problem, is

$$t_{min} = 2.0W \tag{6.80}$$

For a typical $W = 0.75$ μm, a track density $D_t = 670$ tracks per millimeter is easily attained. As the technology evolves, t_{min} can be expected to decrease significantly.

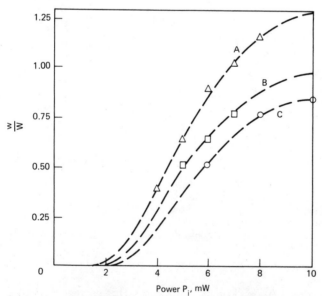

Figure 6.42 Ratio of mark width w to read beam full-width-at-half-maximum W versus laser power P_i incident on the disk, for three disk speeds: (curve A) 4.6 m/s, (curve B) 8.9 m/s, and (curve C) curve 26.3 m/s. Data are written as NRZ 1010 . . . pattern at 5 Mb/s, with 200 ns pulse time for the written marks. Read beam $W = 0.75$ μm.

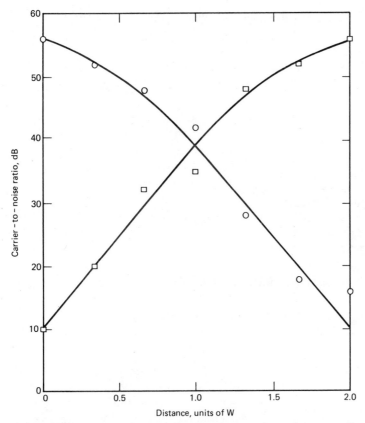

Figure 6.43 Crosstalk measurement of carrier-to-noise ratio versus off-track position of read beam center, in units of W. Tracks written at 2.5 and 3.5 MHz are centered at $x = 0$ and $x = 2W$, respectively. Circles and squares give carrier-to-noise ratios for the 2.5 and 3.5 MHz signals, respectively. For these measurements, $W = 0.75$ μm, and the mark width $w \simeq 0.85W$.

6.5.3.3 Trade-off between capacity and data rate. There are several constraints between the system parameters (disk size, rotation rate, data capacity, and data rate) and the underlying bit length and track spacing. The focus and tracking servos impose an upper limit on the rotation rate of the disk. The decrease in the signal-to-noise ratio at high bit density, as shown by the modulation transfer function, imposes a lower limit on the bit length at the inner track, for a system with a constant number of sectors per track. The minimum track spacing is determined by tracking servo accuracy and the read optics. The data rate is constrained by the inner-track bit length, inner-track radius, and rotation rate. And the data capacity is determined by the disk size, track spacing, and minimum bit length.

The interplay of all these factors can be described by two equations, one for capacity

$$8 \times 10^6 C = \frac{(2\pi r_i/b)(r_o - r_i)}{t} \qquad (6.81)$$

and one for data rate

$$10^6 D = \frac{2\pi r_i}{b} \frac{\Omega}{60} \qquad (6.82)$$

where C = unformatted capacity in MB
  b = inner track bit length
  r_i, r_o = inner and outer track radii
  t = track spacing
  D = data rate in Mb/s
  Ω = rotation rate in rpm.

Substitution of Eq. (6.82) into Eq. (6.81) leads to an equation for inner track radius

$$r_i = r_o - 0.1333 \frac{\Omega C t}{D} \qquad (6.83)$$

and this value can be used in Eq. (6.82) to find the minimum bit length b.

For example, with a data rate $D = 5$ Mb/s, outer radius $r_o = 6$ cm and track spacing $t = 1.5$ μm, Fig. 6.44 shows the minimum bit length b versus rotation rate Ω, for different data capacities. For capacities less than 250 MB, there is a clear trade-off between the servo-limited rotation rate and optics-limited minimum bit length. At higher capacities, an increase in Ω has less effect on b. A similar calculation, for a data rate $D = 10$ Mb/s, demonstrates that if the data rate is chosen to be too large, the minimum bit length will be too small at practical rotation rates and will not be significantly increased by lowering the capacity C.

6.5.4 Data encoding

For simplicity, let us assume that the readback signal is derived from a well-balanced differential channel. The signal is then bipolar, with symmetrical excursions in both directions from the zero level, and a clock defines the channel bit cells. There are two ways to encode and detect message bits. In the first method, known as pulse-position-modulation encoding, the message bit is determined from the signal polarity at the center of the bit cell. The value of the message bit is most easily found by comparing the signal with a threshold voltage level (which is set to zero for this ideal situation). In the second method, known as pulse-duration-modulation encoding, the message bit is determined from the existence (or absence) of a zero-crossing of the signal within the bit cell. This is most directly found by comparing the signal polarities at the cell boundaries. Either encoding method unambiguously determines the message bit to be a 0 or a 1.

In the general encoding process, m data bits are encoded as n message bits (see Chap. 5). The simplest encodings are NRZ and NRZI, where the data and message bits are identical. NRZ is a pulse-position encoding, where 1s and 0s are encoded into regions of opposite magnetization direction. NRZI is a pulse-duration encoding, where 1s and 0s are encoded into the presence or absence of a transition between magnetization directions. Because the domains in (two-pass) magnetooptical recording are always written with the same polarity, opposite to a preerased state (unlike conventional magnetic recording), it is easier to implement a simple polarity

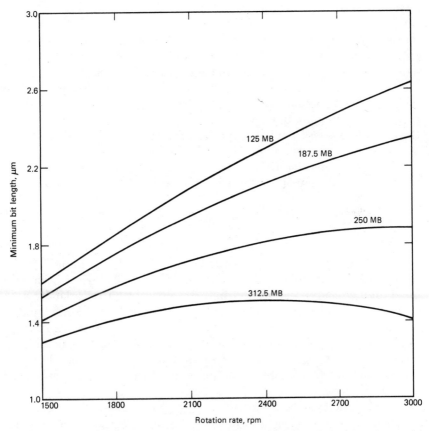

Figure 6.44 Minimum bit length versus rotation rate for a 13-cm disk with various unformatted capacities and with data rate $D = 5$ Mb/s. The track spacing $t = 1.5$ μm, and the number of bits per track is assumed to be constant *(after Bloomberg and Connell, 1985)*.

detection scheme using NRZ. Thus, a region of reversed magnetization is written to store a 1 message bit, and the erased state is left undisturbed for a 0. Implementation of an NRZI encoding with zero crossing detection is more difficult, because two zero crossings will be detected from each isolated region of reversed magnetization. The length of the reversed region must be adjusted so that the detected crossings of a 111 . . . message pattern are equally spaced. Local changes in medium write sensitivity will alter the length of the reversed regions, shifting adjacent zero crossings in the opposite direction. Such shifts cannot be compensated by self-clocking techniques.

Run-length-limited codes with (d,k) constraints (see Chap. 5) are useful both for self-clocking and for packing data at a higher density on the disk. The (d,k) constraints apply to the message bits, which are encoded on the disk as if they were NRZ. The d and k constraints specify that there be at least d and not more than k 0s between adjacent 1s in the message bit stream. The k constraint allows self-clocking, whereas the d constraint can be used to increase the minimum distance between 1s. The penalty for using constraints on the message bits is that each message bit is not entirely random and thus contains less information than a (random) input data bit. Consequently, the m data bits must be encoded into $n > m$ message bits, and the message bit cell is therefore smaller than the data bit cell. Now by definition, for RLL codes there are $d + 1$ message bits within each minimum flux reversal interval. The timing window is always the size of the message bit cell, and for NRZ ($d = 0$), the timing window is also equal to the minimum distance between flux reversals. Therefore, for the same minimum distance between flux reversals, the d constraint reduces the RLL timing window from the NRZ value by

$$T_{\mathrm{RLL}} = \frac{T_{\mathrm{NRZ}}}{d + 1} \tag{6.84}$$

At high density, when the message bit cell is smaller than the optically constrained minimum feature size, regions of reversed magnetization centered on one bit cell will extend into the adjacent bit cells. Under such conditions, which are exacerbated when a $d \geq 1$ RLL code is used, pulse-position encoding may give errors when decoding the adjacent message bits. In contrast, pulse-duration encoding can be used as long as the minimum distance between flux reversals is larger than the optically determined minimum feature size. Thus, pulse-duration encoding must be used at sufficiently high density.

The reduction of the timing window for RLL codes has a significant effect on the required signal-to-noise ratio for magnetooptical systems. Suppose first that additive electronic noise is the only source of noise or timing error. From Eq. (6.84), when using pulse duration modulation encoding, the window around the zero crossing is narrowed to 0.5 of the

NRZ window for $d = 1$ codes and 0.33 for $d = 2$ codes. For a sinusoidal waveform

$$S(t) = S_{\text{MTF}} \sin \frac{\pi t}{T_{\text{NRZ}}} \tag{6.85}$$

shown in Fig. 6.45, the amount of noise that will cause an error is then reduced for these codes from the NRZ value by factors of $\sqrt{2}$ and 2, respectively. Hence, the signal-to-noise requirement is 3 dB (for $d = 1$) and 6 dB (for $d = 2$) greater than when using NRZ with the same minimum feature size.

However, in addition to Gaussian noise there are invariably other sources of noise and timing error that can degrade performance by further reducing the timing window (Treves and Bloomberg, 1986b). The effect of timing jitter is to decrease the window size. Non-Gaussian noise sources (such as laser noise, medium noise, crosstalk from adjacent tracks, and noise from data not entirely overwritten on the same track) typically have an amplitude spectrum without long Gaussian tails, and each acts to decrease the timing window by a fixed amount. Because these amounts are independent of the total window, they will occupy a larger fraction of the window budget for encoding methods with small windows. Consequently, the effective loss of signal-to-noise due to RLL encoding will be larger than the estimate given above.

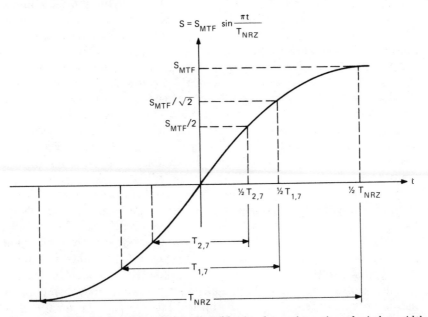

Figure 6.45 Relation between relative allowable signal-to-noise ratio and window width for various pulse duration modulation encodings of a sinusoidal waveform. For a (2,7) code, the window is reduced to one-third that for NRZ encoding, with an effective loss of signal-to-noise-ratio of at least 6 dB and a gain in data density of 50 percent.

These losses in signal-to-noise ratio can be semiquantitatively expressed as follows. Let the time T_{fixed} budgeted to the non-Gaussian noise sources be a fraction γ of the NRZ window:

$$T_{\text{fixed}} = \gamma\, T_{\text{NRZ}} \tag{6.86}$$

Then, again referring to the sinusoidal waveform of Fig. 6.45, the effective fractional loss ΔSNR of signal-to-noise ratio due to window narrowing from both the encoding method and T_{fixed} is

$$\begin{aligned}
\Delta\text{SNR} &= \sin\frac{(\pi/2)(T_{\text{RLL}} - 2T_{\text{fixed}})}{T_{\text{NRZ}}} \\
&= \sin\left[\frac{\pi}{2}\left(\frac{1}{d+1} - 2\gamma\right)\right]
\end{aligned} \tag{6.87}$$

When this is combined with the loss from the modulation transfer function, the normalized effective signal amplitude is

$$S = S_{\text{MTF}} \sin\left[\frac{\pi}{2}\left(\frac{1}{d+1} - 2\gamma\right)\right] \tag{6.88}$$

RLL encoding enables both self-clocking and, in some cases, data compression. For example, use of 1,7 codes gives a 1.33 increase in data density when compared with NRZ, when both have the same minimum distance between flux reversals. Likewise, 2,7 codes give a 1.5 increase in data density. However, as shown by Eq. (6.87), the price in window loss (and, equivalently, loss in signal-to-noise ratio) can be high.

An understanding of intersymbol interference is essential to evaluate the trade-offs involved with the use of RLL codes. At high flux-reversal density, the lowering of the modulation transfer function is accompanied by intersymbol interference, whereby the zero-crossing position of a particular transition depends on adjacent transitions. There are two components to intersymbol interference: a nonlinear effect introduced during writing and a linear part that occurs in readback and is also responsible for the sag effect described earlier. Because the intersymbol interference in readback is linear, it can be greatly reduced, with some loss of signal-to-noise ratio, through equalization of the read channel. The choice between using a $d \geq 1$ RLL code or an enhanced NRZ (with, for example, an extra 1 message bit added after every n data bits for self-clocking) at a higher flux-reversal density depends in part on the ability to remove intersymbol interference with channel equalization. The relatively low signal-to-noise ratio and the absence of high-frequency components because of the sharp cutoff in the optical modulation transfer function, however, will make channel equalization more difficult.

In flying-head conventional magnetic recording, there is a large nonlinear component to intersymbol interference, introduced in the write process by both hysteresis and demagnetization effects, that cannot be equalized. This nonlinear bit shift increases rapidly when adjacent transitions crowd together. Consequently, RLL codes can provide a slight improvement in data bit density for conventional recording by spreading the closest transitions farther apart. In contrast, hysteresis and demagnetizing effects are negligible in magnetooptical recording, and as a result there is minimal nonlinear bit shift. If good channel equalization is achieved, an enhanced NRZ code may be preferable to d $\geq$ 1 RLL codes at moderate flux-reversal densities.

In summary, the choice of a particular code depends on available signal-to-noise ratio, the timing window, the degree of intersymbol interference, the ability to equalize the read channel, and ease of encoding/decoding. Methods can be altered at will. For example, the (d,k = 2,10) EFM code used in digital audio uses an extra "merging" bit that improves the channel by suppressing low frequencies, at a cost of 6 percent overhead (Heemskerk and Immink, 1982). There are multiple trade-offs for any decision, and while designers may agree that several approaches are good, few will agree on a best design.

6.5.5 Error correction and hard errors

Because of the extremely high areal densities attained in optical recording, media currently produced, even under clean room conditions, can be expected to have an uncorrected bit-error rate of up to 10^{-4} associated with disk surface defects. While there will certainly be reductions of the bit-error rate with time, particularly as the substrate technology improves, error correction will always be required to provide a corrected bit-error rate of less than 10^{-12}, as demanded for digital data storage.

To obtain high-bandwidth, real-time operation, the error-correction method chosen must be designed for the expected distribution of hard errors on the disk. An example of the distribution of errors that occur in optical systems is shown in Fig. 6.46. Fortunately, very long burst errors, which can cause loss of clocking in self-clocked systems, are rare. Most of the errors are in the one- to three-bit range, but approximately 10 percent are larger. Thus, the error-correction method must primarily be designed to handle the relatively large number of short-length errors, but must also be capable of correcting the less frequent longer bursts.

Linear block codes, such as Reed-Solomon codes (see Chap. 5), are used because they both meet these requirements and are easily implemented in hardware. In a block code, s message bits form a message symbol, and k message symbols are combined with $p = n - $ k parity symbols to form encoded blocks of n symbols. It can be shown (Clark and Cain, 1981) that using a symbol length $s > 1$ and a block length $n \leq 2^s - 1$, a Reed-Solomon code can be constructed that will correct $t = p/2$ symbols. Further, if the errors occur at not more than p "erasure" positions, all of which are known by some method (e.g., a loss of signal amplitude) external to the decoder, then the code can correct up to $t = p$ errors. Note that error correction applies equally to message and parity symbols.

Theoretically, a Reed-Solomon code with $s = 8$ bits per symbol (i.e., 1 byte symbols), block length $n = 255$ symbols per block, $k = 223$ message symbols per block, and $p = 32$ parity symbols per block, can correct any 16 incorrect symbols. In practice, decoding of such efficient codes has not yet been implemented in hardware. Instead, the relatively easy correction of up to two symbols in each block is made at the data rate. When more than two symbols are in error, decoding is delayed, and correction is performed in software.

Interleaving, whereby consecutive bits are placed in different blocks, can, in two ways, reduce delays in the processing of error-correction algorithms. First, two-way interleaving is particularly useful because of the high probability that a two-bit error will fall across an eight-bit symbol boundary, thus corrupting two symbols in one block. Partitioning such an error into different blocks will decrease the likelihood that it must be decoded in software. Second, for error-correcting architectures that allow

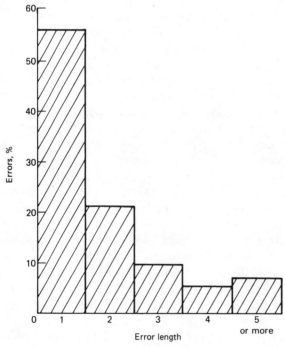

Figure 6.46 Distribution of the size of uncorrectable medium errors for a magnetooptical disk.

decoding of the interleaved blocks in parallel, N-way interleaving will increase the computation time allocated to correction of each block by a factor of N. Therefore it will be possible to implement the efficient codes required for correction of a larger number of incorrect symbols at the data rate. Finally, interleaving also provides the capability to correct the small number of long burst errors that are found in magnetooptical recording media.

These methods of error correction are currently in use in the compact disk read-only memory, digital audio, and write-once systems. Unfortunately, no standard has yet been established by the write-once optical recording industry. In contrast, the industrywide standard of the digital audio disk industry uses two crosswise interleaved Reed-Solomon codes, with $s = 8$ bits per symbol and $p = 8$ parity symbols per block in each block of $n = 32$ symbols (Hoeve et al., 1982). As presently implemented, it can correct errors in any two symbols and up to four errors if the erasure positions are known. Uncorrected error bursts are hidden by interpolation schemes. This is sufficient because the demands of audio technology are met without extremely low corrected error rates.

There is no theoretical impediment to high-data-rate, efficient, real-time error correction. It can be expected that VLSI implementations of interleaved Reed-Solomon codes, capable of correcting worst-case errors at the full data rate, will be operating on magnetooptical systems with corrected error rates well below 10^{-12}.

6.6   Opportunities and Limitations

Magnetooptical recording combines very high storage density, low cost, random access, erasability, and removability. The areal storage densities of 0.5×10^6 b/mm^2 that are easily attainable with magnetooptical recording are one order of magnitude greater than those in currently available magnetic recording products.

The ability to read and write small bits with a large mechanical clearance leads to many favorable attributes of magnetooptical systems. First, it enables second-surface operation, which greatly improves immunity to surface dirt and scratches, and allows good isolation of the active film from the atmosphere. This in turn permits the removal of media, consequently enabling the drive to be used with any medium (erasable, nonerasable, and read-only). Removability requires little additional servo performance, because embedded servos are in any event required at high track densities. Second, there is no possibility of data-destroying head "crashes," and magnetooptical drives should be relatively immune to shock and vibration. A shock large enough to cause loss of focus would only delay disk I/O by a fraction of a second.

The success of magnetooptical recording technology hinges on the ability to make archival-quality erasable media. The primary concern is to protect the magnetooptical thin film from chemical degradation. The problem is particularly severe with the amorphous rare-earth transition-metal films, in which the rare-earth atoms are highly reactive. Nevertheless, such archival-quality films have been produced in various laboratories, and commercialization is proceeding.

A present and continuing limitation of small magnetooptical drives is a relatively low data rate when compared to high performance Winchester drives. This is required because close tracking tolerances and small servo bandwidth limit the rotation rate, and the minimum bit length is limited by the optics.

Agreement on format standards, which are necessary for interchangeability, may take several years because of competing interests. For example, grooves are favored for nonerasable disks because they both help confine the written (ablated) marks and improve track-seeking performance by providing a continuous servo signal. Grooves in magnetooptics, however, are disadvantageous because they lower the medium quality and increase medium noise. A compromise, perhaps made at the expense of continuous tracking, might be to use grooves for both but to write in the (narrow) grooves for nonerasable and on the (wide) lands for magnetooptic media.

The technological push for magnetooptics comes currently from digital audio players (Compact Discs). The commercial success of these drives requires the development of inexpensive diode lasers, optics, actuators, servos, photodiodes, VLSI error correction, and preformatted substrates. Many of these components can be incorporated in magnetooptical systems with small changes. These drives also provide the incentive for investment in manufacturing technology for low-cost components, such as 30-mW lasers, low-noise photodiodes, VLSI circuitry for efficient real-time error correction, high-quality substrates, and preformatting techniques. As a consequence, many companies are actively developing magnetooptical media and drives.

An important aspect of magnetooptical systems is that their capabilities span many applications. At the low end (single-user workstations), they provide inexpensive, random-access replacement and backup for Winchester drives. On multifunctional drives, archival nonerasable and read-only media can also be used.

Magnetooptical drives can also be used with large general-purpose computers. To compensate for the relatively low data rate, two options are possible. For applications where files are being stored or retrieved, an I/O processor (buffer) can match the data rate of the drive to that of the mainframe. For virtual memory systems, the processor needs rapid access to disk memory. To achieve this, the data rate can be improved by using either several magnetooptic drives in parallel through a single data channel, or multiple read beams and read circuitry in one drive. The former has the advantage of simplicity and retains cylinderlike access, whereby the fine tracking actuator gives rapid access to several nearby tracks. With proper staggering of data on adjacent tracks, there should be little loss in peak data rate due to track-seeking or latency. Such a buffered multiple disk configuration will provide tremendous data rates and should be an inexpensive replacement for large magnetic disk drives. Smart controllers will shield the computer from delays due to the two-pass erase-write cycle.

Magnetooptics can also be used to build an on-line mass storage system, which combines removability with random access and terabyte capacity. There are interesting problems relating to the architectural organization of such a system, especially if it is composed of hundreds or thousands of single-disk magnetooptical drives.

Perhaps most important, magnetooptical technology will enable a significant class of new computer-based applications. The processing of graphics, image, video, and audio data makes enormous demands on the storage requirements of computer systems of the 1980s. Computer workstations of the 1990s will enable such applications by using a matched combination of powerful processors, high resolution color monitors, large semiconductor random-access memory, high-bandwidth network communications, and inexpensive random-access magnetooptical storage.

References

Alben, R., J. J. Becker, and M. C. Chi, "Random Anisotropy in Amorphous Ferromagnets," *J. Appl. Phys.*, **49**, 1653 (1978).

Allen, R., and G. A. N. Connell, "Magneto-optic Properties of Amorphous Terbium-Iron," *J. Appl. Phys.*, **53**, 2353 (1982).

Aoki, Y., T. Ihashi, N. Sato, and S. Miyaoka, "A Magneto-optic Recording System Using TbFeCo," *IEEE Trans. Magn.*, **MAG-21**, 1624 (1985).

Aratani, K., T. Kobayashi, S. Tsunashima, and S. Uchiyama, "Magnetic and Magneto-optic Properties of Tb-Fe-Co-Al Films," *J. Appl. Phys.*, **57**, 3903 (1985).

Argyle, B. E., R. J. Gambino, and K. Y. Ahn, "Polar Kerr Rotation and Sublattice Magnetization in GdCoMo Bubble Films," *AIP Conf. Proc.*, **24**, 564 (1974).

Arimoto, A., and M. Ojima, "Diode Laser Noise at Control Frequencies in Optical Videodisc Players," *Appl. Optics*, **23**, 2913 (1984).

Bell, A. E., "Critical Issues in High Density Magnetic and Optical Data Storage," *Proc. SPIE*, **382**, 2 (1983).

Bell, A. E., and F. W. Spong, "Antireflection Structures for Optical Recording," *IEEE J. Quantum Electr.*, **QE-14**, 487 (1978).

Bennett, H. S., and E. A. Stern, "Faraday Effect in Solids," *Phys. Rev. A.*, **137**, 448 (1965).

Bernstein, P., and C. Gueugnon, "Properties of Amorphous Rare Earth-Transition Metal Thin Films for Magneto-optical Recording," *IEEE Trans. Magn.*, **MAG-21**, 1613 (1985).

Biesterbos, J. W. M., A. G. Dirks, M. A. J. P. Farla, and P. J. Grundy, "Microstructure and Magnetic Properties of Amorphous Tb-Fe(O_2) Thin Films," *Thin Sol. Films*, **58**, 259 (1979).

Biesterbos, J. W. M., A. J. D. Boef, W. Linders, and G. A. Acket, "Low-Frequency Mode-

Hopping Optical Noise in AlGaAs Channeled Substrate Lasers Induced by Optical Feedback," *IEEE J. Quantum Elect.*, **OE-19**, 986 (1983).

Bloomberg, D. S., and G. A. N. Connell, "Prospects for Magneto-optic Recording," *Proc. IEEE Compcon Spring*, 32 (1985).

Bouwhuis, G., and J. Braat, "Recording and Reading of Information on Optical Disks," in R. R. Shannon and J. C. Wyant (eds.), *Applied Optics and Optical Engineering*, vol. 9, Academic Press, New York, 1983.

Bouwhuis, G., J. Braat, A. Huijser, J. Pasman, G. van Rosmalen, and K. Immink, *Principles of Optical Disk Systems*, Adam Hilger Ltd., Bristol, England, 1985, p. 49*ff*.

Bozorth, R. M., *Ferromagnetism*, van Nostrand, Princeton, N.J., 1951.

Braat, J. J. M., and G. Bouwhuis, "Position Sensing in Video Disk Readout," *Appl. Optics*, **17**, 2013 (1978).

Carasso, M. G., J. B. H. Peek, and J. P. Sinjou, "The Compact Disc Digital Audio System," *Philips Tech. Rev.*, **40**(6), 151 (1982).

Cargill, G. S., "Structure of Metallic Alloy Glasses," in H. Ehrenreich, F. Seitz, and D. Turnbull (eds.), *Solid State Physics*, vol. 30, Academic Press, New York, 1975.

Carslaw, H. S., and J. C. Jaeger, *Conduction of Heat in Solids*, Clarendon, Oxford, 1959.

Chang, J. T., J. F. Dillon, and U. F. Gianola, "Magneto-optical Variable Memory Based upon the Properties of a Transparent Ferrimagnetic Garnet at its Compensation Temperature," *J. Appl. Phys.*, **36**, 1110 (1965).

Chaudhari, P., J. J. Cuomo, and R. J. Gambino, "Amorphous Metallic Films for Magneto-optic Applications," *Appl. Phys. Lett.*, **22**, 337 (1973).

Clark, G. C., and J. B. Cain, *Error-Correction Coding for Digital Communications*, Plenum, New York, 1981.

Coey, J. M. D., "Amorphous Magnetic Order," *J. Appl. Phys.*, **49**, 1646 (1978).

Cohen, D. K., W. H. Gee, M. Ludeke, and J. Lewkowics, "Automatic Focus Control: The Astigmatic Lens Approach," *Appl. Optics*, **23**, 565 (1984).

Connell, G. A. N., "Interference Enhanced Kerr Spectroscopy for Very Thin Absorbing Films," *Appl. Phys. Lett.*, **40**, 212 (1982).

Connell, G. A. N., "Measurement of the Magneto-optical Constants of Reactive Metals," *Appl. Opt.*, **22**, 3155 (1983).

Connell, G. A. N., "Magneto-optics and Amorphous Metals: An Optical Storage Revolution," *J. Magn. and Mag. Mat.*, **54–57**, 1561 (1986).

Connell, G. A. N., R. J. Nemanich, and C. C. Tsai, "Interference Enhanced Raman Scattering from Very Thin Absorbing Films," *Appl. Phys. Lett.*, **36**, 31 (1980).

Connell, G. A. N., and R. Allen, "Magnetization Reversal in Amorphous Rare Earth-Transition Metal Alloys: TbFe," *Proc. 4th Int. Conf. on Rapidly Quenched Metals, 1981*, Sendai, Japan, 1981.

Connell, G. A. N., R. Allen, and M. Mansuripur, "Magneto-optical Properties of Amorphous Terbium-Iron Alloys," *J. Appl. Phys.*, **53**, 7759 (1982).

Connell, G. A. N., and R. Allen, "Amorphous Terbium-Iron Based Alloys," *J. Magn. and Mag. Mat.*, **31–34**, 1516 (1983).

Connell, G. A. N., D. Treves, R. Allen, and M. Mansuripur, "Signal-to-Noise Ratio for Magneto-optic Readout from Quadrilayer Structures," *Appl. Phys. Lett.*, **42**, 742 (1983).

Connell, G. A. N., S. J. Oh, J. Allen, and R. Allen, "The Electronic Density of States of Amorphous Y-Fe and Tb-Fe Alloys," *J. Non-Cryst. Solids*, **61 & 62**, 1061 (1984).

Connell, G. A. N., and D. S. Bloomberg, "Amorphous Rare-Earth Transition-Metal Alloys," in D. Adler, H. Fritzsche, and S. R. Ovshinsky (eds.), *Physics of Disordered Materials*, Plenum, New York, 1985, p. 739.

Corcoran, J., and H. Ferrier, "Melting Holes in Metal Films for Real-Time High-Density Permanent Digital Data Storage," *Proc. SPIE*, **123**, 17 (1977).

D'Antonio, P., J. H. Konnert, J. J. Rhyne, and C. R. Hubbard, "Structural Ordering in Amorphous TbFe$_2$ and YFe$_2$," *J. Appl. Cryst.*, **15**, 452 (1982).

Deguchi, T., H. Katayama, A. Takahashi, K. Ohta, S. Kobayashi, and T. Okamoto, "Digital Magneto-optic Disk Drive," *Appl. Optics*, **23**, 3972 (1984).

Dekker, P., "Manganese Bismuth and Other Magnetic Materials for Beam Addressable Memories," *IEEE Trans. Magn.*, **MAG-12**, 311 (1976).

Egami, T., "Theory of Intrinsic Magnetic After-Effect," *Phys. Stat. Sol., (a)* **19,** 747 (1973).

Freese, R. P., R. N. Gardner, T. A. Rinehart, D. W. Siitari, and L. H. Johnson, "An Environmentally Stable, High Performance, High Data Rate Magneto-optic Media," *Proc. SPIE,* **529,** 6 (1985).

Freiser, M. J., "A Survey of Magneto-optic Effects," *IEEE Trans. Magn.,* **MAG-4,** 152 (1968).

Fukunishi, S., M. Miyagi, and N. Funakoshi, "Amorphous Magneto-optical Disk with Multi-layered Structure," *Conf. Lasers and Electro-Optics,* 132 (1985).

Gambino, R. J., P. Chaudhari, and J. J. Cuomo, "Amorphous Magnetic Materials," *AIP Conf. Proc.,* **18,** 578 (1974).

Gambino, R. J., and T. R. McGuire, "Enhanced Magneto-optic Properties of Light Rare-Earth Transition Metal Amorphous Alloys," *J. Magn. and Mag. Mat.,* **54–57,** 1365 (1986).

Gangulee, A., and R. J. Kobliska, "Mean Field Analysis of the Magnetic Properties of Amorphous Transition Metal-Rare Earth Alloys," *J. Appl. Phys.,* **49,** 4896 (1978).

Goldberg, N., "A High Density Magneto-optic Memory," *IEEE Trans. Magn.,* **MAG-3,** 605 (1967).

Goodman, J. W., *Introduction to Fourier Optics,* McGraw-Hill, San Francisco, 1968, p. 120.

Hansen, P., and M. Urner-Wille, "Magnetic and Magneto-optic Properties of Amorphous GdFeBi-Films," *J. Appl. Phys.,* **50,** 7471 (1979).

Harper, J. M. E., and R. J. Gambino, "Combined Ion Beam Deposition and Etching for Thin Film Studies," *J. Vac. Sci. Tech.,* **16,** 1901 (1979).

Harris, R. "Metastable States in the Random Anisotropy Model for Amorphous Ferromagnets: The Absence of Ferromagnetism," *J. Phys. F: Metal Phys.,* **10,** 2545 (1980).

Harris, R., S. H. Sung, and M. J. Zuckermann, "Hysteresis Effects in Amorphous Magnetic Alloys Containing Rare Earths," *IEEE Trans. Magn.,* **MAG-14,** 725 (1978).

Hartman, M., J. Braat, and B. Jacobs, "Erasable Magneto-optical Recording Media," *IEEE Trans. Magn.,* **MAG-20,** 1013 (1984).

Hasegawa, R., and R. C. Taylor, "Magnetization of Amorphous Gd-Co-Ni Films," *J. Appl. Phys.,* **46,** 3606 (1975).

Heavens, O. S., *Optical Properties of Thin Films,* Dover, New York, 1965.

Heemskerk, J. P. J., and K. A. S. Immink, "Compact Disc: System Aspects and Modulation," *Philips Tech. Rev.,* **40**(6), 157 (1982).

Heiman, N., K. Lee, and R. I. Potter, "Exchange Coupling in Amorphous Rare Earth-Iron Alloys," *AIP Conf. Proc.,* **29,** 130 (1975).

Heitmann, H., M. Hartmann, M. Rosenkranz, and H. J. Tolle, "Amorphous Rare Earth-Transition Metal Films for Magneto-optical Storage," *J. Phys. Colloq (France),* **46,** 9 (1985).

Hindley, N. K., "Random Phase Model of Amorphous Semiconductors," *J. Non-Cryst. Sol.,* **5,** 17 (1970).

Hoeve, H., J. Timmermans, and L. B. Vries, "Error Correction and Concealment in the Compact Disc System," *Philips Tech. Rev.,* **40**(6), 166 (1982).

Hong, M., D. D. Bacon, R. B. van Dover, E. M. Gyorgy, J. F. Dillon, and S. D. Albiston, "Aging Effects on Amorphous Tb-Transition Metal Films Prepared by Diode and Magnetron Sputtering," *J. Appl. Phys.,* **57,** 3900 (1985).

Horowitz, P., and W. Hill, *The Art of Electronics,* Cambridge University Press, Cambridge, 1980, pp. 288–290.

Hunt, R. P., "Magnetooptics, Lasers and Memory Systems," *IEEE Trans. Magn.,* **MAG-5,** 700 (1969).

Huth, B. G., "Calculations of Stable Domain Radii Produced by Thermomagnetic Writing," *IBM J. Res. Dev.,* **18,** 100 (1974).

Ichihara, K., K. Taira, Y. Terashima, S. Shimanuki, and N. Yasuda, "Highly Reliable TbCo Film for Erasable Optical Disk Memory," *Proc. of Topical Meeting on Optical Data Storage,* Laser and Electro-Optic Society of IEEE, Washington, D.C., 1985, p. WAA2-1.

Imamura, N., S. Tanaka, F. Tanaka, and Y. Nagao, "Magneto-optical Recording on Amorphous Films," *IEEE Trans. Magn.,* **MAG-21,** 1607 (1985).

Jackson, J. D., *Classical Electrodynamics,* Wiley, New York, 1967.

Jackson, W. B., S. M. Kelso, C. C. Tsai, J. W. Allen, and S. J. Oh, "Energy Dependence of the Optical Matrix Element in the Hydrogenated Amorphous and Crystalline Silicon," *Phys. Rev.,* **B31,** 5187 (1985).

Jacobs, B., J. Braat, and M. Hartman, "Aging Characteristics of Amorphous Magneto-optic Recording Media," *Appl. Optics,* **23,** 3979 (1984).

Jipson, V. B., and C. C. Williams, "Two-Dimensional Modeling of an Optical Disk Readout," *Appl. Optics,* **22,** 2202 (1983).

Kanada, T., "Theoretical Study of Noise Reduction Effects by Superimposed Pulse Modulation," *Trans. IECE Japan,* **E68,** 180 (1985).

Kivits, P., R. de Bont, and P. Zalm, "Superheating of Thin Films for Optical Recording," *Appl. Phys.,* **24,** 273 (1981).

Kloosterboer, J. G., G. J. M. Lippits, and H. C. Meinders, "Photopolymerizable Lacquers for LaserVision Video Disks," *Philips Tech. Rev.,* **40,** 298 (1982).

Kocher, D. G., "Automated Foucault Test for Focus Sensing," *Appl. Optics,* **22,** 1887 (1983).

Kryder, M. H., "Magneto-optic Recording Technology," *J. Appl. Phys.,* **57,** 3913 (1985).

Kryder, M. H., W. H. Meiklejohn, and R. E. Skoda, "Stability of Perpendicular Domains in Thermomagnetic Recording Materials," *Proc. SPIE,* **420,** 236 (1983).

Landau, L. D., and E. M. Lifshitz, *Statistical Physics,* Pergamon, London, 1958, pp. 379–406.

Lang, J. K., Y. Baer, and P. A. Cox, "Study of 4f and Valence Density of States in Rare Earth Metals," *J. Phys. F: Metal Phys.,* **11,** 121 (1981).

Loudon, R., "Theory of Resonance Raman Effect in Crystals," *J. Physique,* **26,** 677 (1965).

Luborsky, F. E., J. T. Furey, R. E. Skoda, and B. C. Wagner, "Stability of Amorphous Transition Metal-Rare Earth Films for Magneto-optic Recording," *IEEE Trans. Magn.,* **MAG-21,** 1618 (1985).

Malozemoff, A. P., A. R. Williams, K. Terakura, V. L. Moruzzi, and K. Fukamichi, "Magnetism of Amorphous Metal-Metal Alloys," *J. Magn. and Mag. Mat.,* **35,** 192 (1983).

Mansuripur, M., G. A. N. Connell, and J. W. Goodman, "Signal and Noise in Magneto-optic Readout," *J. Appl. Phys.,* **53,** 4485 (1982).

Mansuripur, M., and G. A. N. Connell, "Magneto-optical Recording," *Proc. SPIE,* **420,** 222 (1983a).

Mansuripur, M., and G. A. N. Connell, "Thermal Aspects of Magneto-optical Recording," *J. Appl. Phys.,* **54,** 4794 (1983b).

Mansuripur, M., and G. A. N. Connell, "Laser-induced Local Heating of Moving Multilayer Media," *Appl. Optics,* **21,** 666 (1983c).

Mansuripur, M., and G. A. N. Connell, "Energetics of Domain Formation in Thermomagnetic Recording," *J. Appl. Phys.,* **55,** 3049 (1984).

Mansuripur, M., M. F. Ruane, and M. N. Horenstein, "Magnetic Properties of Amorphous Rare Earth-Transition Metal Alloys for Erasable Optical Storage," *Proc. SPIE,* **529,** 25 (1985).

Masui, S., T. Kobayashi, S. Tsunashima, S. Uchiyama, K. Sumiyama, and Y. Nakamura, "Magnetic and Magneto-optic Properties of Amorphous GdFeCo and GeFeCoBi Films," *IEEE Trans. Magn.,* **MAG-20,** 1036 (1984).

Mayer, L., "Curie-Point Writing on Magnetic Films," *J. Appl. Phys.,* **29,** 1003 (1958).

Mimura, Y., N. Imamura, and T. Kobayashi, "Magnetic Properties and Curie Point Writing in Amorphous Metallic Films," *IEEE Trans. Magn.,* **MAG-12,** 779 (1976).

Mimura, Y., N. Imamura, T. Kobayashi, A. Okada, and Y. Kushiro, "Magnetic Properties of Amorphous Alloy Films of Fe with Gd, Tb, Dy, Ho, or Er," *J. Appl. Phys.,* **49,** 1208 (1978).

Mizoguchi, T., and G. S. Cargill, "Magnetic Anisotropy from Dipolar Interactions in Amorphous Ferrimagnetic Alloys," *J. Appl. Phys.,* **50,** 3570 (1979).

Nagao, Y., S. Tanaka, F. Tanaka, and N. Imamura, "Improvement of TbFeCo Disks for Magneto-optical Recording," *Intermag. Conf. 1986,* Paper No. FC-07, 1986.

Nishihara, Y., T. Katayama, Y. Yamaguchi, S. Ogawa, and T. Tsushima, "Anisotropic Distribution of Atomic Pairs Induced by the Preferential Resputtering Effect in Amorphous Gd-Fe and Gd-Co Films," *Jap. J. Appl. Phys.,* **17,** 1083 (1978).

Nomura, T., "A New Magneto-optic Readout Head Using Bi-Substituted Magnetic Garnet Film," *IEEE Trans. Magn.*, **MAG-21**, 1545 (1985).

Ohashi, K., H. Takagi, S. Tsunashima, S. Uchiyama, and T. Fujii, "Magnetic Aftereffect Due to Domain Wall Motion in Amorphous TbFe Sputtered Films," *J. Appl. Phys.*, **50**, 1611 (1979).

Ohishi, A., N. Chinone, M. Ojima, and A. Arimoto, "Noise Characteristics of High-Frequency Superposed Laser Diodes for Optical Disc Systems," *Elect. Lett.*, **20**, 821 (1984).

Okada, K., T. Kubo, W. Susaki, and T. Sato, "A New PCM Audio Disk Pickup Employing a Laser Diode," *J. Audio Eng. Soc.*, **28**, 429 (1980).

Okamine, S., N. Ohta, and Y. Sugita, "Perpendicular Anisotropy in Rare Earth-Transition Metal Amorphous Films Prepared by Dual Ion Beam Sputtering," *IEEE Trans. Magn.*, **MAG-21**, 1641 (1985).

Pasman, J. H. T., H. F. Olijhoek, and B. Verkaik, "Developments in Optical Disk Mastering," *Proc. SPIE*, **529**, 62 (1985).

Rhyne, J. J., "Amorphous Magnetic Rare Earth Alloys," in K. A. Gschneider and L. Eyring (eds.), *Handbook on the Physics and Chemistry of Rare Earths*, North-Holland, Amsterdam, 1977.

Saito, J., M. Sato, H. Matsumoto, and H. Akasaka, "Direct Overwrite by Light Power Modulation on Magnetooptic Multilayer Media," *Digest of Int. Symp. on Optical Memory*, Sept. (1987).

Sakurai, Y., and K. Onishi, "Preparation and Properties of RE-TM Amorphous Films," *J. Magn. and Mag. Mat.*, **35**, 183 (1983).

Sato, M., N. Tsukane, S. Tokuhara, and H. Toba, "Magneto-optical Memory Disk Using Plastic Substrate," *Proc. SPIE*, **529**, 33 (1985).

Sato, R., N. Saito, and Y. Togami, "Magnetic Anisotropy of Amorphous Gd-R-Co (R = Tb, Dy, Ho, Er) Films," *Jap. J. Appl. Phys.*, **24**, L266 (1985).

Smith, D. O., "Magneto-optical Scattering from Multilayer Magnetic and Dielectric Films. Pt. I. General Theory," *Optica Acta*, **12**, 13 (1965a).

Smith, D. O., "Magneto-optical Scattering from Multilayer Magnetic and Dielectric Films. Pt. II. Applications of the General Theory," *Optica Acta*, **12**, 193 (1965b).

Smith, D. O., "Magnetic Films and Optics in Computer Memories," *IEEE Trans. Magn.*, **MAG-3**, 433 (1967).

Stubkjaer, K. E., and M. B. Small, "Noise Properties of Semiconductor Lasers Due to Optical Feedback," *IEEE J. Quantum Elect.*, **QE-20**, 472 (1984).

Sunago, K., S. Matsushita, and Y. Sakurai, "Thermomagnetic Writing in TbFe Films," *IEEE Trans. Magn.*, **MAG-12**, 776 (1976).

Takagi, H., S. Tsunashima, S. Uchiyama, and T. Fujii, "Stress Induced Anisotropy in Amorphous Gd-Fe and Tb-Fe Sputtered Films," *J. Appl. Phys.*, **50**, 1642 (1979).

Taylor, R. C., and A. Gangulee, "Magnetic Properties of Amorphous GdFeB and GdCoB Alloys," *J. Appl. Phys.*, **53**, 2341 (1982).

Theuerer, H. C., E. A. Nesbitt, and D. D. Bacon, "High Coercive Force Rare-Earth Alloy Films by Getter Sputtering," *J. Appl. Phys.*, **40**, 2994 (1969).

Thiele, A. A., "Theory of the Static Stability of Cylindrical Domains in Uniaxial Platelets," *J. Appl. Phys.*, **41**, 1139 (1970).

Togami, Y., R. Sato, N. Saito, and M. Shibayama, "Magneto-optical Kerr Effect in Amorphous RE-Co-Fe Films," *Jap. J. Appl. Phys.*, **24**, 106 (1985).

Treves, D., and D. S. Bloomberg, "Effect of Birefringence on Optical Memory Systems," *Proc. SPIE*, **695**, 262 (1986a).

Treves, D., and D. S. Bloomberg, "Signal, Noise and Codes in Optical Memories," *Optical Engineering*, **25**, 881 (1986b).

Tsunashima, S., S. Masui, T. Kobayashi, and S. Uchiyama, "Magneto-optic Kerr Effect of Amorphous Gd-Fe-Co Films," *J. Appl. Phys.*, **53**, 8175 (1982).

Tsunoda, Y., S. Horigome, and Z. Tsutsumi, "Optical Digital Data Storage Technologies with Semiconductor Laser Head," *Proc. SPIE*, **382**, 24 (1983).

Uchiyama, S., "Magneto-optic Properties of Ternary and Quaternary Rare Earth-Transition Metal Amorphous Films," in Y. Sakurai (ed.), *Jap. Annu. Rev. Electron., Comput., Telecommun.: Recent Magnetics For Electronics*, N. Holland, Amsterdam, 1986.

van Dover, R. B., M. Hong, E. M. Gyorgy, J. F. Dillon, and S. D. Albiston, "Intrinsic Anisotropy on Tb-Fe Films Prepared by Magnetron Co Sputtering," *J. Appl. Phys.,* **57,** 3897 (1985).

van Dover, R. B., E. M. Gyorgy, R. P. Frankenthal, M. Hong, and D. J. Siconolfi, "Effect of Oxidation on the Magnetic Properties of Unprotected TbFe Thin Films," *J. Appl. Phys.,* **59,** 1291 (1986).

van Rijsewijk, H. C. H., P. E. J. Legierse, and G. E. Thomas, "Manufacture of Laser-Vision Video Discs by Photopolymerization Process," *Philips Tech. Rev.,* **40,** 287 (1982).

Index

AC-biased audio recording, 13–14
  bias adjustment in, 157, 158
  components for, 153–155
  demagnetization in, 161
  equalization in, 161–162
  erasure in, 170
  gap loss in, 160
  high-frequency losses in, 159–161
  noise in, 162–166
  nonlinearity and distortion in, 155–157
  print-through in, 166–169
  signal-to-noise ratio in (see Signal-to-noise ratio)
  spacing loss in, 160
AC-biased instrumentation recording, 225
Access time in optical recording, 310
Actuators in magnetooptical disk files, 367–370
Aliasing in digital audio recording, 185–187
Analog recording (see AC-biased audio recording; AC-biased instrumentation recording; Video recording)
Analog-to-digital conversion in digital audio recording, 416–419
Anhysteresis (see AC-biased audio recording; Video duplication)
Anhysteretic duplication, 95–96, 98–101
  bifilar-winding method, 99
  double-transfer method, 103
  master and slave tape for, 99–101
  parallel-running method, 98–99
Anhysteretic magnetization curve:
  relation to ac-biased audio recording, 155–157
  relation to anhysteretic duplication, 99–100

Anisotropy constant of magnetooptical materials, 325–328
Areal density:
  historical increases in, 5
  in instrumentation recorders, 226, 245
  in magnetooptical recording, 310, 380–387, 397
Audio cassette recorders:
  equalization in, 178–179
  heads in, 178–179
  mechanical aspects of, 175–177
  noise reduction in, 180
  tapes for, 179–180
Audio duplication at high speed, 180–181
Audio open-reel recorders:
  equalization in, 172–173
  heads in, 171–172
  noise reduction in, 174–175
  tapes for, 173–174
  transports for, 170–171
Audio recording:
  history of, 152–153
  systems of, 13–14
  in video: double-component, 67–68
    FM, 65–68
    frequency division, 65–67
    linear track, 63–65
    with pulse-code modulation (PCM), 68–70
    track division, 68–69
  (See also AC-biased audio recording; Digital audio recording)
Audio signals:
  dynamic range of, 150
  recorder requirements for, 151
  response of human ear to, 149–151
  and unwanted signals, 150–151
Audio tapes:
  coercivity of, 155, 160, 179
  in compact-cassette recorders, 179–180

ABOUT THE EDITORS

DENIS MEE is an IBM Fellow at the IBM General Products Division, San Jose, California. He is the author of *The Physics of Magnetic Recording* (North Holland Publishing Co., Amsterdam, 1964). ERIC DANIEL, consultant, was a Fellow of the Memorex Corporation, Santa Clara, California, until he retired in 1983. Each editor has had more than thirty years of experience in magnetic recording, including both computer data and analog applications, and has published twenty or more papers in scientific journals.